REGULATING HEALTH AND ENVIRONMENTAL RISKS UNDER WTO LAW

Regulating Health and Environmental Risks under WTO Law

A Critical Analysis of the SPS Agreement

LUKASZ GRUSZCZYNSKI

OXFORD
UNIVERSITY PRESS

This book has been printed digitally and produced in a standard specification in order to ensure its continuing availability

OXFORD
UNIVERSITY PRESS

Great Clarendon Street, Oxford OX2 6DP
United Kingdom

Oxford University Press is a department of the University of Oxford. It furthers the University's objective of excellence in research, scholarship, and education by publishing worldwide. Oxford is a registered trade mark of Oxford University Press in the UK and in certain other countries

Reprinted 2012

British Library Cataloguing in Publication Data
Data available

Library of Congress Cataloging in Publication Data
Data available

ISBN 978-0-19-957892-4

For Olga.

General Editor's Preface

The world trading system continues to engage with perplexing and relatively new problems. Often these problems push international regulating norms deeply into the 'sovereign territory' of nation states. Unsurprisingly some national authorities resist the 'globalization' of their turf.

The subject of this new book, which Oxford University Press welcomes into its international economic law series, is a prototypical example of the constant tensions between allocation of power and authority which globalization and its related implications create as the policy necessities of such allocation become increasingly apparent.

The World Trade Organisation is at the center of these policy clashes as the thousands of pages of its intricate jurisprudence attest. The WTO is increasingly forced to regulate threats to animal or plant life and health which often arise from the risks posed by modern additives and also by modern processes of mass production and worldwide trade and transport. To better face up to these risks, the Uruguay Round trade negotiators created a new treaty test, namely the Agreement on the Application of Sanitary and Phytosanitary measures, the now famous "SPS" agreement.

This agreement and its role in mediating between powerful national and global governmental agencies is the subject of this book. The ambiguities and gaps in the treaty text mean that the WTO and its dispute settlement system have come to play a crucial role in resolving controversies. In particular, the WTO system has found itself deeply involved in analysing the concept of risk and the degree to which nations should retain the power to judge the tolerable levels of risk to their citizens without allowing this autonomy to amount to the imposition of unnecessary trade restraints. This is a conundrum that will continue to confront the regulation of expanding international norms that are designed to constrain national authority.

This new book is a remarkably analytical work which should assist policy makers on one of the difficult applications of one of the most perplexing subjects of the WTO. It is an ideal addition to the OUP IEL series.

JOHN H. JACKSON

Foreword

This revised monograph, based on the successful defence of the doctoral thesis by Lukasz Gruszczynski at the European University Institute at Florence in October 2008, offers an excellent legal analysis of diverse risk regulation and risk analysis approaches (Chapter 1), the WTO legal disciplines for sanitary and phyto-sanitary (SPS) and related measures (Chapter 2), the harmonization disciplines of the SPS Agreement and of other relevant international organizations (Chapter 3), the regulation of risk assessments and their scientific justification (Chapter 4), the precautionary principle and related rules (Chapter 5), as well as the 'risk management' rules of the SPS Agreement (Chapter 6). Chapter 7 concludes that the disciplines established by the SPS Agreement, in spite of some defects in the rules and related dispute settlement practices of WTO dispute settlement bodies, provide an effective framework for the supervision and management of national (phyto)sanitary measures and their legal consistency with international trade law.

The legal arguments are thorough and convincing. The proposed distinctions between risk assessment, risk management, and risk communication are today widely accepted. Chapter 2 notes potential problems relating to determination of applicable rules (GATT, SPS, or TBT) for assessment of national risk regulatory measures. In this context, the chapter identifies some deficiencies in the current approach of WTO dispute settlement bodies and argues for greater institutional sensitivity in the interpretation of the SPS Agreement. Chapter 3 explains the three different policy options available to WTO Members: to implement SPS measures that conform to international standards; to only 'base' their SPS measures on international standards; or to completely deviate from international standards by less or more demanding SPS measures. The overview of other international standard setting bodies and of their regulatory problems complements the analyses of the harmonization disciplines of the SPS Agreement. The 'scientific disciplines' and risk assessment obligations of the SPS Agreement, as well as the role of science in the TBT Agreement and GATT 1994, are explained with due regard to the relevant WTO dispute settlement reports. The remaining problems of regulation—such as the tension between the scientific requirements of the SPS Agreement and the right to establish a level of protection which is deemed appropriate by a particular WTO Member—are acknowledged and discussed.

The fact that the SPS Agreement requires WTO Members neither to change their 'appropriate level of protection' with respect to human, animal, or plant life or health, nor to maintain a minimum standard of (phyto)sanitary protection, reflects the limited focus of WTO law on the regulation of trade policy instruments and related policies, such as promoting 'public reason' through scientific justification and non-discriminatory application of SPS measures. The increasing importance of health and environmental risk regulation in international trade, and the regulatory 'gaps' and regulatory problems identified in this book, make it likely that SPS disputes will remain frequent at national and international levels. This book will help lawyers and judges in resolving such disputes over the interpretative and legal problems in the application of WTO law and dispute settlement procedures.

Prof Dr Ernst-Ulrich Petersmann
European University Institute
Florence (Italy)

Acknowledgments

There are many people whom I wish to thank for their contributions to this book. This includes in the first place the supervisor of my PhD thesis Professor Ernst-Ulrich Petersmann whose support, encouragement and advice was indispensable for completing this work.

My special thanks are also reserved for Professors Joanne Scott, Robert Howse and Thomas Cottier, Christian Joerges and Francesco Francioni for taking the time to comment on my ideas and helping me to understand international trade rules and their impact on domestic risk regulations. Other people, which need to be mentioned here, are Professors Laurence Boisson de Chazournes, Gregory Shaffer, John J. Barceló, Gretchen Stanton and two anonymous reviewers from the OUP.

The final thanks go to Olga and my family. Without them nothing would be possible.

Warsaw, 3 November 2009

Contents

List of Abbreviations

ADI	acceptable daily intake
ALOP	appropriate level of protection
Anti-Dumping Agreement	Agreement on the Implementation of Article VI of the General Agreement on Tariffs and Trade 1994
BSE	Bovine Spongiform Encephalopathy
CBD	Convention on Biological Diversity
Codex	Codex Alimentarius Commission
CPB	Cartagena Protocol on Biosafety
CPM	Commission on Phytosanitary Measures
DSB	dispute settlement body
DSU	Understanding on Rules and Procedures Governing the Settlement of Disputes
ECSCP	EC Scientific Committee on Plants
EIA	environmental impact assessment
FAO	Food and Agriculture Organization of the United Nations
GATT 1947	General Agreement on Tariffs and Trade 1947
GATT 1994	General Agreement on Tariffs and Trade 1994
GM	genetically modified
GMO	genetically modified organism
GVP	good veterinary practice
HACCP	hazard analysis and critical control points
IPCS	International Program on Chemical Safety
IPPC	International Plant Protection Convention
ISO	International Organization for Standardization
ISPM	International Standards for Phytosanitary Measures
JECFA	Joint FAO/WHO Expert Committee on Food Additives
JEMRA	Joint FAO/WHO Expert Meetings on Microbiological Risk Assessment
JMPR	Joint FAO/WHO Meetings on Pesticide Residues
MRL	maximum residue level
NGO	non-governmental organization
NOAEL	no-observed-adverse-effect level
NOEL	no-observed-effect level
NRC	National Research Council
NTB	non-tariff barrier
OIE	World Organization for Animal Health
RNQP	regulated non-quarantine pest
RPPO	regional plant protection organization
SCVPH	Scientific Committee on Veterinary Measures Relating to Public Health

SPS	sanitary and phytosanitary
SPS Agreement	Agreement on the Application of Sanitary and Phytosanitary Measures
Standards Code	Agreement on Technical Barriers to Trade 1979
TBT Agreement	Agreement on Technical Barriers to Trade
WHO	World Health Organization
WTO	World Trade Organization
WTO Agreement	Marrakesh Agreement Establishing the World Trade Organization
VCLT	Vienna Convention on the Law of Treaties

Table of Cases

Short title	Full case title and citation	Page reference
GATT/WTO Panel Reports		
Australia – Salmon	Panel Report, *Australia – Measures Affecting Importation of Salmon*, WT/DS18/R and Corr.1, adopted 6 November 1998, modified by Appellate Body Report, WT/DS18/AB/R, DSR 1998:VIII, 3407	69, 111, 112, 117, 121, 135, 178, 240–41, 248–9, 250, 269
Australia – Salmon (Article 21.5 – Canada)	Panel Report, *Australia – Measures Affecting Importation of Salmon – Recourse to Article 21.5 of the DSU by Canada (Article 21.5 – Canada)*, WT/DS18/RW, adopted 20 March 2000, DSR 2000:IV, 2031	108, 128, 134, 228, 231–2, 247
Brazil – Retreaded Tyres	Panel Report, *Brazil – Measures Affecting Imports of Retreaded Tyres*, WT/DS332/R, adopted 17 December 2007, as modified by Appellate Body Report, WT/DS332/AB/R	59, 239, 245
Canada – Continued Suspension	Panel Report, *Canada – Continued Suspension of Obligations in the EC – Hormones Dispute*, WT/DS321/R, adopted 14 November 2008, as modified by Appellate Body Report WT/DS321/AB/R	*See US – Continued Suspension*
EC – Asbestos	Panel Report, *European Communities – Measures Affecting Asbestos and Asbestos-Containing Products*, WT/DS135/R and Add.1, adopted 5 April 2001, modified by Appellate Body Report, WT/DS135/AB/R, DSR 2001:VIII, 3305	237, 239
EC – Bananas III (US)	Panel Report, *European Communities – Regime for the Importation, Sale and Distribution of Bananas, Complaint by the United States*, WT/DS27/R/USA, adopted 25 September 1997, modified by Appellate Body Report, WT/DS27/AB/R, DSR 1997:II, 943	52
EC – Biotech Products	Panel Report, *European Communities – Measures Affecting the Approval and Marketing of Biotech Products*, WT/DS291/R, WT/DS292/R, WT/DS293/R, Corr.1 and Add.1, 2, 3, 4, 5, 6, 7, 8 and 9, adopted 21 November 2006	53–62, 65–6, 67–8, 72, 108, 109, 112, 116, 121, 122, 133, 135, 168–74, 178, 179–84, 185–6, 187, 189, 191, 193, 194–6, 197–8, 206–7, 209, 211–12, 214, 244, 250

(*continued*)

Short title	Full case title and citation	Page reference
EC – Chicken Cuts (Brazil)	Panel Report, *European Communities – Customs Classification of Frozen Boneless Chicken Cuts, Complaint by Brazil*, WT/DS269/R, adopted 27 September 2005, as modified by Appellate Body Report WT/DS269/AB/R, WT/DS286/AB/R, DSR 2005:XIX, 9295	40
EC – Hormones (Canada)	Panel Report, *EC Measures Concerning Meat and Meat Products (Hormones), Complaint by Canada*, WT/DS48/R/CAN, adopted 13 February 1998, as modified by Appellate Body Report, WT/DS26/AB/R, WT/DS48/AB/R, DSR 1998:II, 235	*See EC – Hormones (US)*
EC – Hormones (US)	Panel Report, *European Communities - Measures Concerning Meat and Meat Products (Hormones), Complaint by the United States*, WT/DS26/R/USA, adopted 13 February 1998, modified by Appellate Body Report, WT/DS26/AB/R, WT/DS48/AB/R, DSR 1998:III, 699	63, 69, 70, 88, 90, 97, 103, 108, 116, 120, 129, 133, 167, 221, 229, 232
EC – Tariff Preferences	Panel Report, *European Communities – Conditions for the Granting of Tariff Preferences to Developing Countries*, WT/DS246/R, adopted 20 April 2004, modified by Appellate Body Report, WT/DS/246/AB/R, DSR 2004:III, 1009	182
Japan – Agricultural Products	Panel Report, *Japan – Measures Affecting Agricultural Products*, WT/DS76/R, adopted 19 March 1999, modified by Appellate Body Report, WT/DS76/AB/R, DSR 1999:I, 315	141, 178, 182, 210–11
Japan – Alcoholic Beverages II	Panel Report, *Japan – Taxes on Alcoholic Beverages*, WT/DS8/R, WT/DS10/R, WT/DS11/R, adopted 1 November 1996, as modified by Appellate Body Report WT/DS8/AB/R, WT/DS10/AB/R, WT/DS11/AB/R, DSR 1996:I, 125	40
Japan – Apples	Panel Report, *Japan – Measures Affecting the Importation of Apples*, WT/DS245/R, adopted 10 December 2003, upheld by Appellate Body Report, WT/DS245/AB/R, DSR 2003:IX, 4481	109, 110, 120, 121, 125–6, 127, 183, 187, 192, 199, 211, 242, 252
US – Continued Suspension	Panel Report, *United States – Continued Suspension of Obligations in the EC – Hormones Dispute*, WT/DS320/R, adopted 14 November 2008, as modified by Appellate Body Report WT/DS320/AB/R	48, 51, 116, 117, 120, 122–4, 125, 127, 130, 141, 146, 188–9, 191, 193, 199–202, 212, 213, 214, 222, 252
US – Gambling	Panel Report, *United States – Measures Affecting the Cross-Border Supply of Gambling and Betting Services*, WT/DS285/R, adopted 20 April 2005, as modified by Appellate Body Report WT/DS285/AB/R, DSR 2005:XII, 5797	40

Short title	**Full case title and citation**	**Page reference**
US – Section 301 Trade Act	Panel Report, *United States – Sections 301–310 of the Trade Act of 1974*, WT/DS152/R, adopted 27 January 2000, DSR 2000:II, 815	66
US – Shrimp (Article 21.5 – Malaysia)	Panel Report, *United States – Import Prohibition of Certain Shrimp and Shrimp Products – Recourse to Article 21.5 of the DSU by Malaysia*, WT/DS58/RW, adopted 21 November 2001, upheld by Appellate Body Report, WT/DS58/AB/RW, DSR 2001:XIII, 6529,	239
US – Tuna/ Dolphin II	GATT Panel Report, *United States – Restriction on Imports of Tuna*, issued 16 June 1994, DS29/R	171

Appellate Body Reports

Australia – Salmon	Appellate Body Report, *Australia – Measures Affecting Importation of Salmon*, WT/DS18/AB/R, adopted 6 November 1998, DSR 1998:VIII, 3327	111, 116, 118, 119, 188, 226–7, 231, 232–3, 240, 241, 249
Canada – Continued Suspension	Appellate Body Report, *Canada – Continued Suspension of Obligations in the EC – Hormones Dispute*, WT/DS321/AB/R, adopted 14 November 2008	*See US – Continued Suspension*
EC – Asbestos	Appellate Body Report, *European Communities – Measures Affecting Asbestos and Asbestos-Containing Products*, WT/DS135/AB/R, adopted 5 April 2001, DSR 2001:VII, 3243	136, 245, 249
EC – Chicken Cuts	Appellate Body Report, *European Communities – Customs Classification of Frozen Boneless Chicken Cuts*, WT/DS269/AB/R, WT/DS286/AB/R, and Corr.1, adopted 27 September 2005, DSR 2005:XIX, 9157	39
EC – Hormones	Appellate Body Report, *EC Measures Concerning Meat and Meat Products*, WT/DS26/AB/R, WT/DS48/AB/R, adopted 13 February 1998, DSR 1998:I, 135	3, 38, 40, 49, 50, 51, 52, 70, 91–2, 95, 96, 97, 99, 100, 104, 105, 109 110, 112, 118, 119, 120, 122, 127, 128, 129, 133, 134–5, 136, 139–40, 149, 163, 167, 168, 182, 188, 205, 214, 222–3, 226, 230, 231, 235, 237–8, 239–40, 248, 268
EC – Sardines	Appellate Body Report, *European Communities – Trade Description of Sardines*, WT/DS231/AB/R, adopted 23 October 2002, DSR 2002:VIII, 3359, para 278	97

(*continued*)

Short title	Full case title and citation	Page reference
EC – Tariff Preferences	Appellate Body Report, *European Communities – Conditions for the Granting of Tariff Preferences to Developing Countries*, WT/DS246/AB/R, adopted 20 April 2004, DSR 2004:III, 925	180, 183
Guatemala – Cement I	Appellate Body Report, *Guatemala – Anti-Dumping Investigation Regarding Portland Cement from Mexico*, WT/DS60/AB/R, adopted 25 November 1998, DSR 1998:IX, 3767	71
India – Patents (EC)	Appellate Body Report, *India – Patent Protection for Pharmaceutical and Agricultural Chemical Products*, WT/DS50/AB/R, adopted 16 January 1998, DSR 1998:I, 9	166
Japan – Agricultural Products	Appellate Body Report, *Japan – Measures Affecting Agricultural Products*, WT/DS76/AB/R, adopted 19 March 1999, DSR 1999:I, 277	50, 103, 109, 110, 111, 174, 178, 183, 185, 206, 208, 268
Japan – Alcoholic Beverages II	Appellate Body Report, *Japan – Taxes on Alcoholic Beverages*, WT/DS8/AB/R, WT/DS10/AB/R, WT/DS11/AB/R, adopted 1 November 1996, DSR 1996:I, 97	39, 40, 239
Japan – Apples	Appellate Body Report, *Japan – Measures Affecting the Importation of Apples*, WT/DS245/AB/R, adopted 10 December 2003, DSR 2003:IX, 4391	110, 119, 122, 125, 141, 174–5, 179, 187–8, 190, 191, 195, 211
Korea – Dairy	Appellate Body Report, *Korea – Definitive Safeguard Measure on Imports of Certain Dairy Products*, WT/DS98/AB/R, adopted 12 January 2000, DSR 2000:I, 3	69
Korea – Various Measures on Beef	Appellate Body Report, *Korea – Measures Affecting Imports of Fresh, Chilled and Frozen Beef*, WT/DS161/AB/R, WT/DS169/AB/R, adopted 10 January 2001, DSR 2001:I, 5	245
US – Carbon Steel	Appellate Body Report, *United States – Countervailing Duties on Certain Corrosion-Resistant Carbon Steel Flat Products from Germany*, WT/DS213/AB/R and Corr.1, adopted 19 December 2002, DSR 2002:IX, 3779	66
US – Continued Suspension	Appellate Body Report, *United States – Continued Suspension of Obligations in the EC – Hormones Dispute*, WT/DS320/AB/R, adopted 14 November 2008	48, 96, 103, 111, 118, 119, 121, 124, 128, 130, 136, 137, 144, 145, 178, 185, 191, 196, 199, 202–3, 205, 206, 207, 213, 216, 217, 222–4, 230

Decisions/Awards of Arbitrator

(*continued*)

Short title	Full case title and citation	Page reference
Korea – Alcoholic Beverages	Award of the Arbitrator, *Korea – Taxes on Alcoholic Beverages – Arbitration under Article 21.3(c) of the DSU*, WT/DS75/16, WT/DS84/14, 4 June 1999, DSR 1999:II, 937	209
US – Hot-Rolled Steel	Award of the Arbitrator, *United States – Anti-Dumping Measures on Certain Hot-Rolled Steel Products from Japan – Arbitration under Article 21.3(c) of the DSU*, WT/DS184/13, 19 February 2002, DSR 2002:IV, 1389	208

Introduction

1 Introduction

In April 2009 a new strain of influenza A virus subtype H1N1, commonly known as swine flu or swine influenza, was identified for the first time. The virus was in fact a mixture of four different strains, one of them being endemic in humans, one in birds, and two in pigs. The new virus soon proved to pose great danger leading to thousands of deaths. The disease, which was initially identified only in Mexico and the United States, was very soon detected in many European and Asian countries. The World Health Organization (WHO) concerned about this quick spread declared in June 2009 the outbreak of a pandemic disease. However, at the same time, the WHO emphasized that in most cases swine flu was only of moderate severity that was comparable to more traditional types of influenza. Not surprisingly, the outbreak received wide media attention around the globe. Television and newspapers were constantly supplying the general public with information on the increasing number of casualties. The experts in countless interviews compared the new disease to Spanish or Asian flu, predicting a pandemic on an unprecedented scale. Public hysteria was growing.

Although there seemed to be an agreement between scientists that the disease was transmitted only between humans and could not be acquired from swine, people became increasingly concerned with the safety of pork products, especially those coming from the United States and Mexico. The wide media coverage combined with the use of the common name for the disease that suggested its connection with pigs resulted in a severe decrease in the international pork market. Some countries contemplated and finally introduced import bans or other trade restrictions directed against pigs and processed pork products from those geographical destinations that were considered to be the initial sources of disease. For instance, on 21 April 2009, Russia introduced an importation ban, which still remains partially in force, on fresh and frozen pork meat from animals raised or slaughtered in the United States. Surprisingly, the Russian ban also covered poultry meat and unprocessed beef products. South Korea was less strict and only banned the import of live swine from North America, while a trade in pork products was still permitted.[1] The reaction of the United States was immediate and a number of high officials, including the US Trade Representative, warned that the continuation of the ban may result in a delay in the accession of Russia to the World Trade Organization (WTO).[2] All of them stressed that that Russian measure was unfair and clearly not based on existing international standards.

This growing tension between the countries provoked a reaction from the World Organization for Animal Health (OIE), a body responsible for the international supervision of the animal health situation. In an official statement dated 28 April 2009, the OIE articulated its position on the safety of the international trade in pigs and products of pig origin. The document stated that 'given there is no case of infection in animals confirmed in the zones where cases of human infection have been detected, it is not necessary to introduce specific measures for international trade in swine or their products nor to consider that consumers of pork products are at risk of infection'.[3] In the same document, the OIE also proposed changing the name of the disease, as the current term 'swine influenza' could be misleading for consumers and provok unnecessary distress. Since the response of the international community was quite moderate, a new joint statement was soon issued by the Food and Agriculture Organization of the United Nations (FAO), the OIE, the WHO, and the WTO. The document repeated that 'pork and pork products, handled in accordance with good hygienic practices recommended by the WHO, FAO, Codex Alimentarius Commission and the OIE, will not be a source of infection'.[4] The organizations again stressed that there was no scientific evidence which would show that the virus could be transmitted by food. Nevertheless some of the trade restrictions are still maintained causing considerable losses to the pork industry in Canada, the United States, and Mexico. At the same time, a number of public opinion polls that were conducted in various countries show a relatively high level of belief among domestic consumers that eating pork may constitute one of the potential sources of the infection. Who is right about the safety of those products? How conclusive is scientific evidence? Is there any uncertainty as to the possible pathways for the transmission of disease? Is it legal for a country within the existing international legal framework to institute a trade ban even without possessing sufficient scientific information? Do health concerns indeed constitute a real purpose of these newly introduced trade measures?

[1] Renée Johnson, 'Potential Farm Sector Effects of 2009 H1N1 "Swine Flu": Questions and Answers', CRS Report for Congress (2009) 5–6.

[2] Information available on <http://crisiscrunch.pbndc.com/?p=862>, last visited 5 September 2009.

[3] OIE Position on Safety of International Trade of Pigs and Products of Pig Origin, available at the webpage of the OIE <http://www.oie.int/eng/press/en_090428.htm>, last visited 5 September 2009.

[4] Joint FAO, OIE, WHO, and WTO statement on A(H1N1) virus, available at the webpage of the WTO <http://www.wto.org/english/news_e/news09_e/jt_stat_02may09_e.htm>, last visited 5 September 2009.

These are very complex questions that cannot be easily answered. WTO law attempts to address this type of problem with its Agreement on the Application of Sanitary and Phytosanitary Measures (SPS Agreement)[5] that provides an elaborate set of rules disciplining the imposition and maintenance of measures directed against sanitary and phytosanitary (SPS) risks. As noted by the Appellate Body in one of its reports, the Agreement reflects a 'delicate and carefully negotiated balance... between the shared, but sometimes competing, interests of promoting international trade and of protecting the life and health of human beings'.[6] More specifically, the SPS Agreement while recognizing the right of each WTO Member to establish its own level of SPS protection, also requires the observance of a number of standards such as scientific justification, consistency among different risk measures, or transparency of domestic SPS systems.

2 Objectives

The subject of this book is precisely directed at some of the above issues. The central problem that it attempts to tackle is whether the system established by the SPS Agreement is capable of addressing existing and potential SPS challenges of the new interdependent world. This general question may be translated into a number of more specific issues. Does the SPS Agreement strike an optimal balance between competing objectives of international free trade and protection against environmental and human health risks?[7] How wide is the regulatory discretion of WTO Members? What is the role of science and scientific evidence in the process of adoption of national SPS measures? How might WTO Members respond to those risks which are uncertain? Does the precautionary principle have any relevance under the SPS Agreement?

These questions are addressed within three broader problematic issues that are discussed in book: the appropriateness of the disciplines established by the SPS Agreement, the consistency of their interpretation in the case law, and the normative content of those requirements, which have not yet been interpreted by SPS jurisprudence. Due to the limited size of the book as well as the complexity of potential problems, the scope of this monograph is necessarily restricted. Instead of discussing the whole SPS Agreement, I concentrate only on those provisions which set substantive standards for national risk regulators (for the sake of simplicity those provisions are referred to as substantive requirements). As a consequence of this restriction, the book examines only those aspects of the SPS

[5] Agreement on the Application of Sanitary and Phytosanitary Measures, opened for signature 15 April 1994, 1867 UNTS 493 (entered into force 1 January 1995).

[6] Appellate Body Report, *EC – Hormones*, para 177.

[7] For the sake of simplicity, the book understands the term 'environmental health risks' as risks for animals and plant life and health. At the same time, the book recognizes that the notion of environment is not limited to the categories of fauna and flora (cf Chapter 2, section 2.3).

Agreement which are related to harmonization requirements and science-based obligations, including a special regime for measures adopted in situations that are characterized by insufficiency of scientific evidence and risk management disciplines. This discussion is supplemented with the analysis of some preliminary matters such as the applicability of the Agreement and its relationship with other WTO instruments. This in turn means that the book generally does not address procedural obligations relating to the notification, transparency, or enforcement of SPS measures (eg control, inspection, and approval procedures). The operation of the Committee on Sanitary and Phytosanitary Measures (SPS Committee) in principle also remains outside of my interests.[8] Although such an approach may appear to be reductionist, it also permits me to concentrate on the selected issues and analyse them in a more detailed manner than would otherwise be possible.

3 Methodology

The book predominantly relies on the analysis of the text of the SPS Agreement and corresponding case law. This exegetic approach to the Agreement seems to be justified for three reasons. First, it is consistent with the practice of the WTO dispute settlement bodies. Both panels and the Appellate Body stress that interpretation of WTO rules requires in the first place recourse to a text of relevant provisions. Secondly, and probably more importantly, since one of the aims of this book is to identify inconsistencies in the interpretation of the SPS Agreement, this task can only be successfully accomplished through reliance on the same interpretative methods that are applied by WTO jurisprudence. Thirdly, the book also seeks to clarify the content of those obligations which still remain unaddressed by the case law. Again any meaningful determination of their normative content seems to be possible only with the same approach as used by the SPS jurisprudence. On the other hand, the book also goes beyond pure textual analysis. In line with the interpretative prescriptions of the Vienna Convention on the Law of Treaties (VCLT), it considers the context of a particular provision taken together with the purpose of the SPS Agreement. Special attention is given to other elements, such as subsequent agreement or practice as well as relevant rules of international law applicable in the relations between the parties, which are all identified by the VCLT as relevant elements in the process of interpretation. Moreover, assessing the correctness of SPS disciplines obviously requires recourse to some external benchmarks. In particular, this is provided by legal scholarship and risk science, including the part of research that relates to the problem of risk perception.

[8] Note, however, that the book discusses a number of different documents that were adopted by the SPS Committee whenever they directly related to the substantive obligations of the SPS Agreement.

4 Plan of the book

The book proceeds as follows. Chapter 1 begins with the general discussion on risk and the methods of its regulation. The aim of this is to provide the reader with some useful background for the subsequent analysis of the relevant provisions of the SPS Agreement. After a brief review of different definitions of risk and historical changes in its conceptualization, the chapter concentrates on the problem of risk regulation. Again, a short historical account is combined with the more extensive discussion on contemporary regulatory approaches to risk. In this context, a model of risk analysis is indentified as a dominant paradigm and is addressed in more detail.

Chapter 2 turns to WTO law. It starts with a brief description of the different legal standards that are established by the SPS Agreement, including procedural issues that are relevant for the settlement of SPS disputes. The second part examines two preliminary issues relating to the applicability of the SPS Agreement and its relationship to other WTO instruments that are potentially relevant for the assessment of national risk regulatory actions.

Chapter 3 focuses on the harmonization disciplines of the SPS Agreement. The first section sketches different arguments, which are advanced in a scholarly discussion, on the potential advantages and drawbacks of the international harmonization process. Against this background, the second part provides brief information on those international standard-setting bodies that are explicitly recognized by the Agreement. This particularly includes the Codex Alimentarius Commission, the World Organization for Animal Health and the International Plant Protection Convention. The third section analyses the specific harmonization requirements of the SPS Agreement. Following the approach of the Appellate Body, the chapter distinguishes between three different options that are available to WTO Members: implementing a measure that conforms to an international standard, adopting a measure that is only based on this standard, or completely deviating from it (either upwards or downwards).

Chapter 4 addresses the science-based provisions of the SPS Agreement. The main part of the discussion is concerned with different standards that were elaborated by the case law with respect to risk-assessment disciplines, including the specificity requirement, reliance on scientific minority opinions, the concept of theoretical risks, and the standard of review applicable to adjudication of scientifically complex disputes. On the basis of this analysis, the chapter attempts to identify the ultimate function that is performed by science under the SPS Agreement and to propose some critical observations on the capacity of science to perform this role.

Chapter 5 concentrates on those situations that are characterized by insufficiency of scientific evidence. The problem of the relationship between the precautionary principle and the SPS disciplines serves as a starting point for the subsequent evaluation of the relevant provisions of the Agreement. Article 5.7, being a direct reflection of the precautionary principle in the SPS Agreement, obviously constitutes the core of the discussion, which is further divided into analysis of systemic issues such as applicability of the provision and its substantive requirements. In the concluding remarks, the chapter briefly compares the regime established by Article 5.7 with the precautionary principle and summarizes the direction taken by the SPS case law.

Chapter 6 takes up the problem of the distinction between risk assessment and risk management in the context of the SPS Agreement. After finding that it is legitimate to speak about the risk management dimension of the agreement, it turns to the specific SPS requirements that are relevant for this phase of the risk analysis model. This particularly includes those provisions that pertain to the determination of an appropriate level of protection, options evaluation, implementation, and maintenance of SPS measures, as well as their monitoring and review.

The last chapter draws some overall conclusions on the substantive disciplines of the SPS Agreement, summarizing the previous discussion alongside the three objectives that are served by the book: (i) appraisal of the standards established by the SPS Agreement; (ii) assessment of consistency of the SPS case law; and (iii) identification of the normative content of those provisions of the agreement that have not been yet addressed in the SPS jurisprudence. The above analysis leads me to conclude that despite some defects of the SPS Agreement and corresponding failures of SPS case law, the whole system seems to provide an effective solution for the supervision and management of domestic SPS measures.

1

Risk and Risk Regulation

1.1 Introduction

Risk is an indispensable part of our world. It is a 'necessary evil, inherent to social life, something that must be controlled and whose sources must be regulated'.[1] As human beings we are constantly confronted with risk. Every substance or activity has a certain potential to cause harm: driving a car, walking, eating, or using home appliances are all related to some degree of risk. But what exactly is meant by risk? How can we measure it? What is the role of the modern state in managing and controlling different types of risk?

This chapter attempts to address the above questions. It starts with a brief analysis of the various conceptions of risk. The chapter then looks at risk and its measurement from a historical perspective. The third part considers the role of public authorities in the management of risk. After providing a short historical account, it concentrates on the current regulatory approaches with special attention given to the concept of risk analysis. This section also addresses the problem of uncertainty, which is connected with any probabilistic estimation of risk. The general discussion presented here intends to provide the reader with useful background for the subsequent analysis of the obligations of the SPS Agreement.

1.2 Notion of risk

The concept of risk appeared for the first time in the Middle Ages in the context of maritime ventures.[2] The word 'risk' derives from the early Italian word *riscare*, meaning 'to dare', which denotes potential harm or negative consequence but also a chance, a reward for undertaking risky activity. Thus, risk initially was understood as having both 'good' (winning the game) and 'bad' ('perils that

[1] Sonya Boudia and Nathalie Jas, 'Introduction: Risk and Risk Society in Historical Perspective' (2007) 23(4) History and Technology 317, 325.

[2] A good introduction to the subject of risk may be found in Jakob Arnoldi, *Risk. An Introduction* (2009).

could compromise a voyage'[3]) dimensions. It was the possibility of something happening rather than the indication of potential harmful consequences. This distinction was somehow lost in the course of the last century[4] and the notion of risk started to be used only in the context of negative effects.[5]

As a preliminary remark one also needs to distinguish risk from the related notion of 'hazard'. In everyday language these two terms are frequently considered to be interchangeable. However, in the literature of the subject they have completely different meanings. Hazard is the potential to cause harm which is inherent in a particular object or activity, while risk is the likelihood of that harm. In other words, hazard is the way in which an object or situation may cause harm. A hazard exists when the object, substance, or situation has a built-in ability to cause an adverse effect.[6] Whether hazard will transform into risk depends on many factors such as the degree of exposure (eg low exposure to hazard may result in low risk, no exposure results in no risk), the effects of such exposure, and its likelihood.

In contemporary research, risk is described from various different perspectives that can be broadly labelled as technical, economic, psychological, and sociological approaches.[7] Clearly the summary presented below does not aspire to evaluate the validity or the plausibility of these accounts. It is rather provided as a background for subsequent analysis of the relevant provisions of the SPS Agreement.

The technical paradigm,[8] which is also referred to in the literature as the objectivist school, defines risk as the 'combination of the likelihood (probability) and the harm (adverse outcome, eg mortality, morbidity, ecological damage) resulting from exposure to an activity or substance (hazard)'.[9] To put it a different way, risk in the technical perspective is a combination of the 'expected number of fatalities or injuries likely to arise in the event the risk materialized in harm'.[10] Risk is a real observable phenomenon existing independently from an observer or a bearer. The paradigm also assumes that risk generally can be identified, calculated, and controlled by appropriate risk management tools. Such conceptualization necessarily determines the methods for its assessment, which rely heavily on statistical and actuarial data, scientific information (eg

[3] Deborah Lupton, *Risk* (1999) 5.

[4] Ibid, 8.

[5] Stephen R Graubard, 'Preface to the Issue "Risk"', (1990) 119(4) Daedalus i, v–vi. As noted by Douglas during the twentieth century, the word risk 'has been pre-empted to mean bad risks': Mary Douglas, 'Risk as a Forensic Resource' (1990) 119(4) Daedalus 1, 3.

[6] Lupton (*Risk*), 18; the term 'hazard' can be also equated with 'danger'.

[7] The subsequent discussion is heavily based on Ortwin Renn, 'Concepts of Risk: A Classification' in Sheldon Krimsky and Dominic Golding (eds), *Social Theories of Risk* (1992) 53–5.

[8] Renn distinguishes within the technical paradigm three different approaches: the actuarial approach, assessment of health and environmental risks, and probabilistic risk assessment (Renn ('Concepts of Risk') 58–9).

[9] Jonathan B Wiener and Michael D Rogers, 'Comparing Precautions in the US and Europe' (2002) 5(4) Journal of Risk Research 317, 320.

[10] Jeremy Fraiberg and Michael J Trebilcock, 'Risk Regulation: Technocratic and Democratic Tools for Regulatory Reform' (1998) 43 McGill Law Journal 835, 863.

epidemiological, toxicological data), and calculus of probability. The technical paradigm opposes 'objectively' existing risks with so-called 'subjective' or perceived risks, which are merely psychological or sociological phenomena experienced by individuals or social groups. According to the objectivists, perceived risks, which do not find support in the technical calculations, are clearly erroneous.[11] This also shapes the relationship between risk assessors/managers and the general public. The task of the first group is to educate laypersons about the 'true' nature of risks in order to eradicate irrational fears.

The criticism of the technical paradigm goes in different directions. On the general level, critics argue that risk, rather than being a physical phenomenon, is a social construct. Consequently, technical risk assessment is a 'mental construction that can be checked at best against standards of consistency, cohesion, and internal conventions of logical deduction'[12] and cannot be considered as superior to any other convention (eg risk as perceived by a layperson). Less radical opponents, although they do not question the objective aspects of risk, stress that technical one-dimensional conceptualization is misleading. Risk has a dual nature that combines external (physical) and internal (social and psychological) aspects. The technical paradigm, by disregarding the internal dimension of risk, is simply unable to comprehend its true nature.[13] Critics also question the objectivity of technical risk assessment. Measurability problems, uncertainties in the probability calculations and cause/effect relationship, unreliable historical records, and limitations of science as such, require the application of a number of non-scientific approximations in the process of risk assessment. This in turn greatly reduces any objective value of the final results.[14]

The economic accounts also conceptualize risk as the combination of probability and harm.[15] However, harm is defined not as an absolute adverse effect but rather as a subjective utility (ie satisfaction and dissatisfaction of individuals). As pointed out by Renn, under the economic paradigm, 'the relevant criteria are the subjective satisfaction with the potential consequences rather than predefined list of undesirable effects'.[16] Thus, the aim of risk strategies is to maximize the aggregated utility of society through the reduction of those risks which are considered to produce the highest level of dissatisfaction. Such formulation of

[11] Marjolein BA Van Asselt, *Perspectives in Uncertainty and Risk. PRIMA Approach to Decision Support* (2000) 152.

[12] Andreas Klinke and Ortwin Renn, 'A New Approach to Risk Evaluation and Management, Risk-Based, Precaution-Based, and Discourse-Based Strategies' (2002) 22 Risk Analysis 1071, 1072.

[13] The enumeration is based on the instructive summary presented by Van Asselt, see Van Asselt (*Perspectives*) 153–4.

[14] See generally, Sheila Jasanoff, 'The Songlines of Risk' (1999) 8(2) Environmental Values, Special Issue: Risk 135.

[15] For a general discussion on economic approach to risk, see V Kerry Smith, 'A Conceptual Overview of the Foundations of Benefit-Cost Analysis' in Judith D Bentkover, Vincent T Covello, and Jeryl Mumpower (eds), *Benefit Assessment: A State of the Art* (1986) 13–34.

[16] Renn ('Concepts of Risk') 62.

harm allows integrating societal preferences towards risks and, to some extent, eases the tension in the debate between radical objectivists and constructivists. Critics of this approach, however, highlight difficulties that are inherent in this type of calculation. Particular methodological problems range from estimation of aggregated utility into a single societal welfare function and calculation of delayed costs, through assignment of the monetary value to deaths and injuries, to the evaluation of the distributional effects.[17] Moreover, it is also noted that if laypersons are wrong about the utility of a particular risk strategy, the application of an economic approach may actually result in the decrease of aggregated utility rather that its increase.

The third group of definitions, which can be labelled as a psychological paradigm, defines risk as a function of individual perception. Consequently, risk is not absolute and varies from person to person. As noted by Slovic, risk 'does not exist "out there", independent of our minds and cultures, waiting to be measured.... Although... dangers [of life] are real, there is no such thing as "real risk" or "objective risk" as structure [of quantitative estimates] is subjective and assumption-laden'.[18] Thus, risk is a multidimensional concept, which besides the probabilities, also contains a psychological aspect. In other words, risk as understood under the technical approach, is always mediated through a psychological process of particular individuals (eg risk assessors or laypersons). These theories also stress that the public perception of risk may legitimately differ from the experts' assessment. Risks that are highly ranked by experts can be disregarded by laypersons, while risks which, according to the technical experts, are unlikely, can generate great public concerns. The literature has identified a number of factors which determine the assessment of risk by laypersons:

- heuristic availability (or cognitive availability), which can be defined as the ease with which an instance is brought to mind; events that can be easily recalled by a person (eg due to personal experience, familiarity, media coverage) are considered to be more likely than those that are harder to recall;[19]
- perceptibility—risks that are perceptible by human senses are considered to be less serious than risks that are not (eg chemical substances);
- voluntarity—risks that are voluntarily taken (eg skiing) are perceived to be more acceptable and less risky than involuntary risks (eg risks posed by food preservatives);[20]
- controllability—risks that are perceived as controllable (eg driving a car) are perceived as less serious and less likely to materialize than risks with

[17] Robert Baldwin, Bridget Hutter, and Henry Rothstein, 'Risk Regulation, Management and Compliance. A Report to the BRI Inquiry' (available at <http://www.bristol-inquiry.org.uk, last visted 20 August 2009).

[18] Paul Slovic, 'Trust, Emotion, Sex, Politics, and Science: Surveying the Risk-Assessment Battlefield' (1999) 19(4) Risk Analysis 689, 690.

[19] Lupton (*Risk*) 20.

[20] Paul Slovic, *The Perception of Risk* (2000) 282.

respect to which there is no perceived control (eg potential risks from genetic modifications);

- source of risk—man-made risks (eg environmental pollution) are considered to be more serious than naturally occurring risks (eg floods);[21] this difference arguably results from the catastrophic potential of some man-made risks, their irreversible nature, and asymmetries in the societal distribution of potential costs and benefits related to the activity posing risk;[22]
- catastrophic nature—risks with potential catastrophic consequences (eg aeroplane accidents) are perceived as more serious, irrespective of their probabilities;[23]
- distribution of risks and benefits—risks that are distributed unequally among different groups in society are perceived as more serious (eg if one group benefits while the other bears the costs of risk related to a particular activity); and
- potential benefits which are connected with an activity posing the risks—people tend to downplay the extent of risks connected with those activities which may provide some benefits (eg importers of agricultural products will tend to downplay phytosanitary risks, which may be connected with such importation).[24]

Moreover, laypersons appear to think about risk (and to assess it) in 'black and white' categories rather than in terms of probability. As one scholar put it: 'either the risk is significant (so it is treated as if it were something like 50/50) or it is negligible (so it gets treated as if it were zero)'.[25] This also reflects the more general problem of understanding the probability. The results of research indicate that people have difficulty in conceptualizing risks which are expressed in probabilities (particularly if the chances are low).[26] Moreover, people (including experts) are also subject to confirmation biases, meaning that once an individual reaches a certain view, she has a tendency to ignore additional information that conflicts with her view.[27] Finally, the layperson's distrust of experts and science is also an

[21] Anna-Christina Blomkvist, 'Psychological Aspects of Values and Risks' in Lennart Sjöberg (ed), *Risk and Society* (1987) 107; as it is discussed below, this corresponds closely with the Beck's concept of risk society, the society which is predominantly concerned with technological risks.

[22] Miley W Merkhofer, *Decision, Science and Social Risk Management* (1987) 5.

[23] Paul Slovic, Baruch Fischoff, and Sarah Lichtenstein, 'Facts versus Fears: Understanding Perceived Risk' in Daniel Kahneman, Paul Slovic, and Amos Tversky (eds), *Judgment under Uncertainty: Heuristics and Biases* (1982) 465.

[24] Michael Siegrist and George Cvetovich, 'Perception of Hazards: the Role of Social Trust' (2000) 20(5) Risk Analysis 712, 714.

[25] Howard Margolis, *Dealing with Risk: Why the Public and the Experts Disagree on Environmental Issues* (1996) 85.

[26] Vincent T Covello, Detlof von Winterfeldt, and Paul Slovic, 'Communicating Scientific Information about Health and Environmental Risks: Problems and Opportunities from a Social and Behavioral Perspective' in Vincent T Covello, Lester B Lave, Alan Moghissi, and VR Uppuluri (eds), *Uncertainty in Risk Assessment, Risk Management and Decision-making* (1984) 226.

[27] Paul R Hunter and Lorna Fewtrell, 'Acceptable Risk' in Lorna Fewtrell and Jamie Bartram (eds), *Water Quality: Guidelines, Standards and Health* (2001) 215.

important factor which may explain the frequent divergence between technical and popular assessment of risk.[28]

A number of scholars claim that risk perception is in the first place a cognitive deficiency as laypersons clearly over- or under-estimate risks.[29] This part of scholarship accepts the existence of 'real' risk in the physical world against the risk that the lay person perceives. As summarized by Bradbury, 'in examining the individual's response to risk this research provides a subjectivist interpretation within a realist [objectivist] paradigm'.[30] There is, however, another group, which claims that perception of risk by laypersons is not a psychological deficiency but rather a rival rationality.[31] According to these scholars, 'difference in expert/non-expert risk judgments, if they exist, do not imply that one set of judgments has greater validity than another'.[32] Such understanding is usually combined with an expression of disappointment with the traditional scientific approach to assessment of risk and its management. As noted in the literature 'the faith of previous generations in the ability of science to provide answers has turned to doubt, partly because scientists themselves are not in agreement, partly because the answers science now gives are much more complex and contingent'.[33] Those who believe that risk perception is rival rationality often stress that laypersons may actually conceptualize risk more broadly than experts because they take into account elements which are normally excluded from technical assessment,[34] such as distribution of risk between different social groups or the catastrophic nature of certain risks. Yet another group of scholars emphasizes that experts, like laypersons, are not immune to the epistemological biases. Consequently, one should not unconditionally favour their assessment over the popular perception of risk.

The fourth group, covered by the socio-cultural paradigm, consists of a great array of different approaches.[35] The common feature is an emphasis on 'the social processing of the risk',[36] meaning that risk is socially defined and subsequently socially filtered. Consequently, risk is not an objective phenomenon but is 'constantly construed and negotiated as a part of the network of social interactions

[28] Mary Douglas and Aaron Wildavsky, *Risk and Culture: An Essay on the Selection of Technical and Environmental Dangers* (1982) 10.

[29] See eg, Cass R Sunstein, *Laws of Fear* (2005).

[30] Judith A Bradbury, 'The Policy Implications of Differing Concepts of Risk' (1989) 14(4) Science, Technology and Human Values 380, 384.

[31] Sunstein (*Laws of Fear*) 35–63.

[32] Gene Rowe and George Wright, 'Differences in Expert and Lay Judgments of Risk: Myth or Reality' (2001) 21(2) Risk Analysis 341, 356.

[33] Christopher Hood, Henry Rothstein, and Robert Baldwin, *The Government of Risk: Understanding Risk Regulation Regimes* (2001) 21.

[34] Margolis (*Dealing with Risk*) 35.

[35] A good overview is provided by Lupton, who distinguishes two typologies of socio-cultural definitions: (i) cultural/symbolic, risk society and governmental perspectives; or (ii) structuralists, postructuralists, phenomologists and psychoanalytic approaches (Lupton (*Risk*) 24–8). See also, Renn ('Concepts of Risk') 67–76.

[36] Renn ('Concepts of Risk') 73.

and the formation of meaning'.[37] Both science, which constitutes a basis for technical risk assessment, as well as experts who perform such assessments, are subject to specific socio-cultural conditions. The socio-cultural approaches stress that the technical paradigm does not recognize that risk is not objective and cannot be correctly described outside the social system of beliefs. On the other hand, psychological theories tend to see human beings outside of the social environment which necessarily mediate their judgments and reactions.[38] Various social theories represent different levels of radicalism—from the very weak, which only ask for the inclusion of social factors in the process of technical risk assessment, to stronger versions, which deny any objective value to scientific risk assessment. Opponents of the socio-cultural approach criticize its relativistic character and the overemphasis on the social elements in the concept of risk. Moreover, the application of extreme versions of the socio-cultural approach, as well as purely technical accounts, also seem to provide an oversimplified description of risk.[39] Last but not least, those theories also fail to explain the obvious success of science and technical evaluation in managing and controlling many types of risk.

Against this background, it is interesting to ask a preliminary question on the conceptualization of risk in the SPS Agreement. The Agreement as such does not provide any definition of risk. That is addressed indirectly through the characterization of risk assessment, which is understood as either the evaluation of the likelihood of the entry, establishment, or spread, of a pest or disease within the territory of an importing WTO Member according to the SPS measures which might be applied, and of the associated potential biological and economic consequences (quarantine risks); or the examination of the potential for adverse effects on human or animal health arising from the presence of additives, contaminants, toxins, or disease-causing organisms in food, beverages, or feedstuffs (food-borne risks).[40] There is no doubt that the SPS Agreement relies heavily on the technical paradigm, under which risk is construed in terms of likelihood and adverse outcome. Note also that similar definitions of risk are provided by the international standard-setting bodies that are explicitly recognized by the Agreement. The Codex Alimentarius Commission (Codex) defines risk as 'a function of the probability of an adverse health effect and the severity of that effect, consequential to a hazard(s) in food'.[41] The World Organization for Animal Health (OIE) speaks about 'the likelihood of the occurrence and the likely magnitude of the

[37] Lupton (*Risk*) 29.

[38] Ibid, 23 (giving the example of members of social groups which are less powerful and more concerned with risks than members of more powerful groups, eg non-Whites *v* Whites, women *v* men).

[39] Andreas Klinke and Ortwin Renn, 'A New Approach to Risk Evaluation and Management, Risk-Based, Precaution Based, and Discourse Based Strategies' (2002) 22 Risk Analysis 1071, 1076; see also generally, Eugene A Rosa, 'Metatheoretical Foundations for Post-Normal Risk' (1998) 1(1) Journal of Risk Research 15–44.

[40] Annex A(4) of the SPS Agreement.

[41] Codex, *Procedural Manual* (18th edn, 2008) 73.

consequences of an adverse event to animal or human health in the importing country during a specified time period'.[42] For the International Plant Protection Convention (IPPC), risk is also a product of likelihood (of the pest's entry, establishment, and spread) and magnitude of the impact (the economic, environmental, and social impacts that would potentially be attributable to the pest).[43] This should not come as a surprise since technical risk assessment is a dominant approach in the field of health and environmental regulation. On the other hand, a closer analysis of the SPS Agreement and the corresponding case law also reveals the influences of other paradigms. As will be discussed in more detail in Chapter 6, they may be identified within those provisions which discipline risk management activities.

1.3 Measurement of risk, scientific developments, and social change

The measurement of risk is a relatively new idea in the history of human kind. For a long time, chance was attributed to mystical forces (gods) while the future was described and understood in terms of fate and *fortuna*. Nevertheless, even without understanding the nature of risk as it is conceptualized today, people were able to employ different strategies for avoiding and mitigating potential negative consequences. Some methods, such as religious sacrifices or special magical ceremonies, may appear to us as rather exotic. Others surprisingly resemble contemporary risk management strategies (eg trial and error methods or observations made on surrogates). Of course, risk was not measurable as humankind simply lacked the necessary tools and methods for making calculations. This only started to change with the rediscovery by Renaissance Europe of a system of Arabic numbers and the subsequent development of a mathematical theory of probability. The roots of the theory can be traced back to the scientific problems posed by games of chance but its first scientific formulation came only in the seventeenth century with Blaise Pascal and Pierre de Fermat[44] and was further elaborated by Jacob Bernoulli[45] and Abraham De Moivre.[46] These developments provided a theoretical foundation for the emergence of a technical conception of risk as a combination of probability and harm (or more broadly occurrence

[42] OIE, *Terrestrial Animal Health Code* (18th edn, 2009) Glossary.

[43] International Standard for Phytosanitary Measures (ISPM) No 05, *Glossary of Phytosanitary Terms* (2007) 13.

[44] For a more detailed discussion on Pascal and the theory of probability, see, Oystein Ore, 'Pascal and the Invention of Probability Theory' (1960) 67(5) The American Mathematical Monthly 409.

[45] Bernoulli formulated the Law of Large Numbers (initially referred to as Bernoulli's Theorem); the law provides that as the numbers of independent repetitions of the experiment grow, the average of the observed outcomes approaches the average of all possible outcomes.

[46] De Moivre introduced the concept of normal distribution (bell curve) and standard deviation.

of unwanted event). Complicated statistical calculations, which became possible with the birth of the probability theory, were rapidly used in different spheres of human activity. This also included the protection of human health and much later the protection of environment.

As far as human health is concerned, equally important were scientific developments relating to identification of causal links between adverse health effects and different types of hazardous substances. Early strategies consisted in trial and error methods where individuals were simply drawing conclusions on the effects of a particular hazard from their personal experience (eg testing strange plants). These uncomplicated techniques were subsequently supplemented with epidemiological observations (ie by observing epidemiological effect in different but related situations) and experiments on surrogates such as animals (eg feeding animals with new foods).[47] Interestingly, people were able to establish links between certain activities or substances and potential adverse health effects even without understanding the underlying processes. The qualitative change came only with the seventeenth and eighteenth centuries, which witnessed a number of ground-breaking discoveries in the fields of medicine and epidemiology as well as in general knowledge of the natural world (eg the existence of bacteria). These developments were also accompanied by the sophistication of instrumentation and experimental techniques. Nevertheless, as noted in the literature, 'it was not until the work of Pasteur in the late nineteenth century that scientists first began to comprehend adequately the concept of infection or the causal relationship between the environment and biological agents of infectious disease'.[48] The advances in science continued and even intensified in the course of the twentieth century. Technological progress enabled the exploration of the world to an extent not possible before (eg on the micro-level of genes). This combined with the further sophistication of health risk management strategies, and techniques resulted, at least in developed countries, in a considerable increase of average human life expectancy and overall better health.

Paradoxically, the last century also witnessed another important change in thinking about risk. While previously people were predominantly concerned with the risks coming from nature, nowadays they seem to concentrate more on human produced risks resulting from industrial progress and technological innovation. This shift in thinking has provoked some scholars to argue for the emergence of a new sociological phenomenon that is conventionally labelled as 'risk society'. The underlying idea is that risks, particularly man-made risks, are a characteristic feature of contemporary society. Ulrich Beck,[49] one of the

[47] Vincent T Covello and Jeryl Mumpower, 'Risk Analysis and Risk Management: An Historical Perspective' (1985) 5(2) Risk Analysis 103, 107.

[48] Ibid, 108.

[49] Ulrich Beck, *Risk Society: Towards a New Modernity* (1992); other important works include: Patrick Lagadec, *Major Technological Risks: An Assessment of Industrial Disasters* (1982); Anthony Giddens, *Modernity and Self Identity: Self and the Society in the Late Modern Age* (1991).

founders of this new theorem, compared these types of risk (referred to by Beck as 'manufactured risks') to the traditional ones that are rooted in nature, and identified the first group as a main preoccupation of modern society. According to Beck, manufactured risks are induced by the process of modernization, which includes scientific, technological, and economic expansion with all its positive and negative consequences for society. Science occupies in this framework an ambiguous position. On the one hand, it is responsible for risk creation through technological progress. On the other hand, manufactured risks require even further research and more science in order to be addressed and managed.[50] This is accompanied by the growing erosion of the public trust in science. Beck himself seems to concur with this general distrust as he also questions the capability of science to address new 'manufactured' risks as being largely unpredictable, incalculable, and therefore unmanageable.

This shift from industrial modernity to a risk society (Beck also uses the term 'reflexive modernity')[51] obviously affects the social structure of contemporary society, as it defines the position of its members, in contrast to traditional industrial society, in terms of risk distribution instead of wealth. Manufactured risks are global, in the sense that they may affect people in any part of the world and belonging to any social class. Risk society transcends, therefore, the traditional boundaries, both geographically (world risk society) and socially, and produces its own inequalities.[52] Interestingly, this new distribution of costs and benefits is also combined with so-called organized irresponsibility. Manufactured risks are not attributable to anyone and in consequence nobody can be accountable for them.

1.4 Risk as a subject of regulation

Different forms of interventions of public authorities in the field of human health and environmental risks have an equally long history. One can find examples of governmental efforts to reduce risks or to ensure their distribution in socially acceptable ways as early as in the ancient Egypt or Babylon. Although this involvement has become more sophisticated over time, the underlining idea still remains the same—there are certain risks which, in order to be properly managed, require the engagement of larger community groups.

[50] Carlo C Jaeger, Thomas Webler, Eugene A Rosa, and Ortwin Renn, *Risk, Uncertainty and Rational Action* (2001) 112.

[51] Reflexive modernity may be defined as modernity, which is self-critical in its assessment (George Ritzer (ed), *The Blackwell Companion to Major Contemporary Social Theorists* (2003) 260).

[52] Beck (*Risk Society*) 21; this does not mean that there are no winners and losers of the new risk distribution; the point which was made by Beck is that the risk transgresses previous social structures.

1.4.1 Risk regulation in the historical context

The interventions of public authorities were initially directed against the natural risks and concentrated on the preventive actions with respect to natural disasters and epidemic diseases.[53] This included different types of action, some of them being irrational from a contemporary point of view (eg public sacrifices of a maiden at the beginning of the year or before important challenges such as war), while others resemble modern strategies (eg construction of dams and canals to prevent floods, quarantine requirements in ancient ports and cities, isolation of infected areas, etc). Besides prevention, public authorities also provided *ex post* assistance to victims, once a particular risk materialized into harm (eg 'tax reductions' in ancient Egypt in the periods of famine).[54] Probably to the surprise of many, one may also find examples of governmental interventions aimed at the mitigation of made-man risks. As early as Roman times, laws regulating disposal of wastes and sewage were adopted, despite the fact that the exact causal relationship between their improper disposal and potential health problems was not really understood. A separate part of public efforts was dedicated to the quality and safety of food. Certain religious requirements with respect to food (eg prohibiting specific food products) may be seen as a simplified counterpart of modern food safety regulations.[55] These early forms remained unchanged for centuries to come (with a noticeable decline after the collapse of the Roman Empire). The sewage and waste management requirements adopted in the eighteenth century are strikingly similar to the risk management strategies of ancient Rome.

The above does not mean that public management of risks was only performed by governmental authorities. As nowadays, societies also relied on other strategies, which were implemented side by side with governmental actions. Insurance is a good example. The institution was predominantly used in the context of loans for commercial maritime transitions.[56] A form of life insurance, which in exchange for premiums, covered the expenses connected with the burial of the insured, was another common type in the ancient world. Insurance contracts reappeared again in medieval ages as a risk management tool in the marine trade. Since then, they have become more and more popular (eg the emergence of insurance industry in seventeenth century England) and cover an expanding number of risks. Liability law was yet another form of risk prevention strategy. As explained by Covello and Mumpower 'people [had to] conform to a standard of reasonable

[53] Covello and Mumpower ('Risk Analysis') 111. [54] Ibid.

[55] See generally, Marsha A Echols, *Food Safety and the WTO. The Interplay of Culture, Science and Technology* (2001) 13–28.

[56] Covello and Mumpower ('Risk Analysis') 109 (giving as an example buttomry (loan) contracts in Babylonia for vessels and cargo, which included a risk premium for the chance of loss and consequent cancellation of debt).

conduct (eg cleaning their chimneys, disposing of waste products) or face the prospect of being liable and successfully sued for damages'.[57]

A qualitative change in the public management of health risks came only in the nineteenth century. The industrial revolution required a more sophisticated regulatory response that would be capable of addressing a number of new types of risk.[58] The introduction of novel production methods and technologies, apart from its positive effects such as increased productivity or lower unit costs of final products, also had negative consequences in terms of the augmentation of different risks (eg use of machines in the workplace was connected with a higher accident rate among employees). The industrial revolution also coincided in time with the emergence of socialist movements, which insisted on the responsibility of the governments for the well-being of the citizens, including assurance of good health conditions for employees. Moreover, the nineteenth century was a period of the formation of the modern regulatory state. One element of this process was the extension of state prerogatives to many new areas that had been previously considered as falling outside the competence of government. At the same time, development of science provided public authorities with necessary tools for more efficient management of risks.[59] The regulatory spectrum indeed became very broad and covered different types of measures, ranging from occupational health and safety regulations and compensation schemes, to consumer protection legislation relating to food, pharmaceuticals, and general product safety.

1.4.2 Contemporary regulatory approaches to risk

The regulatory trends set out during the industrial revolution were followed in the course of the twentieth century. The justification for a greater involvement of public authorities was provided by an accelerated rate of technological change, which in turn resulted in the enormous increase in the physical and temporal scale and complexity of risks (eg risks for human health resulting from exposure to chemicals). Obviously, these types of risks could not be addressed by individuals and required governmental intervention. Governments, possessing greater human and financial resources, superior information, and perhaps broader perspective, were better equipped than individual members of society to manage these industrial risks. Secondly, as Beck explained, one of the features of modern societies is its preoccupation with manufactured risks. Consequently, more expansive regulatory actions might be seen as a governmental response to the increasing demands of the citizens. Thirdly, the government also became an important

[57] Ibid, 110 (giving as an example the Code of Hammurabi and the Old Testament).

[58] One of the first initiatives included the Steamboat Inspection Act adopted by the US Congress in 1838. The Act regulated boiler testing, inspection, and liability and was adopted in response to the series of boiler explosions on steamboats which led to thousands of injuries and fatalities during the early nineteenth century (ibid, 114).

[59] Boudia and Jas ('Introduction') 322.

producer of risks 'through its sponsorship of scientific and technological research and development'.[60] A sense of justice required, therefore, that the government also attempted to mitigate these risks. As a result of these different processes, in the majority of contemporary modern societies the problem of environmental and human health risks is primarily addressed through governmental regulations.[61]

An important feature of modern risk regulation is that it tends to address risks before they materialize into harm. It aims at the prediction and prevention of risk rather than compensation for damages.[62] In other words, risk regulation provides *ex ante* solutions.[63] A good example is provided by pre-market regulations, where potential hazards (eg chemical substances) are regulated before entering the market (eg in the form of pre-market approvals).[64] Government regulatory actions can take different forms, varying from traditional command and control measures (ie setting certain standards of conduct and providing sanctions for their violations) through licensing systems (ie allowing marketing of products subject to premarket approvals) to disclosure requirements (ie labelling). Governments can also attempt to control and manage risks through economic incentives (eg different taxes for leaded and unleaded petrol) or direct involvement through their agencies. In practice, regulators combine different strategies to address a particular risk.[65]

One of the most difficult tasks in the regulatory process is selection of risks that need to be regulated. As observed in the literature 'since there are more risks in the world than society could handle at the same time, risk management always implies the task of setting priorities'.[66] This setting of priorities includes a significant number of non-scientific judgments, which reflect the values important to a particular society. Certain risks may be considered worth addressing in one community, while in the other they may be totally disregarded.[67] Of course, due to the high value that is universally assigned to human life and health (and to a lesser extent, to the life and health of animals and plants), the level of acceptable risk will be partially determined by the outcome of scientific risk assessment, which provides information on likelihood of adverse effects. Equally important are available financial resources that may be dedicated to the mitigation of particular risks. In wealthy Western societies even small risks may trigger regulatory action, whereas in poorer developing countries regulators will be frequently forced to

[60] Ibid, 117. [61] Covello and Mumpower ('Risk Analysis') 116–17.

[62] Jacqueline Peel, 'Risk Regulation under the WTO SPS Agreement: Science as an International Normative Yardstick?' Jean Monnet Working Paper No 2002/04, 5.

[63] Fraiberg and Trebilcock ('Risk Regulation') 841.

[64] Note that regulation can also have post-market character; in this case, the intervention of the authorities occurs after the introduction of a potential hazard to the market.

[65] For a more detailed discussion on different regulatory strategies, see Baldwin et al ('Risk Regulation') 6–10.

[66] Klinke and Renn ('A New Approach to Risk Evaluation') 1074.

[67] For an interesting review of the strategies employed in the process of setting acceptable levels of risk, see Hunter and Fewtrell ('Acceptable Risk') 208–18.

tolerate high risks as they lack the resources necessary for their management. On the other hand, there are also a number of non-scientific factors that influence the final determination of the 'safety' threshold. Cultural and psychological elements are particular important in this context. As already mentioned, risks that are taken voluntarily by individuals (eg climbing) or which are historically conditioned (eg eating some type of traditional foods) have less chance to become a subject of regulation and the establishment of an acceptable level of risk may be left entirely to individuals. In other cases, the level of acceptable risk will be influenced by the benefits that are connected with a particularly risky activity. The opposite is also true. Risks that are not significant in terms of probability and potential adverse effects may still provoke intense regulatory action. In particular, this may result from the need to protect certain disfavoured social groups (ie those who are particularly exposed to risks) or may simply be a response to the demands of the general public, which in turn can be amplified or even manipulated by the media. Some regulatory measures may also be produced as a consequence of pressure from powerful domestic interests groups.

One may distinguish two broad tendencies of modern risk regulation.[68] The early approach was more concerned with hazards than risks (the so-called hazard-based approach).[69] Thus regulators, after identifying hazards, usually proceeded with regulatory actions without really estimating relevant risks. There were several reasons for this. Limited scientific knowledge and a lack of precise instrumentation, which prevented calculation of risk, were important factors. There was also a common belief that risk might and should be reduced to zero level. Since risks were incalculable, while governmental agencies aimed at the complete elimination of potential harm, it was rational to adopt and maintain measures solely on the basis of the hazard identification. However, it soon appeared that such an approach, despite being very costly, did not deliver expected health benefits.

[68] A different classification of risk regulatory models was also proposed by Elisabeth Fisher. She rightly observes that risk regulation is predominantly an administrative enterprise. Based on the general theory of public administration, she distinguishes the rational–instrumental and deliberative–constitutive paradigms that dominate modern risk regulation. The first one is defined as a model where public administration acts as an instrument of the legislature and is constrained in its discretion as much as possible (eg by required methodology such as risk assessment or cost-benefit analysis); risk and uncertainties are perceived as manageable with the use of science and expertise. Under the deliberative–constitutive model, public administration is problem-oriented and enjoys a wider margin of discretion in addressing risks. Risk problems are perceived as complex socio-political disputes in which complexity, uncertainty, and socio-political ambiguity dominates. This in turn requires not only the employment of scientific methods but also the involvement of other actors (ie subjects of the particular regulations) in the process of deliberation, which enables the reaching of final decisions (Elisabeth Fisher, 'Beyond the Science/Democracy Dichotomy: The World Trade Organisation Sanitary and Phytosanitary Agreement and Administrative Constitutionalism' in Christian Joerges and Ernst-Ulrich Petersmann (eds), *Constitutionalism, Multi-Level Trade Governance and Social Regulation* (2006) 344).

[69] For example, the International Agency for Research on Cancer and the US Environmental Protection Agency originally classified carcinogens based only on the finding of hazard (cf Lakshman D Guruswamy, 'Sustainable Agriculture: Do GMOs Imperil Biosafety' (2002–03) 9 Indiana Journal of Global Legal Studies 461, 484).

As a result one could observe a gradual shift towards the second approach that concentrates on actual risks rather than mere hazards (the so-called risk-based approach).[70]

The shift between those two approaches was stimulated by the emergence in early 1970s of the concept of risk analysis. Its first systematic formulation came with the publication by the National Research Council (NRC) of the famous Red Book,[71] which clarified and explained the constitutive elements of the concept. Risk analysis as such may be defined as a systematic process of identifying hazards, and the subsequent collection and evaluation of information leading to actions (or inactions) of risk managers. The process is informed by a continuous information exchange between different actors (ie scientists, risk managers, interest groups, and the general public). In other words, risk analysis may be described as a structuralized regulatory response to risks,[72] consisting of three components: risk assessment, risk management, and risk communication.[73] The aim of the functional distinction between scientific risk assessment, which was understood at the beginning only in quantitative terms, and political risk management[74] is to establish and maintain the integrity of scientific advice. In theory, value judgements can be isolated and contained in the management phase, while risk assessment is concerned (entirely or predominantly) with de-contextualized scientific findings. This, in turn, should guarantee the objectivity of the decision-making process and improve the quality of the regulatory outcomes. I will come back to these assumptions later in this chapter.

It is interesting to note that in parallel to the development of risk-based approaches one may also observe, starting from the 1990s, the rebirth of the hazard paradigm. Some scholars, decision-makers, and environmentalists argue that a risk-based approach is actually an inefficient model for dealing with risks posed by modern technology. The advocates emphasize the limitation of scientific knowledge and irreducible uncertainty which result from inherent complexities of natural systems such as the environment and the human body.[75] This scepticism, accompanied by our inability to foresee all potential effects and consequences

[70] This is not to say that a risk-based approach was not present before, obviously it was. My point here is that one may observe a qualitative change, which took place in the late 1960s and at the beginning of the 1970s from a hazard-based to a risk-based approach.

[71] National Research Council, Committee on the Institutional Means for Assessment of Risks to Public Health, *Risk Assessment in the Federal Government: Managing the Process* (1983).

[72] Daniel M Byrd III and C Richard Cothern, *Introduction to Risk Analysis: A Systematic Approach to Science-Based Decision-Making* (2000) 6.

[73] Vlasta Molak (ed), *Fundamentals of Risk Analysis and Risk Management* (1997) 4–6; see also, Codex (*Procedural Manual*) 68; OIE, Terrestrial Animal Health Code, Article 2.1.1.

[74] Cf European Commission, Health & Consumers Protection Directorate-General, *First Report on the Harmonisation of Risk Assessment Procedures, Part 1: The Report of the Scientific Steering Committee's Working Group on Harmonisation of Risk Assessment Procedures in the Scientific Committees Advising the European Commission in the Area of Human and Environmental Health* (2000) 18.

[75] John S Applegate, 'The Taming of the Precautionary Principle' (2002–03) 27(13) William and Mary Environmental Law & Policy Review 13, 36–8.

(epistemological pessimism),[76] as well as irreversibility of some changes, make them believe that in many cases the most useful strategy of avoiding potential harm is to resign from a particular technology or activity (specifically if technology may potentially produce serious damages). As summarized in the literature, 'hazard-based... [paradigm] accepts error on the side of overprotection as preferable to the reverse'.[77]

The risk literature also distinguishes between the so-called precaution-based approach and risk-based or science-based approaches.[78] As will be discussed in more detail in Chapter 5, the precaution (and more precisely the precautionary) principle provides that regulators, rather than awaiting certainty with respect to the actual existence of risk, should act in anticipation of potential environmental and human health harm, in order to prevent it. The precaution-based approach fits well with the above classification of risk- and hazard-based regulatory models. Since there is no uniform definition of the precaution, its various formulations will be located in different places on the continuum between risk- and hazard-based approaches. The stringent versions are closer to the hazard-based extreme as they also question the ability of science to measure and assess complex environmental and human health risks. Consequently, a mere identification of hazard may be sufficient to trigger the regulatory response. The softer versions of the precaution principle are to be located closer to the risk-based end. In such cases, a regulatory action is preceded by scientific evaluation of hazard combined with an attempt to assess potential risks. Only if such examination identifies irreducible uncertainties confirming at the same time the plausibility of claims as to the potential existence of risks, would decision-makers be allowed to take precautionary actions.

1.4.3 The concept of risk analysis

The concept of risk analysis requires some additional discussion. As the practice shows, the concept constitutes a dominant risk regulatory paradigm in the majority of, if not all, developed countries. Risk analysis is used in different areas, including sanitary, phytosanitary, environmental,[79] and engineering

[76] Note the similarities between such epistemological pessimism and the idea of risk society as proposed by Beck.

[77] Applegate ('Taming') 38.

[78] See eg, Jacqueline Peel, Rebecca Nelson, and Lee Godden, 'GMO Trade Wars: The Submissions in the EC—GMO Dispute in the WTO (2005) 6 Melbourne Journal of International Law 141; see generally, Timothy O'Riordan and James Cameron (eds), *Interpreting the Precautionary Principle* (1994).

[79] In the field of environmental protection another important tool is an environmental impact assessment (EIA), which can be defined as the process of identifying, predicting, evaluating, and mitigating the biophysical, social, and other relevant effects of development proposals before taking major decisions and making commitments (International Association for Impact Assessment, Principle of Environmental Impact Assessment Best Practice (1990), available at <http://www.iaia.org/modx/assets/files/Principles%20of%20IA_web.pdf> last visited 23 July 2009). EIA in many respects is similar to risk assessment.

risks. Interestingly, it is also recognized as a central concept by all international standard-setting bodies that are explicitly referred to by the SPS Agreement.

As already mentioned, risk analysis may be conceptually divided into three phases: risk assessment, risk management, and risk communication. All those terms can be defined in various different ways.[80] The prevalent view describes risk assessment as a process of probabilistic estimation[81] of the potential adverse health or environmental effects of a substance, process, action, or event, determined according to scientifically plausible methods.[82] The goal of risk assessment is to provide risk managers with the scientific estimates necessary in order to make a regulatory decision. Science plays a predominant role in that phase—it 'identifies areas for regulatory action, while limiting the field of possible responses'.[83] Under the traditional risk analysis framework, science is deemed to provide political bodies responsible for risk decisions with neutral information on the expected damage or injury from exposure to a particular risk agent. On the other hand, the political choices are confined in the risk management phase.

The assessment as such is conceptualized as a four-step process, consisting of hazard identification, hazard characterization (dose-response assessment), exposure assessment, and risk characterization.[84] Although the overall methodological structure of risk assessment remains the same across different fields, its specific elements may change between various types of risk.[85] The subsequent discussion is predominantly based on the microbiological and toxicological risk assessments.

The first phase of risk assessment aims to identify whether a particular agent (eg chemical substance or microorganism) has a built-in capacity to cause adverse effects.[86] This also comprises a qualitative description of potential adverse effects and initial determination of probable targets (eg humans). Risk assessors also need to define testing methods for the subsequent evaluation of risk (eg *in vivo* or

[80] See eg, William W Lowrance, *Of Acceptable Risk: Science and the Determination of Safety* (1976) 8; NRC (*Risk Assessment*) 18; Vincent T Covello and Miley W Merkhofer, *Risk Assessment Methods: Approaches for Assessing Health and Environmental Risk* (1993); David A Wirth, 'The Role of Science in the Uruguay Round and NAFTA Trade Disciplines' (1994) 27 Cornell International Law Journal 817, 833–4.

[81] Probability can be defined as an expression of confidence that an event will occur: a probability of one indicates that a particular event is certain to occur; on the other side of continuum there is probability of zero, which indicates that the event cannot occur. Probability may be expressed both quantitatively and qualitatively.

[82] The Presidential/Congressional Commission on Risk Assessment and Risk Management, *Final Report, Volume 2: Risk Assessment and Risk Management in Regulatory Decision-making* (1997) 2.

[83] Jeffery Atik, 'Science and International Regulatory Convergence' (1996–97) 17 Northwestern Journal of International Law and Business 736, 737.

[84] See eg, Codex (*Procedural Manual*) 73.

[85] Ortwin Renn, *Risk Governance. Coping with Uncertainty in a Complex World* (2008) 69 (noting that 'although there are clear differences in structuring the assessment process depending upon risk source and organisational culture, there is an agreement on . . . core components of risk assessment').

[86] Byrd and Cothern (*Introduction*) 5.

in vitro studies, laboratory, or field observations). Depending on the knowledge about an agent, the required research may be minimal, as in case of traditional pathogens, or very demanding, for instance in case of emerging pathogens or a new, undocumented chemical substance.[87]

Hazard characterization concentrates on determining whether the exposure to the agent at any level can cause any adverse effects. This is usually done through a dose-response assessment (both for chemical and microbiological agents). The assumption underlying the dose-response model provides that there are no biological effects below a specific concentration threshold.[88] On that basis, it is possible to calculate exposure for human (or animals) which is safe (in the risk assessment jargon—without significant adverse health effects). In chemical (toxicological) risk assessment, risk assessors frequently use the concept of no-observed-adverse-effect level (NOAEL) which is a 'safe' sub-threshold dose.[89] The data on dose-response relationship in humans may be derived from human epidemiological studies, or more often from animals' studies or experimental models. Moreover, in order to obtain meaningful results, experiments on the animals are usually performed on much higher doses than are actually present. Such data are subsequently extrapolated to lower doses as well as to humans. In the process of extrapolation, risk assessors use special safety factors in order to account for differences between species and different groups within a population (eg children or pregnant women may constitute particularly sensitive groups).[90] On that basis risk assessors derive the level of human exposure that will be without adverse effect (acceptable daily intake (ADI), tolerable daily intake, NOAEL, etc). Note that this phase of risk assessment poses some particular methodological difficulties resulting from physiological differences between species, extrapolations from high to low doses as well as from small sample groups to the general population. As far as microbiological risks are concerned, the potential problems are connected with the assessment of pathogenicity which may change over time, for instance due to mutations or interactions with other bacteria, and great variation in hosts' susceptibility to pathogenic agents.[91] If a dose-response relationship cannot be established, in the sense that there is no threshold or there is not enough information on the relationship, a qualitative assessment of potential effects is performed.[92]

[87] European Commission, Health & Consumers Protection Directorate-General (*Report on the Harmonisation of Risk Assessment Procedures*) 51–63.

[88] With respect to contaminants in food such as aflatoxins, the assumption is that there is no threshold, meaning that any exposures to a substance may result in adverse effects in a subject.

[89] WHO, *Principles for the Assessment of Risks to Human Health from Exposure to Chemicals, IPCS Environmental Health Criteria* (1999) 210. Note that the other term, NOEL (no-observed-effect level), is also used; the problem with this concept consists in the fact that not all observed effects (eg certain changes in metabolism) are necessarily adverse.

[90] WHO/FAO, *Report of the Joint FAO/WHO Expert Consultation, Application of Risk Analysis to Food Standard Issues*, WHO/FNU/FOS/95.3, 15.

[91] Ibid, 22.

[92] European Commission, Health & Consumers Protection Directorate-General (*Report on the Harmonisation of Risk Assessment Procedures*) 78–93.

Exposure assessment provides information on predicted and actual exposure of subjects such as humans or animals to a risk agent. While the hazard characterization describes the inherited ability of an agent to cause harm, exposure assessment relates this ability to risk subjects, in order to make a final estimation of risk. Exposure assessment requires evaluation of different variables, including concentration of an agent in the exposure media, duration, and frequency of the exposure. It is also necessary to identify pathways (eg air, food, or water), changes to the agent along the pathway, modes of exposure, or distribution of sensitive population sub-groups.[93] Moreover, exposure assessment needs to evaluate the uptake of the risk agent into a human or animal body and clarify the extent of its modification resulting from metabolism and excretion (bioavailability) as well as determining the concentration of the risk source at the relevant organ or tissue. These complexities and frequent lack of data mean that exposure assessment involves greater uncertainty than hazard characterization.[94] One of the specific problems is identification of the total exposure, which takes into account all relevant pathways and potential subjects.

Risk characterization is the final step of the risk assessment process. It aims at the integration of all preceding steps into a single estimation of adverse effects likely to occur in the given environment (eg population) from the presence of the risk agent in particular exposure media. Risk as such can be expressed in quantitative figures (eg the number of additional cancer deaths over a lifetime in a particular population) or in qualitative terms (eg a high or low possibility of developing cancer over a lifetime). This stage of risk assessment also requires the description of all uncertainties encountered on each step together with the assessment of their impact on the final estimation.[95]

Risk management is defined as 'a process of identifying, evaluating, selecting and implementing actions to reduce risk'[96] and reflects the preferences of a particular society for an acceptable level of risk exposure. Risk management traditionally follows risk assessment and comprises regulatory decisions on how to deal with a particular identified risk. In the words of one of the WTO panels 'once the risks have been assessed, ie, once the risks and their probability of occurrence [has been] identified, a [WTO] Member will need to decide [what SPS measure to adopt], on the basis of its own value judgments'.[97] Technically speaking risk management is composed of four elements or phases: risk evaluation, options evaluation, implementation of decision, and monitoring and review.[98] The first phase consists of an evaluation of risk assessment results in

[93] Ibid, 65–75. [94] Ibid, 65.

[95] WHO/FAO (*Report of the Joint FAO/WHO Expert Consultation*) 19.

[96] The Presidential/Congressional Commission on Risk Assessment and Risk Management, *Final Report, Volume 1: Framework for Environmental Health Risk Management* (1997) 1.

[97] Panel Report, *EC – Hormones (US)* para 8.160.

[98] Presidential/Congressional Commission on Risk Assessment and Risk Management, *Final Report, vol 1,* 7–39; Codex (*Procedural Manual*) 71.

the light of appropriate level of protection (ALOP). ALOP determines how much risk is acceptable for a particular social group or society. It can therefore be understood as the highest level of risk that a group is prepared to tolerate. ALOP may be pre-existing, for instance if it has been already decided with respect to the same or similar risk, or determined during this stage. The risk managers may also conclude at this stage that although risk mitigation action would be desirable, due to other reasons it is not feasible (eg because of high costs). If the answer is positive (risk is above ALOP and that fact is not acceptable), during the second phase available risk management options, such as sale prohibition of a particular product, establishing a tolerable level of a particular substance in a product, requiring an appropriate label, or commissioning additional research, are identified and evaluated with regard to their efficacy and feasibility. This phase usually requires the involvement of risk assessors who provide data on technical aspects of available options, including information on the extent to which a particular alternative reduces the likelihood and magnitude of particular risks. The task of risk managers is to compare those different alternatives as to their effectiveness, including potential societal, operational, and economic costs and benefits. This may require assessing social, commercial, and environmental impacts of proposed measures as well as conducting a cost-benefit analysis.[99] ALOP may be modified during this phase due to the costs or other problems related to regulatory action. On that basis, risk managers decide on further actions, which may result in the implementation of specific regulatory action (phase 3). In making decisions, risk managers need to account for the reliability of performed risk assessment, including the existence of identified uncertainties (ie a high level of uncertainty may result in a strict regulatory measure if the regulator seeks to achieve a high level of protection). The last phase can be described as the ongoing process of monitoring and reviewing of a regulatory option that was implemented in order to assess its effectiveness and to propose necessary adjustments. Risk managers particularly need to consider whether the strategy has accomplished what was intended, whether it still constitutes a viable option (changes in circumstances and scientific developments may affect the viability of the option) and whether there are possibilities for improvements. As a consequence of this review, risk managers may decide on further actions.

Risk communication,[100] the third and the last element of the risk analysis model, is understood either as a one way information flow from risk managers

[99] Note that costs and benefits of particular options may have a monetary or non-monetary character. These would particularly include costs and benefits of the development of new technology, opportunity costs, increased choice for consumers, or changes in the sense of well-being, etc (Presidential/Congressional Commission on Risk Assessment and Risk Management (*Final Report, vol 1*) 25).

[100] For a very interesting discussion on the complexities involved in the interactions between different actors in the process of risk analysis, see David Winickoff, Sheila Jasanoff, Lawrence Busch,

to the general public[101] or as a multidimensional exchange process 'between academic experts, regulatory practitioners, interests groups, and the general public'.[102] The first model dominated in the 1970s and 1980s and was based on the assumption that the general public is inherently wrong about the assessment of risks. Consequently, the function of risk communication was to eradicate irrational fears by improving the knowledge of citizens. Note that such understanding mirrors a radical version of technical paradigm where the public perception of risk is seen as a psychological deficiency. Modern conceptions of risk communication, however, recognize the importance of all the involved actors, as each group represents certain values and political viewpoints. The general public is not perceived as an irrational object of governmental regulations but as a subject and partner in the risk dialogue.[103]

Risk communication has an objective dimension (information on facts) and a subjective one (information of judgments and decisions). The aims of risk communication, in its multidimensional version, are multiple. It enables the provision of the initial information on potential hazards from the general public and risk managers to risk assessors,[104] allows data exchange between scientists and regulators, together with transmission of the information on the outcomes of risk assessment (both to the general public and governmental authorities) as well as the information on the results of the applied risk management policies (again to the general public and risk assessors). In the context of international relations, to which the SPS Agreement applies, risk communication also includes the transfer of information to trading partners that may be affected by a particular SPS measure.[105] Risk communication is also helpful in building up the trust and support of the general public for regulatory decisions as well as improving the legitimacy of such decisions. It may also assist risk regulators in avoiding defects in agenda-setting by providing information on public preferences for particular risk exposure or possible harm, and so on.

The figure below presents a diagram of the risk analysis process.

Robin Grove-White and Brian Wynne, 'Adjudicating the GM Food Wars: Science, Risk, and Democracy in World Trade Law' (2006) 30 Yale Journal International Law 81, 96–104.

[101] Jan Gutteling and Margot Kuttschreuter, 'The Role of Expertise in Risk Communication: Laypeople's and Expert's Perception of the Millennium Bug Risk in the Netherlands' (2002) 5(1) Journal of Risk Research 35, 36.

[102] William Leiss, 'Three Phases in the Evolution of Risk Communication Practice' (1996) 545 Annals 85, 86.

[103] See eg, Lillian Trettin and Catherine Musham, 'Is Trust a Realistic Goal of Environmental Risk Communication' (2000) 32(3) Environment & Behaviour 410, 416.

[104] Winickoff et al (Adjudicating the GM Food Wars) 101 (noting that 'scientific risk assessment necessarily involves the prior selection of the objects of analytic attention, reflecting what is collectively valued and thus worthy of possible protection').

[105] The notion of risk communication is not used by the SPS Agreement, nevertheless such provisions are easily identifiable in the agreement (eg Article 7, Annex B, some provisions of Annex C).

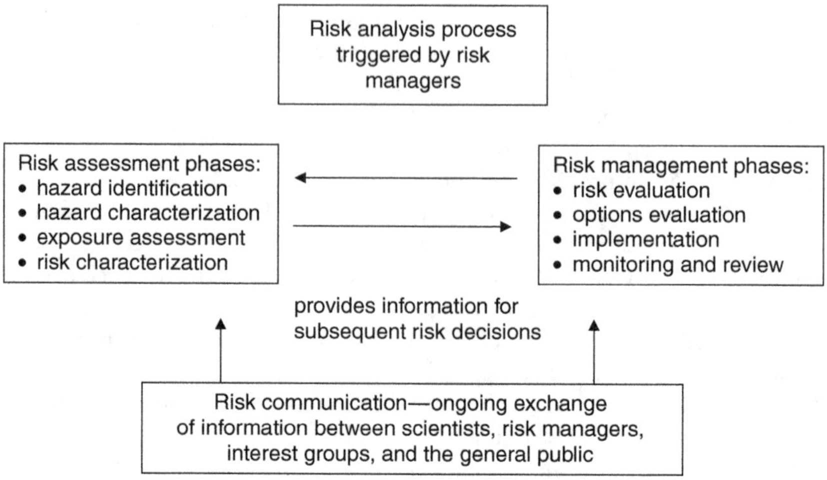

Figure 1. The Risk Analysis Process

The regulatory model of risk analysis has been criticized on different levels. For example, socio-cultural risk scholarship questions the utility of risk analysis on fundamental grounds, submitting that science cannot be regarded as an objective category since it is a social construct that is as valid as any other social process. The more technical criticism goes against the distinction between risk assessment and management.[106] It is particularly submitted that assessment of risk is never value free, while the conceptualization of risk analysis as a process consisting of two separate phases (scientific risk assessment and value driven risk management), creates a false impression that assessment can provide neutral estimates of risks. Due to uncertainties inherited in the evaluation of risks combined with the gaps in knowledge, scientists need to rely in the process of risk assessment on various semi-scientific (in the sense they are not fully scientifically verifiable) and wholly non-scientific assumptions. For instance, the hazard identification may be restricted either to well-characterized outcomes (ie mortality and morbidity associated with known mechanisms of toxicity or disease processes) or may also include less understood, or even unknown or unforeseen, results. Such decisions are non-scientific, and risk assessment, depending on the framing issue, will produce a different set of data.[107] Extrapolating human responses to a particular substance on the basis of animal studies or in extension of observed segments of a dose-response curve to the lower levels of exposure are other examples. These processes rely on certain assumptions (linear or nonlinear dose-response relationship, biochemical

[106] See generally, Ellen K Silbergeld, 'Risk Assessment and Risk Management: An Uneasy Divorce' in Deborah G Mayo and Rachel Hollander (eds), *Acceptable Evidence: Science and Values in Risk Management* (1991); Winickoff et al ('Adjudicating the GM Food Wars').

[107] FAO, *FAO Expert Consultation on Food Safety: Science and Ethics* (2003) 17–18.

and immunological similarities between species, etc), which are only partially scientific. Moreover, for many hazards, one must estimate exposure based on the assumption that the best practice or intended product end use represent a realistic approximation of actual exposure. At the same time, as noted by the expert group of the Food and Agriculture Organization of the United Nations (FAO), 'because of the technical complexity of most scientific risk assessments, value judgments employed in developing the analysis can be very difficult for outsiders to recognize and discern, much less to evaluate in light of their own values and interests'.[108] In other words, although risk assessors may present the outcome of risk assessment as an objective estimate of risks, in reality its result will also reflect the subjective judgements of the assessors.

There are also those who emphasize that the credibility of risk assessment as such may be impaired by the complexities of risk situations. Real-life situations involving environmental and health risks are frequently composed of a large number of variables that may actually prevent the execution of reliable risk assessment.[109] Again, the uncertainty that results from necessary approximations may be not properly accounted for in the final decisions.[110] Last but not least, the credibility of risk assessment may also be compromised by cognitive biases encountered by risk assessors. A number of studies has convincingly shown that experts are as vulnerable to errors and biases when making complex judgments as laypersons.[111]

This criticism was probably responsible for a number of adjustments that were introduced to the traditional model of risk analysis during the 1990s. It was accepted that a strict separation of risk assessment and management is not possible. For instance, the National Research Council (NRC) has reconceptualized its model of risk analysis and now sees the relationship between risk assessment and risk management as a more dynamic and iterative process, where risk managers move back and forth between the parts of the risk analysis using different sciences and cultural factors.[112] Another important change related to the explicit recognition that gaps in knowledge and uncertainties are inevitable in the risk assessment, and reliance on certain default assumptions is not only unavoidable but also desirable. Following the example of the NRC, many national and international risk regulators adopted sets of rules which define and formalize the use of such assumptions in the process of risk assessment (so-called risk assessment policies). As explained by the FAO, those rules can be defined as 'guidelines for value judgment and policy choices which may need to be applied at specific decision points in

[108] Ibid, 20.

[109] Malcolm MacGarvin, 'Precaution, Science and the Sin of Hubris' in O'Riordan and Cameron (eds) (*Interpreting the Precautionary Principle*) 205–6.

[110] Fraiberg and Trebilcock, ('Risk Regulation') 849.

[111] Miller B Spangler, 'Policy Issues Related to Worst Case Risk Analyses and the Establishment of Acceptable Standards of De Minimis Risk' in Covello et al (*Uncertainty in Risk Assessment*) 7.

[112] Paul C Stern and Harvey V Fineberg (eds), *Understanding Risk: Informing Decisions in a Democratic Society* (1996).

the risk assessment process'.[113] In other words, they are scientifically and socially accepted ways of dealing with uncertainties. By their codification, the whole risk analysis process actually becomes more transparent and probably more credible.

Nevertheless, both the NRC and the majority of domestic regulators still recognize the importance of the distinction between risk assessment and its management.[114] The same opinion is shared by a number of international standard-setting bodies and initiatives in the SPS field.[115] As observed in the literature, 'the functional separation . . . reflects a continuing desire to minimise the role of policy in the heavily scientific phase of risk assessment and a continuing belief in the value of separating, as far as possible, the scientific and . . . political aspects of risk regulation'.[116] In fact, their explicit recognition is the first step to control them and assess their impact on the final risk estimation. At the same time, the role of the policy considerations and value judgements in risk assessment should not be overstated (unless one takes a fully constructivist approach toward science). Science remains a central element of this process. Despite the existence of different non-scientific assumptions, and the existence of framing problems or cognitive biases that are encountered by assessors, in practice risk assessment is able to describe a number of potential risks with remarkable precision. This is also reflected in the formidable success of those regulatory schemes that rely on the concept of risk analysis. Although the complexities of real life situations are indeed a serious challenge for risk assessors, they may be remedied (or at least limited) by the application of different strategies, such as reliance on worst case scenarios or application of very conservative assumptions. If a level of uncertainty is so high that it will preclude meaningful risk assessment, risk managers may still undertake preventive actions (eg on the basis of the precautionary principle). A distinction between risk assessment and management is also helpful in making the regulatory process more transparent by providing the possibility of control and review, as well as eliminating some of the political biases.[117]

1.4.4 Assessment of risks and uncertainty

The problem of uncertainty in the risk analysis process, as one of the central issues in the field of risk regulation, requires some further elaboration. Uncertainty may be generally described as 'lack of precise knowledge as to what the truth is,

[113] FAO/WHO, *Report of a Joint FAO/WHO Consultation on Risk Management and Food Safety* (1997) 28.

[114] For example, European Commission, *White Paper on Food Safety*, COM (1999) 719 Final.

[115] FAO/WHO, *Report of the Evaluation of the Codex Alimentarius and Other FAO and WHO Food Standards Work* (2002) 12; Codex (*Procedural Manual*) 69; Articles 15 and 16 of the Cartagena Protocol on Biosafety.

[116] Catherine Button, *The Power to Protect. Trade, Health and Uncertainty in the WTO* (2004) 100.

[117] Fraiberg and Trebilcock ('Risk Regulation') 857–8.

whether quantitative or qualitative'.[118] Sometimes it can be reduced over time as scientific knowledge develops. However, in some cases uncertainty is simply irreducible in the sense that further research is not able to generate knowledge that would explain the nature of a particular phenomenon. Uncertainty results from many different sources such as difficulties associated with the separation of different variables, lack of information, and gaps in scientific theories as well as possible synergies of interactions between those elements (eg synergistic or antagonistic effects). An additional difficulty is added by the imperfection of the scientific methods used in the risk assessment process.[119]

It seems that there is no established classification of uncertainty in the literature. Different authors put forward their own taxonomies, which divide uncertainty according to such criteria as a source, nature, or type of generated methodological challenges.[120] An interesting typology was proposed by Klinke and Renn and it may be used here as a basis for subsequent discussion on the conceptualization of uncertainty under the SPS Agreement.

The authors divide uncertainty *sensu largo* into different categories that may be identified in the context of modern risk analysis. The first type of uncertainty results from the nature of science and the reality of probabilistic risk assessment. As explained by the NRC 'scientific truth is always somewhat uncertain and is subject to revision as new understanding develops'.[121] Even if one believes that science in general is able to describe reality (with greater or lesser accuracy), its particular assertions will be valid only as long as they are not falsified. Consequently, there always remains uncertainty as to the correctness of a particular scientific claim. Moreover, note that risk assessment is only a probabilistic estimation of future events. Probabilities as such 'represent only an approximation to predict uncertain events'.[122] These predictions are based on data relating to past trends, and interferences from experiments or systemic observations. Of course, there is obvious uncertainty in making predictions on the basis of past experience.

The second type of uncertainty results from the variability problem,[123] that is, the 'variation of individual responses to an identical stimulus among the

[118] National Research Council, Committee on Risk Assessment of Hazardous Air Pollutants, Commission on Life Sciences, *Science and Judgment in Risk Assessment* (1994) 161.

[119] Lester B Lave, *Quantitative Risk Assessment in Regulation* (1982) 23.

[120] Van Asselt (*Perspectives*) 84–93; Vern R Walker, 'The Siren Songs of Science: Toward a Taxonomy of Scientific Uncertainty for Decisionmakers' (1990–91) 23 Connecticut Law Review 567. The NRC distinguishes between aleatory uncertainty (random variations and chance outcomes in the physical world), epistemic uncertainty (lack of knowledge), indeterminacy (resulting from the choice of risk generating process) and ignorance (unrecognized uncertainty), see Stern and Fineberg (*Understanding Risk*) 107.

[121] NRC (*Science*) 161.

[122] Klinke and Renn ('A New Approach to Risk Evaluation') 1074.

[123] Variability (aleatory uncertainty) as understood by the NRC and some other authors (eg Van Asselt) is a much broader term as it refers to the inherent randomness of the nature. Van Asselt also adds other sources of variability such as value diversity, behavioural variability, societal variability (social, economic and cultural dynamics) and technological surprises (Marjolein BA

individual targets within the relevant population such as humans, animals, plants'.[124] The variation may be caused by the target's age (eg children and elderly people are usually more sensitive), sex, individual characteristics (eg diet and metabolism), and other factors. The third type of uncertainty is a consequence of systemic and random measurement errors. This category particularly includes imprecision or imperfection of measurement and a need to rely on different forms of extrapolations that are normally used in risk assessment (eg from animal data to humans, from large to small doses, from small statistical samples to larger population groups). The fourth type, which is labelled by Klinke and Renn as indeterminacy, results 'from a genuine stochastic relationship between cause and effect(s), apparently non-casual or noncyclical random events or badly understood nonlinear chaotic relationship'.[125] The last form of uncertainty, which seems to overlap to a certain extent with the previous categories, is connected with a lack of knowledge. That may result from 'ignorance of the deliberate definition of system boundaries and hence exclusion from external influences, measurement impossibilities and other'.[126] This category is very broad and may be also divided into further classes. According to Van Asselt and Rotmans, lack of knowledge (and corresponding uncertainty) may be a consequence of:

- lack of observations/measurements resulting from failures to collect relevant data or to perform measurements;
- practical immeasurability—data can be theoretically collected and measured but such a process is either overcomplicated or too expensive;
- conflicting evidence (with corresponding competing theories), meaning that the same set of data allows different interpretations;
- reducible ignorance, including processes which, although cannot be observed or theoretically imagined at the moment, may be possible in the future; or
- irreducible ignorance—there may be 'processes and interactions between processes that cannot be (or not unambiguously) determined by human capacities and capabilities'.[127]

The first type of uncertainty is normally disregarded in the risk analysis. If the scientific risk assessment is to be an operable tool, one needs to assume that it has the capacity to describe and measure phenomena such as hazards, probabilities, potential damages, or injuries. All the strategies that are conventionally used to deal with the problem of uncertainty are based on this assumption. The second and third types of uncertainty are normally addressed through application

Van Asselt and Jan Rotmans, 'Uncertainty in Integrated Assessment Modelling' (2002) 54(1–2) Climatic Change 75).

[124] Klinke and Renn ('A New Approach to Risk Evaluation') 1079.

[125] Ibid; this is also the category which is closest to the concept of aleatory uncertainty.

[126] Ibid.

[127] Van Asselt and Rotmans ('Uncertainty') 80–81.

in the process of risk assessment of different techniques such as safety factors, conservative assumptions, or worst-case scenario approaches. A safety factor is a figure by which an observed or estimated toxic concentration or dose is divided to arrive at a criterion or standard that is considered to be without effect. The Codex typically applies, in the process of establishing ADI for a particular substance, an overall safety factor of 100, which consists of a safety factor of 10 for diversity within the human population[128] and a safety factor of 10 for extrapolation from animal test results to humans. The conservative assumptions, which are adopted in the process of risk assessment, may consist in assuming a daily intake of a particular substance at much higher levels than statistically supported, ingestion of a substance over a whole lifetime (which in reality is rarely the case), or the absence of a safe threshold in the exposure to particular chemical (ie the chemical substance causes harm at any level of concentration).[129] Other strategies include the employment of uncertainty analysis—the qualitative or quantitative examination of uncertainties and assessment of the probability distribution between different possible outcomes (usually in the form of a continuum).[130] Such data are subsequently transferred to risk managers, who decide on risk strategies on the basis of selected estimates (including the worst-case scenario). Consequently, the existence of these two types of uncertainty in principle does not prevent the performance of risk assessment. Nevertheless, one also needs to admit that, even with the application of the above strategies, some risk of under-inclusion (failing to properly identify risk) always remains.[131]

The fourth type of uncertainty is more problematic. The application of safety factors, conservative assumptions, and uncertainty analysis may address this class of uncertainty only to a limited extent and the risk of under-inclusion may actually be high.[132] If the nature of the relationship between cause and effect, the probability of occurrence, or the extent of possible damage, is unknown, it is not really possible to address uncertainty through the application of traditional tools. Of course, in some cases one may reduce this type of uncertainty through further research. However, it is also true that sometimes the uncertainty may remain over

[128] Sometimes additional safety factors are applied where there is an identifiable sub-group that might reasonably be expected to be more sensitive than the group from which data was obtained.

[129] John D Graham and Lorenz Rhomberg, 'Challenges in Risk Assessment and Risk Management' (1996) 545 Annals of the American Academy of Political and Social Science 15, 20 (giving the example of formaldehyde, which is conventionally considered to be a substance without a safety threshold, despite the fact that there is no conclusive evidence in this respect).

[130] NRC (*Science*) 167.

[131] In practice, the application of these strategies will rather result in over-inclusion (identifying risk where there is none).

[132] As noted by the EC in the *Suspension* disputes 'the genotoxic and carcinogenic potential of oestradiol-17 cannot be adequately addressed by the safety factors applied by JECFA, because the underlying scientific uncertainty about the mechanisms causing cancer are not amenable to quantification so as to be adequately addressed by the safety factors' (Panel Report, *US – Continued Suspension, Annex F, Comments by the Parties on the Replies of the Scientific Experts, Codex, JECFA and IARC to Questions Posed by the Panel and Comments by the Parties on the other Parties' Comments*, F-8).

a long period of time or even be irreducible. In such cases, risk managers may be required to rely on other values and principles, and on that basis decide to adopt a protective action without a real understanding of the underlying process.

The fifth type of uncertainty, depending on the class to which it belongs, can be reduced through further research or simply be tolerated. Lack of observations/measurements and practical immeasurability typically compel risk managers to commission additional scientific studies (which sometimes, due to the additional costs, may remain only a theoretical option). When faced with conflicting evidence, risk managers, depending on the regulatory context in which they act, may be required to base a regulatory measure on the best science or to choose between conflicting scientific views. Reducible and irreducible ignorance, as the first type of uncertainty, is normally disregarded in risk assessment. If the existence of potential risk is not even recognized, there cannot be a regulatory response. Nevertheless, risk managers may adopt a measure in anticipation of potential risks (even without determining what would be the nature of such risks). That option may be particularly used when a new technology is introduced and the amount of relevant knowledge is limited.

1.5 Conclusions

This chapter has shown that risk is a complex phenomenon that may be defined from various perspectives. Against this background, it has identified technical understanding of risk as the one which is adopted by the SPS Agreement. The chapter has also discussed different regulatory models that are used to deal with problems of risk and recognized the concept of risk analysis as dominant regulatory approach. As the subsequent chapters will show, this model may also be used in describing the disciplines provided by the SPS Agreement.

2

The SPS Agreements within the WTO System

2.1 Introduction

The last 60 years have witnessed a massive and unprecedented expansion of international trade. The system created in 1947 by the General Agreement on Tariffs and Trade (GATT 1947) proved to be a very successful instrument for the elimination of trade tariff barriers. By limiting or removing tariffs, countries have gradually gained access to foreign markets at considerably lower costs.[1] However, as discussed in the previous chapter, the process of international trade liberalization has also coincided in time with the increase of national regulatory activism. This was particularly visible in the area of risk regulation where governments, responding to the fears and demands of their domestic constituencies, adopted a wide range of regulatory measures aimed at the protection of the environment and human health and safety. In some cases, such measures had an impact on international trade and could be regarded as non-tariff barriers (NTBs).[2] Although normally, the new regulatory initiatives served fully legitimate goals, it also became clear that they might also become an attractive vehicle for protectionism and take the place that was traditionally occupied by tariffs barriers.[3]

In response to these concerns, the GATT Contracting Parties, seeing the traditional GATT rules as insufficient, decided to develop a new legal instrument that would discipline national regulatory measures. The subsequent efforts resulted in the adoption during the Tokyo Round of multilateral negotiations of the Agreement on Technical Barriers to Trade (the Standards Code)[4] that

[1] On the economic justification of free trade and possible objections, see Michael J Trebilcock and Robert Howse, *The Regulation of International Trade* (2005) 2–20.

[2] A typical example of an NTB is a ban on sale and import of specific products on health or environmental grounds; other examples include measures that establish some positive requirements for such products (eg certification); for details see, Tracey Epps, *International Trade and Health Protection. A Critical Assessment of the WTO's SPS Agreement* (2008) 12–15.

[3] On the historical account on SPS disputes, see Justin Kastner and Douglas Powell, 'The SPS Agreement: Addressing Historical Factors in Trade Dispute Resolution' (2002) 19 Agriculture and Human Values 283.

[4] Agreement on Technical Barriers to Trade, 12 April 1979, 1186 UNTS 276, GATT, BISD, 26th Supp 8 (1980).

was specifically designed for technical regulations, including a sub-category of measures that relate to food safety, and animal and plant life and health.[5] Failing to reach a common consensus, the Standards Code was adopted as a plurilateral agreement that bound only a limited number of countries that accepted the agreement. This feature, combined with the consensus-based dispute settlement process, which allowed blocking of the establishment of the panel or the adoption of a final report, were probably the main reasons for a practical failure of the Standards Code.[6] No single SPS measure was successfully challenged in the GATT dispute settlement, while its disciplines had only very limited impact on the regulatory practice of the signatories. The need for an effective instrument became even more compelling as the new round of trade negotiations was launched in 1986. There was a common fear among the negotiators that liberalization of international trade in the agricultural sector, which was an important item of the Uruguay Round, could be undermined by the increased recourse of the countries to SPS measures.[7] These concerns were also reflected in the Punta del Este Declaration,[8] which explicitly proclaimed a need to minimize 'the adverse effects that sanitary and phytosanitary regulations and barriers can have on trade in agriculture' as a one of the objectives of the negotiations.[9]

Initially, the negotiations on the SPS disciplines were conducted as a part of a general discussion on agriculture. In 1988 an individual group (the Working Group on Sanitary and Phytosanitary Regulations and Barriers) was formed with the aim of producing a separate agreement aimed specifically at SPS measures. The intention was to create a set of rules that would regulate the establishment and maintenance of national measures for protection of human, animal and plant life and health from specific SPS risks, which might have an impact on international trade flow. The draft text was finally produced in November 1990 and reflected an overall consensus among the main actors that were involved in the negotiation process.[10] Following a deadlock in the negotiations, a new draft was tabled (the so-called Dunkel Text) with some minor changes compared to the November version. This latter appeared to be the final formulation of the SPS Agreement.[11]

[5] David A Wirth, 'The Role of Science in the Uruguay Round and NAFTA Trade Disciplines' (1994) 27 Cornell International Law Journal 817, 824.

[6] David Victor, 'The Sanitary and Phytosanitary Agreement of the WTO: An Assessment after Five Years' (2000) 32 New York University Journal of International Law & Policy 865, 874.

[7] For details concerning the negotiation process, see Denise Prévost and Peter Van den Bossche, 'The Agreement on the Application of Sanitary and Phytosanitary Measures' in Patrick FJ Macrory, Arthur E Appleton and Michael G Plummer (eds), *The World Trade Organization: Legal, Economic and Political Analysis* (2005) 239–44.

[8] Ministerial Declaration on the Uruguay Round: Declaration of September 20, 1986, GATT BISD 33S/19 (1987).

[9] Ibid, 20.

[10] The major actors involved in the process of negotiation included the United States, the EC and the Carins Group that was composed of number of developing and developed countries such as Argentina, Australia, Brazil, and Thailand.

[11] Simonetta Zarilli, 'WTO Sanitary and Phytosanitary Agreement: Issues for Developing Countries', TRADE Working Papers No 3 (1999) 4.

The WTO legal framework that came as a result of the Uruguay Round proved to be a complex and multi-layered system. It is composed of different agreements regulating different aspects of international trade. At the top of the system is the Marrakesh Agreement Establishing the World Trade Organization.[12] The other agreements are gathered in annexes attached to the WTO Agreement. Annex 1A consists of those legal instruments that are concerned with trade in goods. Here the central one is the General Agreement on Tariffs and Trade (GATT 1994),[13] which is a successor of the old GATT 1947. Both the Agreement on Technical Barriers to Trade (TBT Agreement)[14] and the SPS Agreement belong to this group as well. Annex 1B and 1C include agreements that regulate other aspects of international trade. A separate set of obligations is dedicated to the dispute settlement process and the policy review mechanism (Annexes 2 and 3 respectively). All those rules reflect the idea of the WTO law as a single legal undertaking[15] and they are binding on all WTO Members. Although some efforts were made in the course of the negotiation to ensure consistency between different agreements, this task proved to be difficult. The countries were rather reluctant to introduce any changes to drafts already adopted within individual negotiating groups. There was also no extensive discussion on the relationship between the agreements and their hierarchy. Altogether this resulted in the adoption of norms that sometimes appear to be overlapping or inconsistent. At the same time, the collision rules remained underdeveloped,[16] while the complexity of the system added an additional layer of difficulty.

This chapter starts with a brief overview of the SPS Agreement. This part is intended as a summary of the SPS Agreement and should give the reader a broader picture of the structure and the basic requirements of the Agreement. The second section concentrates on the issue of applicability and discusses in more detail the conditions that trigger the application of the SPS disciplines. The third part analyses the position of the agreement vis-à-vis other WTO legal instruments that are relevant for the assessment of national risk regulatory measures. This particularly includes the GATT 1994 and the TBT Agreement.

[12] Marrakesh Agreement Establishing the World Trade Organization, opened for signature 15 April 1994, 1867 UNTS 154 (entered into force 1 January 1995) (WTO Agreement).

[13] General Agreement on Tariffs and Trade, opened for signature 15 April 1994, 55 UNTS 194, 1867 UNTS 187 (entered into force 1 January 1995).

[14] Agreement on Technical Barriers to Trade, opened for signature 15 April 1994, 1868 UNTS 120 (entered into force 1 January 1995).

[15] This concept emerged during the negotiation process and was understood as requiring future WTO Members to accept all the agreements under discussion as a single package, constituting one international treaty.

[16] Gabrielle Marceau and Joel P Trachtman, 'The Technical Barriers to Trade Agreement, the Sanitary and Phytosanitary Measures Agreement, and the General Agreement on Tariff and Trade' (2002) 36 Journal of World Trade 811, 813–15.

2.2 Overview of the disciplines of the SPS Agreement

The preamble to the SPS Agreement outlines a number of different objectives that are supposed to be accomplished by the Agreement. Taking into account the overall aim of the WTO to further liberalization of international trade, probably the central goal of the SPS Agreement is to establish a multilateral framework of rules that regulates the development, adoption, and enforcement of SPS measures in order to minimize their negative trade effects. This objective is, however, limited by a second set of considerations. The SPS Agreement emphasizes that its disciplines are not intended to prevent WTO Members from adopting or enforcing measures necessary to protect human, animal, or plant life or health and do not require Members to change their appropriate level of SPS protection.[17] The SPS Agreement also explicitly provides that one of its goals is the improvement of human and animal health as well as the phytosanitary situation in all WTO Members.[18] The existence of these multiple and potentially conflicting objectives is well reflected in the observation of the Appellate Body, which characterized the SPS Agreement as embodying a 'delicate and carefully negotiated balance... between the shared, but sometimes competing, interests of promoting international trade and of protecting the life and health of human beings'.[19]

However, before going into details of the SPS Agreement, it is worth briefly reviewing the interpretative methods that are used by the WTO dispute settlement bodies. As will be seen later, this matter proves to be highly relevant for a number of issues, which are discussed in this book.

2.2.1 Rules of treaty interpretation in the WTO jurisprudence

The WTO jurisprudence has progressively recognized the rules of the Vienna Convention on the Law of Treaties (VCLT) as a source of interpretative guidelines for WTO dispute settlement bodies.[20] Article 3.2 of the Understanding on Rules and Procedures Governing the Settlement of Disputes (DSU),[21] which requires interpreting provisions of the WTO agreements in accordance with customary rules of interpretation of public international law, was conventionally referred to as a relevant legal basis.[22]

[17] Recital no 1, Preamble to the SPS Agreement. [18] Ibid, Recital no 2.

[19] Appellate Body Report, *EC – Hormones*, para 177.

[20] Appellate Body Report, *US – Gasoline*, 17.

[21] Understanding on Rules and Procedures Governing the Settlement of Disputes, 1869 UNTS 401 (entered into force 1 January 1995).

[22] Article 3.2 of the DSU specifically provides that '[t]he dispute settlement system of the WTO is a central element in providing security and predictability to the multilateral trading system. The Members recognize that it serves to preserve the rights and obligations of Members under the covered agreements, and to clarify the existing provisions of those agreements in accordance with

The VCLT distinguishes between the primary and secondary means of interpretation. The first set of rules may be found in Article 31, which provides that a treaty is to be interpreted in good faith and in accordance with the ordinary meaning to be given to its terms in their context and in the light of its object and purpose. A special meaning has to be given to a particular term only if such was an intention of the parties (Article 34 of the VCLT). According to the VCLT, the context comprises elements such as the text of the treaty, including the preamble and annexes to the treaty, as well as any agreement relating to the treaty which was made between all the parties in connection with the conclusion of the treaty, together with any instrument which was made by one or more parties in connection with the conclusion of the treaty and accepted by the other parties as an instrument related to that treaty. Article 31.3 also mandates to take into account, together with the context, some other extrinsic elements:

- any subsequent agreement between the parties regarding the interpretation of the treaty or the application of its provisions;
- any subsequent practice in the application of the treaty which establishes the agreement of the parties regarding its interpretation; and
- any relevant rules of international law applicable in the relationship between the parties.

According to Article 32 of the VCLT, an interpreter may rely on the supplementary means of interpretation in order to confirm the meaning determined on the basis of the primary rules. An interpreter is also instructed to use supplementary means when the interpretation leaves the meaning ambiguous or obscure, or leads to a result which is manifestly absurd or unreasonable. The supplementary means of interpretation include the preparatory work of the treaty and the circumstances of its conclusion, but the list of relevant factors enumerated in the VCLT is not exhaustive.

The approach of the WTO case law is generally compatible with the normative prescriptions of the VCLT. The text of an agreement is always a starting point[23] and is followed by the analysis of the purpose and the object of a treaty as well as the relevant context.[24] The purpose was conceptualized as referring to the treaty as a whole and not to a particular term or provision.[25] It may be looked for not only in the preamble to a particular agreement but also in the Marrakesh Agreement, which is an umbrella treaty for the other WTO agreements. The context, in line

customary rules of interpretation of public international law. Recommendations and rulings of the DSB cannot add to or diminish the rights and obligations provided in the covered agreements'.

[23] Appellate Body Report, *Japan – Alcoholic Beverages II*, 12.

[24] The approach of the case law appears to be sometimes over-mechanical with consideration of different elements prescribed by the VCLT in a rigid step-by-step analysis. The VCLT seems to promote more a holistic approach where different elements are considered in a single analysis; see Federico Ortino, 'Treaty Interpretation and the WTO Appellate Body Report in US-Gambling: A Critique' (2006) 9(1) Journal of International Economic Law 117.

[25] Appellate Body Report, *EC – Chicken Cuts*, para 238.

with the VCLT, comprises the text of a particular agreement, its preamble and annexes[26] as well as agreements or instruments related to the treaty.[27] The WTO case law has also recognized the relevance of the good faith principle,[28] which was translated into the requirement of an effective treaty interpretation that mandates an interpreter to read different provisions in a harmonious way so as to give meaning to all of them.[29] The elements which are enumerated in Article 31.3 were also considered as potentially relevant for interpretation of WTO provisions, although subparagraph (c) proved to be problematic. I will come back to this issue in Chapter 5.

The recognition of Article 31 as a reflection of customary rules of interpretation of public international law was followed by the same approach with respect to Article 32 of the VCLT.[30] The WTO case law, in line with the VCLT, has accepted the recourse to Article 32 in those cases when the application of primary rules leaves the meaning ambiguous or obscure, or leads to a result that is manifestly absurd or unreasonable.[31] In line with the VCLT, the reference to a supplementary means of interpretation is also used to confirm the results obtained on the basis of Article 31. The elements which are most referred to in this process comprise the *travaux preparatoires* and the circumstances of the conclusion of WTO agreements.[32]

The WTO dispute settlement bodies also relied on other interpretative guidelines, which are not specifically provided by the VCLT.[33] As noted by Lennard 'they have been treated in WTO fora as emerging naturally from the expressed principles in the Vienna Convention',[34] particularly the good faith principle. This catalogue is broad and includes, among other things, the principle of effective treaty interpretation; the principle *in dubio mitius* (if the meaning of the term is ambiguous, that meaning is to be preferred which is less onerous to the party assuming an obligation);[35] the presumption of consistency (different terms are taken to have different meanings);[36] and the concept of evolutionary meaning (generic terms may be interpreted in an evolutionary manner).[37]

2.2.2 Substantive obligations of the SPS Agreement

The overview of the SPS Agreement may start from Article 2, which defines the basic rights and obligations of WTO Members. Paragraph 1 confirms the right

[26] For example, Panel Report, *EC – Chicken Cuts (Complaint by Brazil)*, para 7.153.

[27] For example, Panel Report, *United States – Section 110(5) of the US Copyright Act*, para 6.45.

[28] See eg, Appellate Body Report, *Japan – Alcoholic Beverages II*, fn 21.

[29] For example, Panel Report, *US – Gambling*, para 6.49.

[30] For example, Appellate Body Report, *Japan – Alcoholic Beverages II*, 10.

[31] Panel Report, *Japan – Alcoholic Beverages II*, para 87.

[32] Panel Report, *US – Gambling*, para 6.114.

[33] See generally, Isabelle Van Damme, *Treaty Interpretation by the WTO Appellate Body* (2009).

[34] Michael Lennard, 'Navigating by the Stars: Interpreting the WTO Agreement' (2002) 5(1) Journal of International Economic Law 17, 55.

[35] Appellate Body Report, *EC – Hormones*, fn 154.

[36] Ibid, para 164.

[37] Appellate Body Report, *US – Shrimp*, para 153.

of each WTO Member to adopt and maintain SPS measures necessary for the protection of human, animal or plant life or health provided that such measures are not inconsistent with the other provisions of the Agreement. Article 2.2 introduces a traditional necessity test by prohibiting the application of the SPS measure over what is necessary to protect human, animal or plant life or health. This obligation is combined with the requirement of rational decision-making—each measure needs to be based on scientific principles and cannot be maintained without sufficient scientific evidence. Article 2.3 imposes on WTO Members a non-discriminatory obligation and requires them to ensure that their SPS measures do not arbitrarily or unjustifiably discriminate between Members where identical or similar SPS conditions prevail. On the basis of the same article, SPS measures have to be applied in a manner which does not constitute a disguised restriction on international trade.

One way of limiting the potential distortive effects that SPS measures may have on international trade is to promote their harmonization on as wide a basis as possible. Elaborating on this idea, Article 3.1 encourages WTO Members to base their measures on international standards, guidelines, and recommendations, while Article 3.2 establishes a rebuttable presumption of consistency with the SPS Agreement for those measures that conform to such standards. The relevant SPS standards are identified by the SPS Agreement through reference to specific international standard-setting bodies. These include the Codex Alimentarius Commission (for food safety issues), the World Organization for Animal Health (for animal health), and the Secretariat of the International Plant Protection Convention (for plant health). At the same time, the harmonization disciplines are not absolute and WTO Members, aiming for different levels of protection, may decide to adopt and maintain measures that are not based on international standards. However, under this scenario a Member needs to be able to provide appropriate scientific justification that supports its measure. This is an instance where the obligation of Article 2.2 is activated.

Articles 5.1 to 5.3 of the SPS Agreement elaborate on this issue and translate the general obligation of Article 2.2 into very specific duties. Article 5.1 requires WTO Members to base their SPS measures on a scientific assessment of risks to human, animal or plant life or health, while subparagraph 2 enumerates elements that need to be taken into account in such assessment (eg available scientific evidence but also relevant inspection, sampling, and testing methods). Article 5.3, which only applies to risks to animal or plant life or health, supplements that list with economic factors (eg the potential damage in terms of loss of production and sale in the event of the entry of a pest). If scientific evidence is insufficient to perform an adequate risk assessment, a WTO Member may still act on the basis of a provisional measure that is merely supported by pertinent information. While a measure is in force, a Member is obliged to seek additional information in order to perform a more objective assessment of risk (Article 5.7).

Articles 5.4 to 5.6 introduce certain disciplines with respect to the risk management phase of the national decision-making process. In particular, WTO Members, when determining their ALOPs, should take into account the objective of minimizing negative trade effects (Article 5.4). WTO Members are expected to avoid arbitrary or unjustifiable distinctions in the levels of protection that are applied in different but comparable situations, however, only if such distinctions result in discrimination or disguised restriction on international trade (Article 5.5). WTO Members also need to ensure that SPS measures are not more trade-restrictive than is necessary to achieve their appropriate level of SPS protection. They are obliged to take into account the technical and economic feasibility of alternative options in order to achieve this level (Article 5.6). The necessity and the least-restrictive-alternative requirements are reflected in the obligation to adapt SPS measures to regional conditions of a country of import and export (the concept of regionalization).

The SPS Agreement explains one of the possible risk management strategies in more detail. According to Article 4, WTO Members are obliged to accept the SPS measures of other Members as equivalent, even if such measures differ from their own or from those used by other Members trading in the same product. This recognition is conditional on an objective demonstration by the exporting country of equivalence in terms of ALOP.

2.2.3 Risk communication (transparency) and procedural obligations

A separate set of obligations of the SPS Agreement is dedicated to risk communication disciplines. Since the SPS Agreement is an international treaty that creates obligations towards other states, the risk communication aspect of the national decision-making process is regulated only as far as it concerns other WTO Members. In consequence, the Agreement does not establish any requirement on the internal dimension of the risk communication process (eg vis-à-vis the general public of a particular country).

The central obligation in this context is the notification requirement relating to changes in SPS measures (Article 7).[38] Elaborating on that provision, Annex B requires prompt publishing of all SPS regulations in such a manner as to enable interested Members to become acquainted with them. The publication needs to be followed by reasonable *vacatio legis* (subject to the exceptions for urgent matters) in order to allow for necessary adjustments of products and production methods used in the exporting countries. To facilitate the information flow, WTO Members are asked to establish special enquiry points responsible for the

[38] On 30 May 2008, the Committee adopted a revised version of the Recommended Procedure for Implementing the Transparency Obligations of the SPS Agreement G/SPS/7/Rev.3. The new procedure, including the new notification formats, took effect as of 1 December 2008.

provision of relevant documents regarding, amongst other things, information on national SPS measures (adopted and planned), control and inspection procedures, production and quarantine treatment, risk assessment procedures, membership and participation in international and regional SPS organizations, as well as other information.

The SPS Agreement also provides details on a special notification procedure to be followed for those SPS proposals that substantially differ from international standards, guidelines, and recommendations (or if no standard exists) and which may have a significant effect on trade with other Member(s). The notification process takes place through the WTO Secretariat, which is responsible for subsequent diffusion of information. Other Members may comment on the proposed SPS regulation while the informing Member is obliged to take those comments and considerations into account when enacting a final measure. The SPS Agreement permits a deviation from the notification procedure only in cases of urgency. However, even then, a WTO Member is expected to provide *ex post* notice.

The procedural obligations of the SPS Agreement aim at facilitating the operation of control, inspection, and approval procedures in importing countries (Article 8). Members are obliged to ensure, with respect to any such procedure, that they are undertaken and completed without undue delay. Other provisions included in Annex C are intended to guarantee the transparency of such procedures (eg the obligation to inform an applicant on anticipated processing periods). Finally, some obligations simply transpose the traditional WTO principles into a procedural context. This includes the national treatment (eg procedures are undertaken and completed in no less favourable manner than for like domestic products), necessity (eg requirements are limited to what is reasonable and necessary), or the most favoured nation (eg application of the same fees irrespectivel of the country of importation).

2.2.4 Developing countries and the SPS Agreement

The SPS Agreement contains a number of provisions relating to the special and differential treatment of developing countries. Article 10, being a central rule, requires WTO Members to take into account, when preparing and applying SPS measures, the special needs of developing Members, including the least-developed countries (the external dimension). This obligation is reflected in a call for longer time frames for compliance that should be available to developing Members, but only if the ALOP of an importing country allows for the phased introduction of an SPS measure. The internal dimension of the special and differential treatment concerns the compliance of SPS measures enacted by developing countries with the provisions of the SPS Agreement. In this context, the Committee on Sanitary and Phytosanitary Measures (SPS Committee) may grant specific and temporary exceptions with respect to the whole or parts of the Agreement.[39] In addition, WTO Members are expected to encourage and facilitate the active participation of developing countries in the relevant international standard-setting organizations (Article 10.4). All those requirements are formulated in non-mandatory language ('should' or 'take into account') and may be seen as a type of best endeavours obligation.

The above rules are supplemented with the obligations of Article 9, which call for the provision of technical assistance to developing countries.[40] The article in non-exhaustive enumeration identifies some of the areas where assistance is particularly needed, such as processing technologies, research, and infrastructure, including the establishment of national regulatory bodies. The aim of this assistance is to allow such countries to adjust to, and comply with, SPS measures necessary to achieve ALOP of their exporting markets. Again, as in case of Article 10, these provisions may be characterized as best endeavours obligations.[41]

Article 14 allowed the least-developed countries to delay the application of provisions of the SPS Agreement for a period of five years from the date of entry into force of the WTO Agreement (until 2000). Other developing countries also had such a right, although with the exception of some provisions and for a shorter period of time (until 1997). Since the deadlines provided in the article have already expired, the provision needs to be regarded as obsolete.[42]

2.2.5 SPS Committee

The SPS Committee is a special body composed of representatives from all WTO Members,[43] which is responsible for the administration of the SPS Agreement (Article 12).[44] The mandate of the Committee is described very broadly to carry out all the functions necessary for implementation of the provisions of the SPS Agreement and the furtherance of its objectives, in particular with respect to harmonization. These broad objectives are translated into a number of

[39] Note that no request to the SPS Committee under Article 10.3 of the SPS Agreement has ever been made.

[40] Cf the Implementation Decision adopted at the Doha Ministerial Conference, Fourth Session, 9–14 November 2001, Implementation—Related Issues and Concerns, Draft Decision WT/MIN(01)/W/10, para 3.6(ii).

[41] As a part of efforts to increase the technical capacities of developing countries in the SPS area, the special Standards and Trade Development Facility (SDTF) was created in 2002. The aim of the STDF is to assist developing countries to enhance their expertise and implement international SPS standards.

[42] There are also other provisions of the SPS Agreement which explicitly or implicitly touch upon the special treatment of developing countries. This particularly includes Article 5.1, Article 5.7, Annex B(2), and B(9).

[43] Other entities, such as international organizations, may be granted observer status (either on a permanent or ad hoc basis). The full list of entities that have observer status may be found in: SPS Committee Requests for Observer Status in the Committee on Sanitary and Phytosanitary Measures Revision, G/SPS/W/78/Rev.3 (2007).

[44] For extensive discussion of the SPS Committee, see Joanne Scott, *The WTO Agreement on Sanitary and Phytosanitary Measures. A Commentary* (2007) 41–74.

specific functions, which are enumerated in subsequent paragraphs of Article 12. First of all, the Committee is expected to encourage and facilitate consultations and negotiations between its Members on specific SPS issues. This function is principally performed through organizing a series of meetings, which normally take place at least three times a year. During such meetings, WTO Members have an opportunity to raise and subsequently discuss with other Members, specific trade concerns relating to the application and maintenance of SPS measures. If the concerned Members continue such discussion in the form of bilateral negotiations, they are obliged to inform the SPS Committee of its outcome. The Committee is also responsible for the promotion, on the basis of international standards, guidelines and recommendations, of the standardization process in the SPS area. In particular, the SPS Committee monitors the process of international harmonization and the use of such standards by WTO Members (Article 12.4).[45] Thirdly, the Committee drafts different guidelines, which aim to clarify the legal content of specific provisions of the SPS Agreement (eg consistent application of ALOPs, equivalence, notification, regionalization and transparency). Fourthly, some specific duties are conferred on the Committee, such as recognizing international standard-setting organizations as relevant for the purpose of the SPS Agreement (Annex A(3)); or granting time-limited exceptions from the obligations of the Agreement to developing countries (Article 10.3). Fifthly, the Committee is obliged to review the operation and implementation of the SPS Agreement and, where appropriate, it may submit to the Council for Trade in Goods proposals on possible amendments to the Agreement.[46] To date two review processes have been completed, while the third one is currently in progress. Article 12 also contains provisions that are intended to guarantee efficient and mutually advantageous cooperation in the SPS field. The SPS Committee is therefore obliged to maintain close contacts with international standard-setting organizations in order to secure the best available scientific and technical advice for the administration of the SPS Agreement. This should also help to avoid unnecessary duplication of the work that is done by the different bodies, including the Committee itself (Article 12.5).

The normative nature of different documents that are adopted by the SPS Committee is not entirely clear. Those which relate to the internal procedure of the Committee (eg the Rules of Procedures)[47] are undoubtedly binding on all Members. However, the other types, such as guidelines, appear to be more problematic. They use language which indicates their non-binding character (eg 'should' instead of 'shall'). This is also reflected in the explicit reservation

[45] On that basis, the SPS Committee adopted a provisional Procedure to Monitor the Process of International Harmonization and the Use of International Standards, Guidelines or Recommendation, G/SPS/W/200. This procedure was subsequently extended for an indefinite period of time.

[46] Note that to date there has been no such proposal.

[47] SPS Committee, Working Procedures of the Committee, G/SPS/1 (1995).

that is always included in such a document.[48] At the same time, they cannot be regarded as providing a binding interpretation, since this task is reserved for the Ministerial Conference and the General Council.[49] Having said that, it is difficult to accept the position that they are without any legal relevance. Note that their adoption is preceded by a transparent and comprehensive discussion that allows all participants to express their views and concerns, while the final text is approved only when consensus occurs. This obviously guarantees that the outcome reflects the will of WTO Members. Indeed, the literature recognizes the acts of the WTO organs, such as guidelines adopted by the SPS Committee, as constituting a source of WTO law. Albeit that they cannot be enforced through the dispute settlement system, they 'must be taken into account by panels and the Appellate Body'.[50] But how can this be done technically?

The customary rules of interpretation as provided by the VCLT seem to offer a possible solution to this problem. Remember that dispute settlement bodies, when interpreting WTO provisions, are obliged to take into account any subsequent agreement between the parties regarding the interpretation of the treaty or the application of its provisions. The same is true for any subsequent practice in the application of the treaty which establishes the agreement of the parties regarding its interpretation.[51] The decisions of the SPS Committee appear to potentially fall into one of these categories.[52] It is clear that such decisions do not constitute an international treaty with equivalent status to the SPS Agreement. However, there is nothing in the VCLT which would imply the existence of such a threshold as a prerequisite for the application of Article 31.3(a). To the contrary, the jurisprudence of different international tribunals tends to interpret this provision liberally and accepts less formal or even informal arrangements as sufficient.[53] What seems to be more important is a common understanding that emerges between the parties at some point in time after the conclusion of the original treaty. As mentioned above, the adoption of a decision is preceded by intense discussion between all the members of the SPS Committee, while a final decision is adopted on the basis of consensus. This clearly may amount to a common comprehension. Another possibility is to regard the Committee's decisions as evidence of subsequent practice in the application of the SPS Agreement, which establishes the

[48] See eg, SPS Committee, Guidelines to Further the Practical Implementation of Article 5.5, G/SPS/15 (2000) 1, which specifically provides that '[t]hese guidelines do not add to nor detract from the existing rights and obligations of Members under the SPS Agreement nor any other WTO Agreement'.

[49] Article IX:2 of the WTO Agreement; note also that SPS guidelines also stipulate that they do not provide any legal interpretation of or modification to the Agreement itself.

[50] Peter Van den Bossche, *The Law and Policy of the World Trade Organization: Text, Cases, and Materials* (2005) 56. Lack of enforcement results from the fact that decisions of the SPS Committee cannot be considered to constitute 'covered agreement' for purposes of the DSU.

[51] Article 31.3 of the VCLT.

[52] See also, Marceau and Trachtman ('Technical Barriers to Trade Agreement') fn 127.

[53] Richard K Gardiner, *Treaty Interpretation* (2008) 220–21 giving as an example the arbitral award issued in *European Molecular Biology Laboratory Arbitration (EMBL v Germany).*

understanding of the parties as to its meaning. The practical disadvantage of the second approach is the fact that one would probably need to look for additional evidence of such practice, which is external to the decision itself (eg whether a practice is indeed concordant, common, and consistent). This can be deduced, for example, from statements and views that are expressed by the members in the course of adopting a decision.

Note also that conceptualizing decisions of the SPS Committee as a source of interpretative guidelines is entirely compatible with both Article IX(2) of the WTO Agreement as well as the general reservation clauses that are normally included in such documents. The VCLT does not see these elements as creating any binding interpretation, they rather inform, together with other factors, the interpretative process, while the discretion always remains with the interpreter.[54] In the context of the SPS Agreement this would mean that dispute settlement bodies, although obliged to take into account decisions of the SPS Committee, retain a wide margin of discretion when interpreting provisions of the Agreement. Of course, since the issue has never been addressed in the case law, some uncertainty as to the actual practice of the dispute settlement bodies remain. This fact has recently provoked a group of WTO Members to express their desire to seek legal advice from the WTO Secretariat on the question of whether these decisions may constitute a subsequent agreement or practice.[55] At the time of writing this book, the Secretariat has not provided any answer.

2.2.6 Dispute settlement

The process of dispute settlement under the SPS Agreement, including the consultation phase, is governed by Articles XXII and XXIII of GATT 1994 as elaborated and applied by the DSU.[56] Interestingly, the SPS Agreement refers to the right of WTO Members to have recourse to alternative dispute settlement mechanisms available within other international organizations or established under international agreements (Article 11.3). This includes, for instance, procedures provided by the OIE and the IPPC.

Due to the technical character of the SPS disputes, the SPS Agreement explicitly confirms the right of a panel to seek advice from experts on scientific and technical issues or to consult the relevant international organizations. In fact the Agreement encourages panels to do so, by using the expression 'should'

[54] Note that Article 31.3 of the VCLT only mandates an interpreter to take into account the subsequent agreement and practise.

[55] SPS Committee, Guidelines to Further the Practical Implementation of Article 6 of the Agreement on the Application of Sanitary and Phytosanitary Measures, 25 February 2008, G/SPS/W/218, fn 1.

[56] For a general discussion on the WTO dispute settlement system, see David Palmeter and Petros C Mavroidis, *Dispute Settlement in the World Trade Organization: Practice and Procedure* (2004).

as opposed to 'may' that is included in Article 13.2 of the DSU.[57] This can be done either at the request of one of the parties to the dispute or on its own initiative. At the same time, it remains at the discretion of a panel whether to establish an advisory technical experts' group or to consult the relevant international organization. Obviously panels can do both. The choice of expert(s) rests on a panel, although the parties to the dispute should also be consulted.[58] Notwithstanding that the SPS Agreement speaks about the assistance of the technical experts group, the case law preferred consultations with individual experts. This approach, although questioned in the literature,[59] appears to be fully compatible with both Article 11 of the SPS Agreement and relevant provisions of the DSU. Under WTO law panels enjoy a considerable discretion in the facts-finding process, including the way they seek advice on technical matters.[60] On the other hand, the discretion of panels is limited by the need to afford due process to the parties of the dispute. This obligation, for example, requires a guarantee that experts who advise the panel are independent and impartial.[61]

The SPS Agreement does not set out any special procedural rules regarding the appointment of experts and the subsequent consultation process. The applicable provisions for technical expert groups may be found in Appendix 4 of the DSU. As far as the participation of individual experts is concerned, the relevant rules are adopted on an ad hoc basis and formally apply only to particular cases.[62] Again the determination of those rules takes place in consultation with the parties to the dispute. It is also an established practice to request the experts to provide written answers to specific questions drafted by the panel, while parties are given the opportunity to comment on those answers. The procedure normally provides for meeting(s) of the panel with parties and experts where additional clarification or information may be sought.

In discussing the dispute settlement process under the SPS Agreement, two other issues require further elaboration: the allocation of burden of proof between the parties to the dispute, and the applicable standard of review. As will be discussed later in more detail, their proper conceptualization has important consequences for the operation of the SPS Agreement as a whole.

[57] Article 13.2 of the DSU provides that 'panels may seek information from any relevant source and may consult experts to obtain their opinion on certain aspects of the matter. With respect to a factual issue concerning a scientific or other technical matter raised by a party to a dispute, a panel may request an advisory report in writing from an expert review group'.

[58] For details concerning the role of the experts under the SPS Agreement, see Joost Pauwelyn, 'Expert Advice in WTO Dispute Settlement' in George A Bermann and Petros C Mavroidis (eds), *Trade and Human Health and Safety* (2006) 235–56.

[59] Joost Pauwelyn, 'The Use of Experts in WTO Dispute Settlement' (2002) 51(2) International and Comparative Law Quarterly 325.

[60] Panel Report, *US – Continued Suspension*, para 7.74.

[61] Appellate Body Report, *US – Continued Suspension*, para 446.

[62] For example, Panel Report, *US – Continued Suspension*, Annex A5.

(a) Burden of proof

The SPS Agreement does not contain any specific rules on the allocation of the burden of proof between the parties to the dispute and the SPS case law has simply followed the general approach of WTO law.[63] Thus it is for a party who asserts a particular fact, irrespective of whether it is a claimant or a respondent, to provide evidence which would support its assertion (claim or defence). If that party is successful in adducing evidence sufficient to raise a presumption that what is claimed is true (prima facie case), the burden shifts to the other party, who will fail unless it adduces sufficient evidence to rebut this presumption.[64] This means that it is for a complainant to establish a prima facie case of inconsistency of a particular SPS measure with provision(s) of the SPS Agreement (because it a complainant who asserts that a measure is inconsistent with WTO law). When such a prima facie case is made, the burden of proof shifts to the defendant,[65] who must counter or refute the claimed inconsistency.[66] If a particular WTO provision is an exception or an affirmative defence (a form of justification for an otherwise inconsistent measure), the above logic requires a defendant to be assigned with the burden of proof as a party making particular claims (ie that a measure which violates a particular WTO provision is nevertheless justified). This was confirmed by the Appellate Body when it observed, with respect to Article XX of GATT 1994, that 'the burden of establishing... a defence should rest on the party asserting it'.[67] Sometimes, as in the case of the SPS Agreement, it may be disputable whether a particular provision can be characterized as a positive obligation or as an exception. I will discuss this issue in more detail in Chapters 3 and 5. The Appellate Body also noted that what is required to establish a prima facie case is highly context-dependent and 'will necessarily vary from measure to measure,

[63] For a general discussion on burden of proof in WTO law, see Yasuhei Taniguchi, 'Understanding the Concept of Prima Facie Proof in WTO Dispute Settlement' in Merit E Janow, Victoria Donaldson, and Alan Yanovich (eds), *The WTO: Governance, Dispute Settlement, and Developing Countries* (2008) 553: James Headen Pfitzer and Sheila Sabune, 'Burden of Proof in WTO Dispute Settlement: Contemplating Preponderance of the Evidence', ICTSD Dispute Settlement and Legal Aspects of International Trade, Issue Paper No 9 (2009).

[64] Cf Appellate Body Report, *US – Shirts and Blouses*, 14.

[65] Note a certain lack of clarity in WTO case law. The common law distinguishes between the burden of persuasion, which indicates which party must 'bear the risk that after evaluating all the evidence and arguments presented by both sides the adjudicator will in equipoise', whereas the burden of production determines which party must present evidence to support its version of all disputed facts (John J Barceló III, 'Burden of Proof, Prima Facie Case and Presumption in WTO Dispute Settlement' (2009) 42(1) Cornell International Law Journal 23, 27. The burden of proof in civil law jurisdictions is principally understood as burden of persuasion. It is not entirly clear whether WTO law recognizes both aspects of burden of proof (as in common law systems) or only one (as in civil law jurisdictions).

[66] Appellate Body Report, *EC – Hormones*, para 98; the Appellate Body defined prima facie case as the 'one which, in the absence of effective refutation by the defending party, requires a panel, as a matter of law, to rule in favor of the complaining party': ibid, para 105.

[67] Appellate Body Report, *US – Shirts and Blouses*, 16.

provision to provision, and case to case'.[68] Nevertheless, it is argued in the literature that a test of a prima facie case is not a very demanding task under the SPS Agreement and a WTO Member adopting a measure needs to be prepared to present, during the panel proceedings, sufficient scientific backup.[69]

The standard described above is, however, not easily identifiable in practice. Note that the panel makes a decision as to whether a proponent established a prima facie case only after reviewing all the evidence submitted by both parties. This would suggest that panels actually apply a standard of 'weighing up of the evidence' or 'preponderance of evidence' rather than a traditional prima facie approach. Different evidence presented by both parties is simply balanced against each other in order to determine which side prevails.[70]

(b) Standard of review

The standard of review may be defined as 'the level of intensity of the scrutiny that the reviewing body will exert over the decision or regulation being reviewed'.[71] The SPS Agreement does not provide any specific rules on the standard of review that is applicable to the assessment of factual issues. The Appellate Body, in line with general case law, held that the applicable standard of review is defined by Article 11 of the DSU as 'neither *de novo* review as such nor "total deference", but rather the "objective assessment" of the facts'.[72] This rather imprecise clarification left future panels with little guidance as to the degree of required scrutiny. As will be discussed in Chapter 4 in more detail, the applicable standard of review is of a crucial importance in the adjudication of science-based disputes involving complex technical determinations. It defines, alongside the substantive requirement, a margin of discretion that is enjoyed by WTO Members and effectively allocates the power between the international bodies (ie panels and the Appellate Body) and national authorities.[73] The fully deferential standard results, in principle, in the inability of a panel to review the scientific findings made on a national level. Instead, a panel may only examine whether a WTO Member has followed the procedures prescribed by the SPS Agreement (eg whether it performed risk assessment). On the other hand, a *de novo* standard allows a panel to review the domestic determinations and substitute them with its own judgments.

[68] Ibid, 14.

[69] Jacqueline Peel, 'Risk Regulation under the WTO SPS Agreement: Science as an International Normative Yardstick?', Jean Monnet Working Paper No 2002/04, 74. But cf the argument of the US in *Japan – Agriculture Products* which claimed that the required threshold was set excessively high (Appellate Body Report, *Japan – Agricultural Products*, para 133).

[70] Pfitzer and Sabune ('Burden of Proof') 22, citing Edmond McGovern, *International Trade Regulation* (2006).

[71] Catherine Button, *The Power to Protect. Trade, Health and Uncertainty in the WTO* (2004) 164.

[72] Appellate Body Report, *EC – Hormones*, para 117.

[73] For a more detailed discussion on vertical allocation of power, see Matthias Oesch, *Standards of Review in WTO Dispute Resolution* (2003) 25.

That includes assessing the quality of scientific evidence put forward by parties and the accuracy of interferences that were drawn from such evidence, as well as the appropriateness of the domestic decisions on how to address specific risks. Of course, between those two extremes, there is a great array of more or less deferential/*de novo* approaches.

The standard of review, which is applicable to assessment of facts, including the assessment of scientific evidence, has to be conceptually distinguished from the standard of review that is applied to the interpretation of WTO obligations and domestic law. In principle, both the panel and the Appellate Body enjoy a wide margin of discretion when interpreting provisions of WTO law and they are not constrained by the interpretations proposed by the parties.[74] Domestic law is treated in principle as a question of fact.[75] In addition, one may also speak about the standard of the review that is applied by the Appellate Body when reviewing panel reports. As provided by Article 17.6 of the DSU, the review is limited to issues of law covered in a panel report and legal interpretations developed by a panel. At the same time, the case law also recognizes that the standard of review which was applied by a panel in the examination of facts, is to be regarded as a legal issue subject to review by the Appellate Body.[76]

The history of SPS case law may be seen as the constant attempt of dispute settlement bodies to identify, on the axis between *de novo* and deferential extremes, a proper standard of review. The case law, which followed *EC – Hormones,* went in the direction of a *de novo* review, reaching its culminating point with panels' decisions in *EC – Biotech Products* and *US/Canada – Continued Suspension*. The panels have gradually engaged in more and more intrusive assessment of scientific evidence, evaluating its quality, persuasive force, and accuracy of conclusions made on the basis of scientific data. Although the case law has never gone so far as to decide whether an SPS measure was based on the best available science and has always accepted divergent scientific opinions, panels came very close to this position. This discretion of the panel in evaluation and assessment of scientific evidence is well exemplified in one of the rulings, where the panel said that 'it is also generally within [the panel's] discretion to decide which evidence we choose to utilise in making findings'.[77] Consequently, as summarized by one of the commentators, despite the fact that the 'line between an objective standard and a *de novo* review appears blurry...an objective assessment of the facts would arguably entitle the panel to impose its own views on the scientific evidence'.[78] The Appellate Body report in *US/Canada – Continued Suspension* seems to constitute

[74] Ibid, 18.

[75] For an extensive discussion on a treatment of domestic law in the WTO context, see Sharif Bhuiyan, *National Law in WTO Law. Effectiveness and Good Governance in the World Trading System* (2007).

[76] Appellate Body Report, *EC – Hormones*, para 132.

[77] Panel Report, *US – Continued Suspension*, para 7.416.

[78] Nick Covelli and Viktor Hohots, 'The Health Regulation of Biotech Foods under the WTO Agreement' (2003) 6(4) Journal of International Economic Law 773, 783.

a turning point in SPS case law. The standard of review that was prescribed by the Appellate Body is to great extent a deferential one, giving WTO Members a substantial degree of discretion on how to assess scientific evidence and what kind of conclusions to make on its basis.

2.3 Applicability of the SPS Agreement

The scope of the SPS Agreement, at least in theory, is rather narrow, as in principle it only applies to measures intended for the protection, within the territory of the importing WTO Member, of the life and health of humans, animals and plants from certain specific SPS risks. Moreover, according to Article 1.1, measures may become relevant for the purpose of the SPS Agreement only if they directly or indirectly affect international trade. Thus, measures with exclusive internal impact fall outside the scope of the SPS Agreement. However, since Article 1.1 does not introduce any threshold with respect to such impact, one may assume that any kind of interference with international trade (including *de minimis*) will be sufficient to meet a requirement of the SPS Agreement. This conclusion is in line with the approach taken by the *Biotech* panel, which held that an SPS measure did not need to actually affect trade but that potential interference was sufficient.[79] Likewise, the applicability of the SPS Agreement does not seem to be restricted to those measures which affect a trade of a complainant, and measures having an effect only on the trade of other WTO Members may equally satisfy the requirement of Article 1.1.[80] Last but not least, the SPS Agreement applies to all SPS measures irrespectively from the time of their adoption. This means that measures taken before the Agreement came into force are also covered.[81]

The specific SPS risks consist of: (a) risks for animal or plant life or health arising from the entry, establishment, or spread of pests, diseases, disease-carrying organisms, or disease-causing organisms; (b) risks for human or animal life or health arising from additives, contaminants, toxins, or disease-causing organisms in foods, beverages, or feedstuffs; (c) risks for human life or health arising from diseases carried by animals, plants, or products thereof, or from the entry, establishment, or spread of pests; and (d) other risks related to the entry, establishment, or spread of pests.[82] The relevant footnote explains that the notion of animal includes fish and wild fauna, plant includes forests and wild flora, pests

[79] This may be deduced from the use of word 'may', which indicates a potential, rather than actual, effect.

[80] Cf Panel Report, *EC – Bananas III*, para 7.50 ('in our view a Member's potential interest in trade in goods or services and its interest in a determination of rights and obligations under the WTO Agreement are each sufficient to establish a right to pursue a WTO dispute settlement proceeding').

[81] Appellate Body Report, *EC – Hormones*, para 126.

[82] Annex A(1) of the SPS Agreement.

include weeds, while contaminants include pesticide and veterinary drug residues as well as other extraneous matters.[83] The list is exhaustive and if a particular risk does not fit to one of the categories, the SPS Agreement does not apply.

As far as the legal form of a measure is concerned, the SPS Agreement broadly stipulates that an SPS measure embraces all relevant laws, decrees, regulations, requirements and procedures. This broad formulation seems to cover also informal procedures not codified in the law.[84] In addition, Annex A(1) provides a non-exhaustive list of relevant requirements and procedures, such as end product criteria, processes and production methods, testing, inspection, certification and approval procedures, quarantine treatments (eg relevant requirements associated with the transport of animals or plants, or with the materials necessary for their survival during transport), provisions on relevant statistical methods, sampling procedures and methods of risk assessment, and packaging and labelling requirements directly related to food safety.[85]

The comprehensive analysis of the applicability conditions of the SPS Agreement was provided for the first time by the *Biotech* panel. The examination of the report indicates that specific categories of risk enumerated by the SPS Agreement are to be interpreted broadly. First, the panel believed that environmental risks, which can be reduced to life and health of animals and plants, fall within the scope of the SPS Agreement.[86] In particular, this means that risks to biodiversity are covered by the SPS Agreement because any damage to biodiversity implies damage to living organisms.[87] As will be discussed below, other types of environmental risks may also be covered by the agreement, on the basis of subparagraph (d).

Secondly, the panel held that the expression 'animal and plant' includes micro- and macro-flora and fauna, irrespective of whether its components were target or non-target organisms.[88] According to the panel, this broad interpretation was justified by the footnote to Annex A, which mandates to include wild flora and fauna within the expression, indicating that the scope of the phrase was meant to be comprehensive in coverage.

Thirdly, an equally broad interpretation was applied to the expression 'arising from'. The panel, referring to the dictionary definition, interpreted the term as meaning 'to occur as a result of'. This finding, coupled with the observation of its unqualified character,[89] led the panel to conclude that the scope of the notion 'risks

[83] Ibid, fn 4.

[84] But cf Panel Report, *US – Continued Suspension*, para 7.429 (failing to recognize that informal procedures can be also qualified as an SPS measure).

[85] Annex A(1) of the SPS Agreement.

[86] Panel Report, *EC – Biotech Products*, para 7.207.

[87] Ibid, para 7.372.

[88] Ibid, para 7.219 (also adding that 'we note that, textually, the unqualified terms "animal" and "fauna", on the one hand, and "plant" and "flora", on the other, can encompass macro- and micro-fauna, on the one hand, and macro- and micro-flora, on the other').

[89] Ibid, para 7.226 (noting that the expression 'risks arising from' is not qualified by words 'directly' or 'immediately').

arising from' is sufficiently broad to include risks which may only potentially arise in the future.[90] In consequence, according to the panel, the SPS Agreement is not limited to the measures which address risks of immediate and direct character (eg the spread of pests), and equally covers those measures which are concerned with risks that are indirect and more remote in time. In practical terms, this means that the required causality between the entry, establishment, or spread of pests, diseases, disease-carrying organisms, or disease-causing organisms, and risk to animal and plant life and health is not very demanding (eg potential gene transfer to other plants leading to expansion of this species and reduction of biological diversity). At the same time, the panel did not explain what level of causality is required in order to satisfy the condition of 'risks arising from'.

Fourthly, in defining the notion of 'pest', the panel again referred to the dictionary meaning of this term as well as the context of the relevant subparagraphs of Annex A. According to the panel, a pest under the SPS Agreement is to be understood as an 'animal or plant which is destructive, or causes harm to the health of other animals, plants or humans, or other harm, or a troublesome or annoying animal or plant'.[91] An interesting thing about this interpretation is the fact that this description is broader than the general definition provided by the IPPC.[92] This approach seems, however, to be justified since the language of the SPS Agreement indeed indicates a broader understanding of the concept of pest.[93] On that basis, the panel concluded that cultivated plants (ie genetically modified (GM) plants) growing where they are not wanted and causing harm, qualify as weeds, a subcategory of pests.[94]

Fifthly, the panel noted that even without deciding whether a GM plant which cross-breeds with other plants could be viewed as a pest, it was sufficient to establish that cross-breed plants could be classified as pests.[95] In the words of the panel 'there is nothing in the text of Annex A(1) to suggest that the product subject to an SPS measure—in this case, a GM plant to be released into the environment—need itself be the pest which gives rise to the risks from which the measure seeks to protect'.[96] The panel was satisfied with the existence of a rational relationship between controlling the release into the environment of the GM plant which might cross-breed with other plants, and the purpose of protecting animal or

[90] Ibid, para 7.225.

[91] Ibid, para 7.240 (observing that the footnote in Annex A stipulates that pests also include weeds, while the reference in various subparagraphs of Annex A(1) relates pest to different risks: to human, animal and plant life and health as well as other damages).

[92] The IPPC defines pest as '[a]ny species, strain or biotype of plant, animal or pathogenic agent injurious to plants or plant products' (International Plant Protection Convention, Article II.1). Note that this wording excludes harm to animal and human life and health as well as other damages caused by a pest as defining elements.

[93] According to the SPS Agreement a pest may be defined through reference to harm to human health (eg allergenic reactions), animal life and health (eg outcompeting those plants which are vital for fauna in particular region), as well as other damage (eg economic).

[94] Panel Report, *EC – Biotech Products*, para 7.245.

[95] Ibid, para 7.255.

[96] Ibid, para 7.258.

plant life or health from risks arising from the entry, spread, or establishment of cross-breeds with undesired traits.[97] Note also a certain logical trick here. The panel seems to construe the notion of a pest not in absolute terms (whether a GM plant is a pest or not) but with reference to potential risks, which may arise in particular situations. Thus, a GM plant growing where it is undesired would be a pest, while a GM plant as a vector of genetic materials may not be.

Sixthly, when addressing risks arising from the establishment or spread of diseases, disease-carrying organisms, or disease-causing organisms, the panel found again that it was not necessary to take a position on whether a GM plant could have been classified as falling into one of these categories. Instead, the panel was satisfied with the fact that there was a rational relationship between the release of a GM plant containing the antibiotic resistance marker genes and the establishment or spread of a pathogen (understood as 'disease-causing organisms'), which might develop resistance to the antibiotic.[98] Note that this approach closely corresponds with the interpretation adopted by the panel with respect to the notion of 'pest'.

Seventhly, the panel noted that GM crops grown for the explicit purpose of providing food to humans and animals could be classified as food or feedstuff respectively. This rather uncontroversial statement was accompanied by the finding that GM plants grown for other purposes (eg as a material for biofuels) might also be qualified as food, as these plants may be eaten by non-target animals (eg rabbits or birds).[99] After establishing that a GM plant can be regarded as food (or feedstuff), the panel also found that added genes constitute an additive. The panel made two arguments in support of its finding. First, it referred to the definition of the Codex, which defines an additive as an intentional addition in the manufacturing process for technological reasons. According to the panel, this language also covered the development and production phase of seeds of the GM plants. Secondly, the panel also observed that in any event the Codex definition was not dispositive of the meaning of the term,[100] while its ordinary meaning indicated that added genes can be defined as an additive.[101]

Eightly, the panel held that protein produced through the unintended expression of modified genes in agricultural crops may be considered as a 'contaminant',[102] while a poisonous substance, which may be produced during the metabolism or growth of a GM crop, could qualify as a 'toxin'.[103]

Ninthly, the panel found that the language used in the expression 'other damage within the territory of the Member from the entry, establishment or spread of

[97] Ibid. [98] Ibid, paras 7.282–4.
[99] Ibid, para 7.292.
[100] It seems that the panel made the second argument as it felt uncomfortable with qualifying a gene as a substance and development of seeds as a manufacturing process.
[101] Panel Report, *EC – Biotech Products*, paras 7.297–301 (noting that 'Annex A(1)(b) does not suggest that for an added substance to qualify as an "additive" in a food, the substance needs to have been added at a particular stage prior to the consumption of the food in question').
[102] Ibid, para 7.313. [103] Ibid, para 7.323.

pests' also required a broad interpretation.[104] According to the panel, it particularly includes damage to property (eg water intake systems), economic damage (eg damage in terms of sales lost by farmers due to the presence of unwanted genetically modified organisms (GMOs) on the agriculture fields)[105] as well as damage to the environment other than damage to the life or health of animals or plants (eg adverse effects of GMOs to non-living components of the environment, including damage to geochemical processes).[106]

Finally, the panel qualified labelling requirements as an SPS measure, although only to the extent that it was related to health concerns.[107] According to the panel, the existence of a rational relationship between the purpose of the legislation (ie protection of human life and health) and the labelling requirement was sufficient to establish the applicability of the Agreement. On the other hand, the same requirement applied for other purposes, such as prevention of misleading consumers, fell outside the scope of Annex A(1) of the SPS Agreement.[108]

As a result of this expansive interpretation, the *Biotech* panel concluded that almost all risks addressed by the EC measures fell under one of the subparagraphs of Annex A(1). This in turn suggests that in principle any domestic regulatory framework dealing with the approval and control of GMOs will be classified in future as an SPS measure subject to the disciplines of the Agreement.

This broad conceptualization of the applicability of the SPS Agreement was met with intense criticism in the literature. In particular, it is submitted that the SPS Agreement was only designed to address classic SPS measures that aim at protecting 'specific subjects from damages or risks that arise from specific causes listed in Annex A'.[109] This is reflected, the argument goes, both in the negotiating history and previous WTO jurisprudence. The drafting history shows that the negotiators saw the adoption of the SPS Agreement as a response to the liberalization of trade in agriculture products, worrying that the reduction of tariff barriers would be undermined by the excessive recourse to SPS measures. Another concern which negotiators had in mind was the long-standing dispute between the EC and the US over the use of hormones for growth promotion purposes. Consequently, relevant SPS risks appear to be limited to traditional quarantine and food safety risks. As far as the previous case law is concerned, there are indications which may suggest a rather narrow scope of the SPS Agreement. All disputes decided before *EC – Biotech Products* were concerned with risks that were

[104] Ibid, para 7.370 (noting that the category of 'other damage' is not qualified by any other expression).

[105] Ibid (basing its observation on the dictionary meaning of damage, which defines it as 'physical harm impairing the value, usefulness, or normal function of something' and on the context of Annex A(1)(d), particularly Article 5.3 which recognizes economic damage as one of the relevant elements in a risk assessment).

[106] Ibid, para 7.372.

[107] Ibid, para 7.389.

[108] Ibid, para 7.412.

[109] Christiane R Conrad, 'The EC-Biotech Dispute and Applicability of the SPS Agreement: Are the Panel's Findings Built on Shaky Ground?' (2007) 6 World Trade Review 233, 241.

directly and physically related to regulated products (eg residues of hormones in beef or pathogens responsible for fire blight that may be found in apples).[110] On the other hand, potential risks that are posed by GMOs are of a quite different nature. Some of them require intermediation objects (eg a cross-breed plant) or only indirectly relate to animal and plant life and health (eg risks for plant and animal health resulting from changes in biogeochemical cycles, particularly carbon and nitrogen recycling).

Likewise, environmental risks, including risks for biodiversity, are generally far more complex than any of the categories that are enumerated in Annex A(1)[111] and should be excluded from the scope of the SPS Agreement. This conclusion seems to find some support in the language of the TBT Agreement. Note that its Article 2.2 explicitly refers to the protection of the environment as one of legitimate objectives for adoption of technical regulations, whereas the SPS Agreement does not contain any comparable wording. This fact may be relevant in determining what constitutes a relevant risk for the purpose of the SPS Agreement. As noted by the EC 'when the drafter of an international agreement uses a term in one instrument but not in another, the drafter intended to exclude that term from the latter instrument'.[112]

A separate set of concerns relates to the broader implications that may result from the expansive reading proposed by the *Biotech* panel. Remember the panel's conclusion that the regulated product (GM plant) does not have to be a pest or a carrier of a risk agent. It is sufficient for the purpose of applicability of the SPS Agreement that there is rational relationship between a product and a risk agent (cross-breed plant).[113] The same approach was taken with respect to disease, disease-causing, and disease-carrying organisms. The panel did not require that the GMO fall into one of these categories (or be a host for a risk agent), and was satisfied with the existence of a rational relationship between the GMO and that agent (ie pathogen that might develop resistance to antibiotics). Consequently, the SPS Agreement, under the analysis of the *Biotech* panel, equally covers risks that are indirect or more remote in time. What was important for the panel was the existence of a causal relationship between the regulated product and the risk agent.

If one takes this approach to its extreme, the majority of environmental regulations will surely fall into the ambit of the SPS Agreement. Since it is not necessary for the regulated product to be a risk agent or a direct risk transmitter and the existence of some causal relationship is sufficient, in theory every environmental regulation that is reducible to the protection of human, animal or plant health and which involves at some point a pest or disease would qualify as an SPS

[110] Conrad (EC – Biotech Dispute) 240; Jacqueline Peel, 'A GMO by any Other Name... Might be an SPS Risk!: Implications of Expanding the Scope of the WTO Sanitary and Phytosanitary Measures Agreement' (2007) 17(5) European Journal of International Law 1009, 1016.

[111] Conrad (EC – Biotech Dispute) 242.

[112] Panel Report, *EC – Biotech Products*, para 7.198.

[113] Ibid, para 7.258.

measure.[114] The problem with such an unrestrained interpretation of Annex A(1) is the fact that environmental protection is a 'quite different field from the [SPS] Agreement's traditional subject matter'.[115] It frequently relies on the precautionary principle, while the assessment of risks is not limited to scientific considerations but also includes other elements such as broad environmental effects and socio-economic consequences. Applying strict scientific disciplines of the SPS Agreement may undermine the ontological basis of such regulatory policies, both on a national and international level. In addition, it may also have a 'chilling effect over negotiations for new MEAs [multilateral environmental agreements] or supplementary protocols that might overlap with the (extended) ambit of the SPS Agreement'.[116] Since the TBT Agreement provides more permissible disciplines, presumably it would be easier to reconcile the obligations of international environmental law with WTO obligations under this Agreement.[117]

Having said that, one also has to admit that the language of the SPS Agreement is not without ambiguity. It includes an explicit reference to the protection of wild flora and fauna, elements that are traditionally considered as falling within the notion of environmental protection.[118] This may indicate that the founding Members actually intended to create an agreement with a broader scope of application that would not be limited only to traditional SPS risks. One may also find hints of such an intention in the drafting history. As observed by the *Biotech* panel, one of the initial drafts of the SPS Agreement incorporated language that explicitly removed environmental measures from its scope. This clause was, however, dropped out in the final version of the Agreement,[119] which may suggest that environmental measures are not per se excluded from the scope of the SPS Agreement. Further, the inclusion of subparagraph (d) in the definition of relevant SPS risks may be equally seen as an attempt by the drafters to extend the reach of the Agreement. The residual category of other damage arising from the entry, establishment, or spread of pests is very broad and potentially covers a number of environmental risks. Consequently, it appears rational to accept that some environmental regulations should be regarded as SPS measures. Besides that, there are also pragmatic reasons which would speak against the general and unconditional exclusion of environmental risks from the scope of the SPS Agreement. Note that this would allow WTO Members to escape the scrutiny of the Agreement by simply labelling measures as environmental. Accordingly, whether a particular measure aims to address environmental rather

[114] Peel ('GMO by any Other Name') 1026 (giving the example of qualifying regulations designed to protect bio-diverse marine ecosystems from the adverse effects of pesticide run-off as an SPS measure on the basis that risks arise indirectly from the introduction of a pest through trade).

[115] Ibid. [116] Ibid, 1029.

[117] Conrad ('EC – Biotech Dispute') 235.

[118] Cf Panel Report, *EC – Biotech Products*, para 7.204.

[119] Ibid, para 7.211 (the relevant clause provided '[r]equirements concerning quality, composition, grading, [consumer preferences, ... the environment or ethical and moral considerations] are *not* included in the definition of sanitary or phytosanitary measures').

than traditional SPS risks cannot be decisive to determine the applicability of the SPS Agreement.

An important part of the *Biotech* panel's conclusions on the applicability of the SPS Agreement was based on the concept of a rational relationship. Remember that it was not necessary for a GMO to be a pest or pest risk transmitter, and a causal relationship between the product (a GMO) and the establishment and spread of pests was sufficient. It is true that an overly broad conceptualization of this term may expand the scope of the SPS Agreement beyond what may be properly considered as an SPS risk. Although the *Biotech* panel did not establish any clear standard that would explain the required level of causality, its language actually indicates the existence of some limits. First, the panel asked for a *rational* relationship rather than merely causal association. While the rational threshold in itself does not predetermine the level of required connection, it arguably imposes some restraints that prevent the applicability of the SPS Agreement to every imaginable environmental measure. The limits which are implied in the rational relationship were actually explicitly recognized by the *Biotech* panel in another section of its report. The panel particularly gave the example of an environmental measure aimed at the reduction of air pollution that was applied to protect the life or health of plants.[120] Despite the fact that such a measure is concerned with one of the subjects of protection enumerated in Annex A(1) and may be reduced to a risk posed by disease,[121] it will nevertheless be considered as falling outside the scope of the SPS Agreement because the causal relationship is too remote. Interestingly, even the complainants in *EC – Biotech Products* were aware of the limited coverage of the SPS Agreement. For instance, Canada identified some general types of environmental measures, such as those relating to air and water quality or waste management, as falling outside the scope of the Agreement.[122]

One should also not be surprised by the reluctance of the panel to formulate any general standard here. Although such an approach would improve the predictability of the SPS system (and from this viewpoint can be criticized), it simply follows the normal practice of the WTO dispute settlement bodies to decide a particular case rather than to establish blueprint criteria. At the same time, the more recent WTO case law may also suggest that the scenario under which the SPS Agreement overshadows the whole field of environmental protection is somehow exaggerated. In *Brazil – Retreaded Tyres*, Brazil introduced a ban on the importation of retreaded tyres, claiming that those tyres, having a shorter lifespan, contribute to the faster accumulation of waste. This, in turn, increases

[120] As explained by the panel 'to the extent that high levels of air pollution could result in certain plant species lacking sufficient sunlight for them to exist and survive' (Panel Report, *EC – Biotech Products*, para 7.210).

[121] As noted by Conrad, air pollution may lead to biological interactions that impact the spread of diseases and disease-causing organisms as well as corresponding risks to plant health (Conrad ('EC – Biotech Dispute') 239).

[122] Panel Report, *EC – Biotech Products*, para 7.205.

the population of mosquitoes, which frequently act as vector for malaria and pose direct risks to human health. If one follows the extreme interpretation of the *Biotech* decision, it is not difficult to find such a requirement to be an SPS measure. There is a causal relationship between the importation of retreaded tyres and amplification of risk for human health (the increase of the mosquito population being the mediating element). Mosquitoes as such can be easily qualified as disease-carrying animals within the meaning provided by Annex A(1)(c) of the SPS Agreement. Note, however, that the EC never argued that such a measure could be regarded as an SPS regulation. Of course, the fact that the EC did not present its claims under the SPS Agreement does not prove that its disciplines would not be applicable to the contested measure. The EC could have had other reasons for not wishing to argue in favour of a broad reach for the SPS Agreement. Nevertheless, the approach of the EC may also indicate a recognition among the WTO Members of the inherent limits in the applicability of the SPS Agreement, or at least contribute to the development of litigation practice to be used in the future for similar disputes. Remember also that Canada took essentially the same position in *EC – Biotech Products* when it identified a waste management regulatory system as falling outside the scope of the SPS Agreement.

Yet another indication is provided by the practice of the SPS Committee. A brief review of the SPS notifications shows that they relate to classic SPS risks (with the exception for GMO risks). WTO Members do not notify any environmental measures that are only loosely connected with an SPS element. This, in turn, may suggest that the GMO-related risks are just a special case under the SPS Agreement rather than the first step in the expansive development of the SPS disciplines. Last but not least, as Peel herself admits, the Appellate Body when deciding controversial issues usually displays considerable 'political astuteness'.[123] The recent ruling in *US/Canada – Continued Suspension* is a clear example of this sensitivity. One can rationally expect that the Appellate Body will be very careful when deciding that an environmental measure with only a remote SPS aspect is actually covered by the SPS Agreement.

Of course the above does not mean that that the criticism of the panel's approach is without merit. It may be reasonably questioned whether unconditional inclusion of all indirect risks posed by GMOs into the scope of the SPS Agreement is a proper approach. Similarly, one may legitimately worry about a reductionist interpretation of the concept of biodiversity. Biodiversity indeed appears to constitute an autonomous value, which exceeds the value of its components and which cannot be reduced to particular units.[124] What also seems to be disappointing in the panel report is the reasoning itself. For example, when discussing the meaning of the terms 'animal or plant', the panel observed that these notions also encompass soil and aquatic microorganisms. The panel finding was

[123] Peel ('GMO by any Other Name') 1030, citing Victor ('Sanitary and Phytosanitary Agreement').
[124] Conrad ('EC – Biotech Dispute') 242–3.

based on the observation that the SPS Agreement explicitly provides that animals and plants comprise wild fauna and flora. This meant to the panel that the scope of the phrase 'animal or plant' was comprehensive in coverage and included microorganisms.[125] This teleological interpretation differs, however, from the interpretation which was applied with respect to the term 'additive', where the panel preferred to rely on the textual methods.[126] Note, however, that the application of such an approach in the context of 'animal' and 'fauna' would lead to entirely different outcomes than those which were reached by the panel. According to the dictionary, 'an animal' means 'a multicellular organism...differing from plants in certain typical characteristics such as capacity for locomotion' and so on, while 'fauna' is defined as 'animals, especially the animals of a specific region'. On the other hand, a dictionary definition of a microorganism provides that it is 'an organism of microscopic or submicroscopic size, especially bacterium or protozoan'.[127] These definitions clearly exclude microorganism from the scope of animal category, since bacteria are unicellular creatures. Note that the point of my concern is not whether risks for microorganisms should be included or not in the SPS Agreement, but rather the consistency of panel's approach. The report as it stands now gives the impression that the panel decided beforehand to apply the SPS Agreement and simply produced the rational *ex post*. What are the criteria or justifications of applying one method of interpretation in one case and another in the other case? The panel report does not provide an answer.

Another problem in the panel analysis relates to its over-reliance on the dictionary meaning of the terms. Remember the finding of the panel that crops grown for purposes other than for human or animal consumption (eg cotton or plants grown as a material for biofuels) can be regarded as 'food' and consequently fall into the ambit of the SPS Agreement. The panel referred to the Oxford English Dictionary and found that 'food' is a substance taken into the body to maintain life and growth. This allowed the panel to classify a GM crop that was grown for a different purpose, but which could be eaten by animals (including wild fauna), as 'food' for those animals. According to the panel, this included a GM crop that was consumed by insects and GM plants eaten by non-target insects, deer, rabbits, or other wild fauna. The panel reached the same conclusion with respect to GM seeds used for sowing purposes because it found that these seeds could be spilled next to a field or on a farm and subsequently eaten by birds. As an effect of such equilibristic interpretation, the panel came up with a definition which is completely unrelated to its understanding in the SPS field, where food is always defined through reference to human consumption.

A similar problem arose with the panel's interpretation of the term 'additive'. The panel again referred to the dictionary meaning of the term. It noted that

[125] Panel Report, *EC – Biotech Products*, para 7.219.
[126] Ibid, paras 7.297–7.301.
[127] *The American Heritage Dictionary of the English Language* (2000).

an additive was defined as a substance added to another so as to give it specific qualities.[128] Then, using the same method, it concluded that a gene may be considered as a substance.[129] These two findings allowed the panel to come up with the conclusion that genes, intentionally added for a technological purpose to GM plants, were additives.[130] At the same time, the panel recognized that its definition of an additive was different from the one provided by the Codex. However, according to the panel, the Codex understanding of additive was not dispositive for the purpose of WTO law. Principally, the same approach was also adopted with respect to other terms used in Annex A(1), such as 'disease' or 'pest'. Again, the outcome of panel's interpretation could hardly be seen as satisfactory, as both additive and pest have specific meanings in the SPS field that cannot really be captured when using a dictionary.

The SPS Agreement is a legal instrument which regulates highly technical issues and the meaning of some of its terms cannot simply be established on the basis of dictionaries. Such terms have their own specific meaning, which can be ascertained only within a particular technical or scientific discipline where they are used. Consequently, it seems that the *Biotech* panel in its quest for the ordinary meaning of the words actually failed to identify it. The panel should rather look into the documents of sister organizations or other international entities such as the FAO or the WHO specializing in SPS issues.[131] Such documents may be regarded as a kind of glossary which reveals the meaning ascribed to specific technical terms by the international community.[132] Note also that this kind of approach is not alien to WTO dispute settlement bodies which sometimes use the extraneous normative materials for identification of the ordinary meaning of words used in WTO law.[133] Although it may look as if the *Biotech* panel followed this approach,[134] the ultimate outcome indicates that the meaning of terms used in Annex A(1) was determined in principle on the basis of dictionaries.

[128] Panel Report, *EC – Biotech Products*, para 7.297.

[129] Ibid, para 7.298. [130] Ibid, para 7.301.

[131] Contrast Joe McMahon and Margaret A Young, 'The WTO's Use of Relevant Rules of International Law: An Analysis of the Biotech Case' (2007) 56(4) International and Comparative Law Quarterly 907, who criticize such an approach as being incompatible with the VCLT. In particular, they argue that the *Biotech* panel actually followed this approach and considered extraneous normative materials as a context. However, according to the authors, the context in the VCLT is defined very narrowly and cannot include such materials. As a matter of principle, the arguments may be defended. At the same time, the use of extraneous normative materials by the *Biotech* panel seems to be more apparent than real.

[132] Cf Robert Howse, 'The Use and Abuse of Other "Relevant Rules of International Law" in Treaty Interpretation: Insight from WTO Trade/Environment Litigations', IILJ Working Paper 2007/1, 15.

[133] Eg Decision by the Arbitrators, *Brazil – Aircraft (Article 22.6 – Brazil)* fn 48; Appellate Body Report, *US – Shrimp*, para 130; also cf Campbell McLachlan, 'The Principle of Systemic Integration in Treaty Interpretation and Article 31(3)(c) of the Vienna Convention' (2005) 54 International and Comparative Law Quarterly 279, 315.

[134] The *Biotech* panel made a number of reference to such materials (see eg, Panel Report, *EC – Biotech Products*, paras 7.299, 7.321, 7.334).

2.4 The relationship between the SPS Agreement, the TBT Agreement, and GATT 1994

2.4.1 The SPS and TBT Agreements

The SPS and TBT Agreements are both concerned with internal measures that constitute NTBs to international trade. How should one navigate between those two agreements? What features of national regulations are decisive in qualifying a measure as falling under the SPS or TBT Agreement? The TBT Agreement broadly states that it covers all types of technical regulations, product standards, and conformity assessment procedures. In principle, the applicability of the TBT Agreement is not limited by the objective of a measure. Although the Agreement enumerates some examples, such as national security, protection of human health or safety, animal or plant life, and health or environment, this catalogue is not exhaustive. The scope of the SPS Agreement is much more limited as it only applies to specific types of technical regulations (product standards and conformity assessment procedures), which are intended to protect the life and health of humans, animals, and plants from certain specific SPS risks. Thus, the distinction between the two agreements is not based on the nature of regulated product but on the purpose underlying a measure and a type of risk which it seeks to address. If the purpose of the measure is to protect human, animal, and plant health or life from risks enumerated in Annex A(1) of the SPS Agreement, such a measure, despite being a technical regulation, will only be qualified as an SPS measure.

The relationship between the SPS and TBT Agreements seems to be clear. Article 1.5 of the TBT Agreement explicitly provides that its provisions do not apply to SPS measures as defined in the SPS Agreement. Similarly, Article 1.4 of the SPS Agreement confirms a mutually exclusive relationship between the two agreements by stressing that the SPS Agreement does not affect the rights of WTO Member under the TBT Agreement.[135] However, this clear-cut distinction is not always easy to operate. Although in the majority of cases, it should be rather easy to identify which agreement is applicable, there are also more problematic situations. The first group consists of measures that appear on their face to be technical regulations covered by the TBT Agreement but which were actually adopted due to SPS concerns.[136] The second group includes multi-purpose measures that serve both SPS and TBT objectives (eg measures which address certain aspects of food safety as well as information on nutritional value).

[135] See also, Panel Report, *EC–Hormones (US)*, para 8.29.

[136] Pauwelyn also notes these difficulties; see Joost Pauwelyn, 'The WTO Agreement on Sanitary and Phytosanitary (SPS) Measures as Applied in the First Three Disputes' (1999) 4 Journal of International Economic Law 641, 643.

The determination of the applicable agreement is not only an academic exercise without any practical consequences. The disciplines established by each agreement differ in many important aspects. The SPS Agreement provides rather rigid requirements of scientific risk assessment, while the TBT Agreement is more liberal, permitting consideration of different factors in the process of adopting a measure. Although the TBT Agreement may also require a scientific justification for a measure that is aimed at protecting human, animal, or plant life or health, this obligation is not translated into demanding obligations of formal risk assessment. In a specific case, this requirement may constitute a more demanding threshold than that provided by the necessity test in the TBT Agreement.[137] As a consequence, it may be tempting for an individual WTO Member to formulate its regulation in a way which will trigger the applicability of one or other agreement. From the perspective of the WTO, such forum shopping between agreements is highly undesirable as it would allow WTO Members to evade their obligations.

(a) Purpose of a measure

The first category of measures which requires more detailed analysis consists of those regulations which, although adopted in order to address SPS concerns, 'pretend' to be TBT measures. Should a panel defer to a purpose of a measure articulated by a defendant? Or, rather, should it carry out its own assessment of a measure? If a panel is expected to make its own assessment, what is the appropriate level of scrutiny?[138] Theoretically, one can imagine at least four different approaches. First, a panel may be fully deferential and base its findings on the declaration of a defendant made in the course of the proceedings, irrespective of the real nature of the contested measure. A more moderate approach would require a panel to concentrate on the purpose explicitly stated in a measure rather than to rely on the declaration of a defendant. Another approach is to look at the structure and substance of a particular measure so as to ascertain its purpose. In such a case, an analysis of a panel is not limited to the purpose(s) expressly stated in a regulatory measure (ie in the preamble). This factor is given consideration, but constitutes only part of the examination process. A panel may also evaluate other elements of a measure, such as the content of the substantive requirements, prescribed procedures and possible exceptions, and so on. As a fourth option, a panel may seek out the subjective intent of a domestic regulator.

[137] Norbert Wilson, 'Clarifying the Alphabet Soup of the TBT and the SPS in the WTO' (Fall 2003) 8 Drake Journal of Agricultural Law 703, 723; Scott also notes that a measure may breach the SPS Agreement and be TBT-consistent, see Joanne Scott, 'European Regulation of GMOs: Thinking about Judicial Review in the WTO' in Jane Holder, Colm O'Cinneide and Michael Freeman (eds) *Current Legal Problems 2004: Vol 57* (2005) 125.

[138] Scott articulates this problem as whether the underlying objective is to be ascertained by reference to considerations of subjective intent, or on the basis of objective factors embedded in regulatory design or text, see Scott (European Regulation of GMOs) 123.

The early SPS case law did not have a chance to address this issue. One could only speculate, on the basis of the approach of the Appellate Body to the determination of ALOP, that the applicable standard was a highly deferential one. Since ALOP was conceptualized as an objective, it was rational to assume that the same model of examination would apply to a determination of the purpose of a measure. It was only the *Biotech* panel that provided more detailed guidance. Although the panel in its analysis of the purpose predominantly relied on the text of the EC measures, it also held that 'nothing in the *SPS Agreement*...bar a panel from considering purposes which were not articulated by the Member States'.[139] In the other section of the report, the panel went even further and proclaimed that it did not 'need to accept at face value assertions of purposes which are implausible in the light of all relevant circumstances'.[140] Consequently, a purpose of a measure is to be determined on the basis of objective considerations (eg the text and structure of a relevant measure). The panel also distinguished between probative values of the purpose that is articulated in a measure as opposed to the purpose which is merely asserted in the proceedings. With respect to the first, there seems to be a rebuttable presumption which requires the panel to accept such purpose as genuine unless there is sufficient indication to the contrary.[141] Thus, it is for a complainant to present adequate evidence which will rebut the presumption. In the second situation, it is the responsibility of a defendant to substantiate its assertions since a mere pronouncement is not sufficient.[142]

The analysis in *EC–Biotech Products* follows the third approach, which requires a panel to ascertain the 'real' nature of the measure on the basis of 'objective factors embedded in regulatory design or text'.[143] This appears to be the correct interpretation. A purpose that is expressly stated in a measure is given a proper consideration (ie it is refutably presumed to represent a genuine purpose). At the same time, the panel is not barred from further examination of a measure if there are doubts as to its real purpose. This seems to be a reasonable compromise between the sovereign right of WTO Members to adopt any regulatory measure in order to achieve any legitimate objective and a need to provide WTO dispute settlement bodies with the meaningful tools for assessment of such a measure. The panel's approach is also compatible with the general practice of the WTO jurisprudence on the standard of review applicable to the examination of domestic law. Note

[139] Panel Report, *EC – Biotech Products*, para 7.2558. [140] Ibid, fn 1691.

[141] Ibid, para 7.385 (noting 'in the absence of sufficient indications of a different or additional purpose, we may and should presume that the labelling requirement is intended to serve the purpose articulated in the Directive').

[142] Ibid, para 7.2573 (finding that 'the European Communities asserts that Austria's safeguard measure on T25 maize is also applied in view of concerns about labelling. This concern was not articulated by Austria in the documents discussed by us above. Furthermore, the European Communities has neither substantiated nor explained its assertion. In the light of this, we are not persuaded that Austria is applying its safeguard measure also to address the additional concern identified by the European Communities').

[143] Scott ('European Regulation of GMOs') 124.

that regulatory purpose is actually part of a measure. A purpose may be provided either directly (eg in the preamble) or indirectly, in the structure and the content of the measure. The examination of the purpose is therefore an inquiry into the meaning of a measure. In *US – Section 301 Trade Act*, the panel regarded a question of domestic law as a question of fact[144] and described its mandate as being to 'establish the meaning of Sections 301–310 as factual elements'.[145] Consequently, the panel is not bound by the interpretation of the domestic law by national governments. The elements that can be taken into account when ascertaining the meaning of the domestic law are the 'text of the relevant legislation or legal instruments, which may be supported, as appropriate, by evidence of the consistent application of such laws, the pronouncements of domestic courts on the meaning of such laws, the opinions of legal experts and the writings of recognized scholars'.[146] Clearly, a panel is authorized to go beyond the language of a domestic act. At the same time, a degree of deference has to be granted to WTO Members. In the same case, the panel stated that 'any Member can reasonably expect that considerable deference be given to its views on the meaning of its own law'.[147] This corresponds with the finding of the *Biotech* panel that the purpose expressly stated in a measure constitutes a form of a rebuttable presumption.

(b) Multi-purpose measures

The second category of measures is also problematic. How is the applicable agreement for multi-purpose measures established? One approach would be to accept the absolute priority of the SPS Agreement. This would mean that any single domestic measure, which has both an SPS and a TBT dimension, should always trigger the applicability of the SPS Agreement to the exclusion of the TBT Agreement.[148] Possible support for this position may be provided by Article 1.5 of the TBT Agreement, which declares a preference for the SPS Agreement.[149] A broad interpretation of the provision would automatically remove all measures with some SPS element from the scope of the TBT Agreement. Another approach would be to concentrate on the primary–ancillary relationship.[150] A panel, after identifying the primary purpose of the measure, should disregard any other objective. This approach, as with the first approach, secures the mutual exclusivity of the agreements, however, without providing automatic priority of the

[144] For a more detailed discussion, see Oesch (*Standards of Review*) 188–205.

[145] Panel Report, *US – Section 301 Trade Act*, para 7.18.

[146] Appellate Body Report, *US – Carbon Steel*, para 157.

[147] Panel Report, *US – Section 301 Trade Act*, para 7.19.

[148] Cf Argentina's argument in Panel Report, *EC – Biotech Products*, para 7.161.

[149] Article 1.5 of the TBT Agreement specifically provides that '[t]he provisions of this Agreement do not apply to sanitary and phytosanitary measures as defined in Annex A of the Agreement on the Application of Sanitary and Phytosanitary Measures'.

[150] As noted by Scott, the European Court of Justice prefers this method (Scott 'European Regulation of GMOs' 123).

SPS disciplines. Under the third approach, a single measure that is motivated by both SPS and TBT considerations may be regarded as 'partly an SPS measure and partly a TBT measure, and subject to both agreements'.[151] In consequence, the examination of different parts of the measure under different agreements would depend on the objectives underlying a particular part. This approach requires a panel to divide an initial measure into 'sub-measures', and subsequently to classify each of these sub-measures under the appropriate agreement. Such extracted sub-measures may reflect either different parts of the initial measure or simply reproduce the entire measure (ie if the initial measure as a whole is motivated by different concerns).

Again it was only the *Biotech* panel which systematically analysed the above problem. The panel first referred to the hypothetical situation where a WTO Member decides to consolidate two identical requirements (measures), one of them being SPS and the second non-SPS, into a single law. According to the panel, this purely technical endeavour cannot remove such consolidated act from the scope of a particular WTO agreement (ie the SPS Agreement).[152] The panel also added that requiring WTO Members to enact two separate measures containing identical provisions, each of them being adopted due to different concerns (purposes), would go against the rational legislative practice.[153]

Consequently, to the extent that the requirement is applied for one of the purposes enumerated in the SPS Agreement, it falls under this Agreement. On the other hand, to the extent that the same requirement 'is applied for a purpose which is not covered by Annex A(1), it may be viewed as a separate measure which falls to be assessed under a WTO agreement other than the SPS Agreement'.[154] The panel also believed that its interpretation was compatible with the language of Article 1.5 of the TBT Agreement. Referring to the same hypothetical situation, it noted that Article 1.5 would only call for excluding from the scope of the TBT Agreement those extracted requirements (measures) that had qualified as SPS.[155] Consequently, Article 1.5 does not affect the applicability of the TBT Agreement to those sub-measures which pursue non-SPS objectives. The panel also observed that its analysis was based on the assumption that a single requirement (measure) could be divided into two separate requirements (measures) 'which would have an autonomous raison d'être, i.e., a different purpose which would provide an independent basis for imposing the requirement'.[156] This observation remained, however, somehow theoretical. As was discussed in section 2.3, the panel found that almost all the objectives, which were purportedly pursued by the EC legislation, were of SPS character.[157]

151 Marceau and Trachtman ('Technical Barriers to Trade Agreement') 865.
152 Panel Report, *EC – Biotech Products*, para 7.166.
153 Ibid, paras 7.168–70.
154 Ibid, para 7.165.
155 Ibid, para 7.167.
156 Ibid, para 7.165.
157 Except from the objective of ensuring that novel foods do not mislead the consumers (ibid, para 7.412).

The approach of the *Biotech* panel seems to be plausible. It respects the autonomy of WTO Members to regulate for whatever purpose they wish. Its advantages become visible if one compares it with the available alternatives.[158] Although the first approach may look attractive as it provides a clear blueprint criterion, the risk of over-simplification is high. The automatic classification of a measure that has some SPS elements as an SPS cannot capture the specificity of truly multi-facet measures. If the SPS consideration is only a minor one, this approach will diminish the value of genuine and legitimate municipal objectives.[159] On the other hand, the second approach requires a rather artificial simplification of the rationale underlying a measure and anticipates the possibility of identifying the principal purpose. This is not always the case as a measure may equally serve two different objectives, neither of them being principal. Moreover, since a decision on the importance of specific objectives requires a certain degree of subjectivity, it may be difficult to be applied in practice.

At the same time, the *Biotech* panel did not answer the very important question of how to proceed with those measures which are indivisible. Probably no perfect answer is possible here. It seems that the primary–ancillary approach, as the second best option, should be applied. Thus, a panel would need to establish which purpose underlying a particular measure is the most important. The strong predominance of one objective may justify the application of only one agreement, as the interference with WTO Member freedom to regulate would be minimal in such a case. In those cases where it is not possible to divide a measure into sub-measures or to identify the strong principal objective, arguably the priority of the SPS Agreement should be presumed. This will provide a safety exit in all those cases where two first methods failed, guaranteeing the mutual exclusivity of both agreements as required by Article 1.5 of the TBT Agreement.

2.4.2 The SPS Agreement and GATT 1994

The relationship between the SPS Agreement and GATT 1994 seems to be less complicated than the TBT Agreement. The SPS Agreement establishes a special and additional regime for the assessment of specific domestic regulations, which are concerned with the protection of human, animal, and plant life and health against SPS risks. It elaborates upon existing GATT 1994 rules,[160] but it also adds obligations that are not found in the basic agreement for trade in goods

[158] Note, however, that division of an initial measure in order to analyse extracted sub-measures under different agreements may fail to appreciate the overall importance of values underlying the initial measure. In practical terms, this may affect the assessment of conformity of a measure with the disciplines of Article 5.6 of the SPS Agreement (Scott (*Commentary*) 20).

[159] Cf Panel Report, *EC – Biotech Products*, para 7.156.

[160] The preamble to the SPS Agreement provides as follows: '[d]esiring therefore to elaborate rules for the application of the provisions of GATT 1994 which relate to the use of sanitary or phytosanitary measures, in particular the provisions of Article XX(b)'.

(ie requirement of risk assessment, obligation to base measures on international standards).[161]

In contrast to the TBT and SPS relationship, the SPS Agreement does not exclude the applicability of GATT 1994 and both treaties equally apply to given factual situations.[162] This reflects the general position of WTO law, which considers different agreements as 'a "Single Undertaking". All WTO obligations are generally cumulative and Members must comply with all of them simultaneously.'[163] The SPS case law also confirmed that there is no requirement of prior violation of the GATT provisions in order to trigger the applicability of the SPS Agreement.[164] In practice, it is the decision of the complainant whether to advance its claims under one or other agreement. In the majority of cases, it will opt for the SPS disciplines as they provide more demanding requirements than GATT 1994 rules. The WTO dispute settlement bodies are bound by this choice and a panel cannot refuse to consider a measure under the provision of the particular agreement (eg GATT 1994) simply because it believes that the other set of rules is more suitable for the examination. However, if a party raises in one proceeding two concurrent claims under both agreements, a panel will first assess the compatibility of a measure under the SPS Agreement and only afterwards may examine it under GATT 1994. As explained by the panel in *EC – Hormones*:

> [i]f we were to examine GATT first, we would in any event need to revert to the SPS Agreement: if a violation of GATT were found, we would need to consider whether Article XX(b) could be invoked and would necessarily need to examine the SPS Agreement, if, on the other hand, no GATT violation were found, we would still need to examine the consistency with the SPS Agreement. For these reasons, and in order to conduct our consideration of this dispute in the most efficient manner, we shall first examine the claims raised under the SPS Agreement.[165]

In *Australia – Salmon* this even meant that despite the fact that Canada first addressed its claims under Article XI of GATT 1994 and argued the violation of the SPS Agreement in the second place, the panel nevertheless decided to begin with the analysis under the SPS provisions. The panel again referred to the need to guarantee the efficiency of the proceedings.[166]

In practice, once a measure is found to be incompatible with the SPS Agreement, panels, relying on the principle of judicial economy, do not continue the examination under GATT 1994. In line with the instruction of the Appellate

[161] Cf also Panel Report, *EC – Hormones (US)*, para 8.38.

[162] Ibid, para. 8.40; Panel Report, *Australia – Salmon*, para 8.38.

[163] Appellate Body Report, *Korea – Dairy*, para 74.

[164] Panel Report, *EC – Hormones (US)*, para 8.41.

[165] Ibid, para 8.42.

[166] Panel Report, *Australia – Salmon*, para 8.39. Note also that this approach corresponds closely with the soft version of the *lex specialis* principle, which determines the sequence of examination of different provisions (agreements). Accordingly, the more specific rules are to be examined before the general provisions (Pauwelyn, *Conflict of Norms in Public International Law: How WTO Law Relates to Other Rules of Intenational Law* (2003) 411).

Body that '[a] panel need only address those claims which must be addressed in order to resolve the matter in issue in the dispute',[167] all SPS panels limited their findings to the violation of the SPS Agreement.[168] The approach of the *Hormones* panel may also indicate that the judicial economy is not the only obstacle for further analysing an SPS-inconsistent measure under GATT 1994. The panel particularly noted that even if it found a measure to be incompatible with Articles I and III of GATT 1994, it would 'then need to examine whether this inconsistency could be justified . . . under Article XX(b) of GATT and would thus necessarily need to revert to the SPS Agreement under which . . . [it has] already found inconsistencies'.[169] Thus, a measure that violates the SPS Agreement cannot be saved under Article XX(b) of GATT 1994. Note that subparagraph (b) actually exhausts all defence options that are available under Article XX. In order to fall into the ambit of the SPS Agreement, a measure needs to be adopted for the protection of human, animal, or plant life and health from certain specific risks. If a measure is adopted due to other concerns, such as the protection of public morals as provided by Article XX(a), it will not qualify as an SPS measure in the first place.

Those measures which comply with the SPS Agreement are presumed to conform to the relevant provisions of GATT 1994.[170] The character of this presumption is not completely clear.[171] The text of Article 2.4 suggests that the presumption is rebuttable, as it simply uses the expression 'to presume' without introducing any qualification.[172] That would correspond with the approach under Article 3.2, where the same language was described by the Appellate Body in terms of rebuttable presumption.[173] The principle of cumulative application also seems to support this conclusion. If both agreements are applicable cumulatively, there is no need to disturb this application with irrefutable presumption.

In practice, however, the presumption will rather operate as an irrefutable one. Note that the SPS Agreement provides a more sophisticated and developed regime for those measures which aim at the protection of human, animal, and plant

[167] Appellate Body Report, *US – Shirts and Blouses*, 19.

[168] Cf Panel Report, *Australia – Salmon*, para 8.185; Panel Report, *Japan – Apples*, para 8.328; Panel Report, *EC – Biotech Products*, para 7.3422. But cf the approach of the *Biotech* panel which, after finding that product-specific measures violated Annex C(1)(a) first sentence, proceeded with the analysis of the very same measures under Article III:4 of GATT 1994 (Panel Report, *EC – Biotech Products*, paras 7.2509–7.2516).

[169] Panel Report, *EC – Hormones (US)*, para 8.273.

[170] Article 2.4 provides that '[s]anitary or phytosanitary measures which conform to the relevant provisions of this Agreement shall be presumed to be in accordance with the obligations of the Members under the provisions of GATT 1994 which relate to the use of sanitary or phytosanitary measures, in particular the provisions of Article XX (b)'.

[171] Cf Scott (*Commentary*) 28–9; Marceau and Trachtman ('Technical Barriers to Trade Agreement') 871.

[172] Marceau and Trachtman ('Technical Barriers to Trade Agreement') fn 238 (observing that 'if the intent were to deem such measures to comply with GATT, the treaty could have said so, or could have stated that the presumption is irrebuttable').

[173] Appellate Body Report, *EC – Hormones*, para 170.

life and health from certain specific risks. Its obligations go well beyond what is required under GATT 1994. In consequence, it is difficult to imagine a situation where a measure which has passed various demanding tests provided by the SPS Agreement will be found GATT-incompatible. Even if such an option exists, in the sense that Articles I, III, and XI would be interpreted as providing some additional obligations that are not reflected in the SPS Agreement, Article XX(b) will nevertheless cure any possible violation. Since the SPS Agreement is the elaboration of Article XX(b), explaining its disciplines for specific types of measures, it would be counter-intuitive to demand more under the latter than is required by the SPS Agreement itself.[174]

Although any conflict between SPS and GATT rules is highly improbable, it is worth noting that WTO law contains an explicit provision to regulate such a situation. A general interpretative note to Annex A instructs that 'in the event of conflict between a provision of the General Agreement on Tariffs and Trade 1994 and a provision of another agreement in Annex 1A to the Agreement Establishing the World Trade Organization . . . the provision of the other agreement shall prevail to the extent of the conflict'. This gives an automatic precedence to any conflicting rule of the SPS Agreement over the GATT provision. However, since the conflicts are conceptualized in the context of WTO law narrowly ('adherence to one provision will lead to a violation of the other provision'[175]), while the Appellate Body generally assumes that all provisions may be read in harmony, the space for any potential clash between the GATT 1994 and SPS provisions is very limited, if it exists at all.[176]

2.5 Conclusions

The SPS Agreement provides a special regime for the assessment of a specific category of national measures which aim for the protection of human, animal, and plant life and health from certain specific risks. As a consequence of the broad interpretation of the conditions pertaining to its applicability, the Agreement appears to cover different regulatory measures, some of them exceeding traditional SPS risks. In particular, indirect risks as well as some environmental risks, which are reducible to animal and plant life and health and which result from the entry, establishment, and spread of pests, proved to constitute an SPS risk. While such a broad interpretation that was proposed in the case law may cause legitimate concerns, part of it seems to be overstated. In this context, the chapter notes that the explicit language of Annex A(1) actually requires WTO dispute

[174] Cf Scott (*Commentary*) 28 (noting that 'any measure passing muster under the SPS Agreement must surely be consistent with the GATT, and in particular Article XX(b)').

[175] Appellate Body Report, *Guatemala – Cement I*, para 65.

[176] Marceau and Trachtman ('Technical Barriers to Trade Agreement') 869 (noting 'it is difficult to imagine circumstances where one requires what the other forbids').

settlement bodies to qualify some environmental risk as relevant for the purpose of the Agreement.

The test of a rational relationship that was introduced by the *Biotech* panel appears to be a key issue in defining the scope of SPS disciplines. Extensively broad conceptualization may indeed result in expansive application of the SPS Agreement beyond what may properly be considered within its scope. However, drawing any final conclusions on the basis of the *Biotech* report seems to be premature. Although the panel did not establish any clear standard that would explain the required level of causality, the language that it used implies the existence of certain limits. The practise of WTO Members, both within the WTO dispute settlement system as well as in the SPS Committee, also shows that the scenario under which the SPS Agreement overshadows the whole field of environmental law seems to be exaggerated. At the same time, I agree that it may be legitimately questioned whether unconditional inclusion of all indirect risks posed by GMOs in the scope of the SPS Agreement is a proper approach. The same problem is posed by the panel's reductionist approach to the concept of biodiversity.

A closer analysis of the *Biotech* report reveals, however, a number of other deficiencies. The panel analysis particularly appears to lack overall interpretative consistency, sometimes preferring a functional approach, while in other cases relying heavily on purely textual methods. This gives an impression that the panel, instead of interpreting provisions of the SPS Agreement, was rather looking for *ex post* justification for its prior normative choices. The other important failure relates to the panel's interpretative approach to some technical terms, which appear in the text of the SPS Agreement. It seems that defining those terms on the basis of the *Oxford English Dictionary* cannot produce a satisfactory outcome, as their meaning can only properly be established within a relevant technical or scientific discipline. A preferable approach would be to look for an interpretative guidance into documents produced by recognized international bodies that specialize in the SPS field (including sister organizations). Such documents may be regarded as a kind of glossary, which reveals the common meaning ascribed to specific technical terms by the international community. Although one may detect traces of such an approach in the *Biotech* report, the ultimate outcome shows that the interpretative process was entirely dominated by textual methods, without proper regard for the technical character of the SPS Agreement.

WTO law does not provide any clear rules which would precisely limit the scope of application of different legal instruments. In the context of the SPS Agreement, measures that serve multiple objectives and those which on their face appear to be TBT regulations but in fact were adopted due to SPS concerns, are particularly problematic. In both cases, the SPS jurisprudence succeeded in adopting an approach that provides a workable solution. Recognizing that a single measure may serve different objectives, and accepting that the examination of sub-measures under both agreements, not only respects the regulatory autonomy of WTO Members but also guarantees the mutual exclusivity of TBT and SPS disciplines. I also concur with panel's holding on the standard of review applicable for assessment of a purpose of a measure, which seems to provide a reasonable compromise between the sovereign right of WTO Members to adopt any regulatory measures and a need to provide WTO dispute settlement bodies with the necessary competences. It respects a purpose that is expressly stated in a measure but also permits further inquiry if the objective circumstances indicate the existence of other undisclosed objectives.

In contrast to the TBT and SPS relationship, the SPS Agreement does not exclude the applicability of GATT 1994 and both treaties equally apply to a given factual situation. At the same time, the SPS Agreement establishes a rebuttable presumption of consistency with GATT 1994 for those measures which comply with its disciplines. Although the nature of this presumption is not without ambiguity, in practice it seems to operate as an irrefutable one. Since the SPS Agreement provides a more stringent regime for assessment of SPS measures compared to GATT rules, a measure that passes examination under the former, logically needs to be compatible with less demanding rules. Even if two agreements are seen as providing complementary standards, Article XX(b) will ensure GATT-compatibility for all measures which comply with SPS disciplines.

3

Harmonization Disciplines of the SPS Agreement

3.1 Introduction

One of the objectives of the SPS Agreement is to promote harmonization of national SPS requirements. Harmonization is not an end in itself, but rather a means of facilitating international trade through the establishment of common standards. This goal is explicitly pronounced in the Preamble to the SPS Agreement, which expresses its desire 'to further the use of harmonized sanitary and phytosanitary measures between Members, on the basis of international standards, guidelines and recommendations developed by relevant international organizations'.[1]

The central provision in this context is Article 3. It requires WTO Members to base their SPS measures on international standards, guidelines, or recommendations.[2] Measures that conform to such standards enjoy a presumption of consistency with relevant provisions of WTO law. A Member may still adopt and maintain SPS measures which are not based on international standards, but only if they are supported by scientific justification in the form of risk assessment. The SPS Agreement also requires WTO Members to play a full part, within the limits of their resources, in the operation of relevant international organizations and their subsidiary bodies (Article 3.4). At the same time, the SPS Agreement provides a non-exhaustive list of such organizations (Annex A(3)) and instructs the SPS Committee to develop a procedure to monitor the process of harmonization and to coordinate efforts in this regard with the appropriate international organizations (Article 3.5).[3]

[1] Recital no 6, Preamble to the SPS Agreement.

[2] For the purpose of simplicity, where possible, I refer to standards, guidelines, and recommendations taken together as standards. This appears to be a justified approach due to the uniform treatment of these documents under the SPS Agreement.

[3] The Committee adopted the first provisional procedures to monitor the process of international harmonization and the use of international standards, guidelines, and recommendations in 1997 (revised in 2004). In 2006 the procedure was extended indefinitely; for details see, SPS Committee, Procedure to Monitor the Process of International Harmonization, Ninth Annual Report, 10 July 2007, G/SPS/45.

The relevance of international standards and standard-setting bodies is, however, not exhausted by Article 3. WTO Members, when assessing risks, are obliged to take into account risk assessment techniques developed by appropriate international organizations (Article 5.1). If scientific evidence is insufficient, WTO Members may provisionally adopt SPS measures on the basis of available pertinent information, including that from the relevant standard-setting bodies (Article 5.7). Each WTO Member is also required, upon a request from another Member, to provide explanations of the reasons underlying an SPS measure that constrains the import. This obligation only applies if a measure is not based on the relevant international standard or if there is no such standard (Article 5.8). Last but not least, when assessing the SPS status of a region in the exporting country, one has to take into account appropriate criteria or guidelines which may be developed by the international organizations (Article 6).

This chapter starts with a short discussion on the concept of harmonization, including a brief review of arguments for and against harmonization. This section is intended as a useful background for subsequent assessment of the disciplines established by the SPS Agreement. The second part concentrates on international standard-setting bodies that are relevant in the context of the Agreement, while the third part examines the legal meaning of the expression 'international standards, guidelines or recommendations'. The fourth part discusses in more detail each of the three options that is available to WTO Members under Article 3: a situation when a measure conforms to an international standard, when it is merely based on such a standard, as well as when it deviates from it. The last section summarizes the previous finding and draws overall conclusions on the harmonization requirements of the SPS Agreement. The scope of this chapter is limited to the obligations provided by Article 3, while the other SPS provisions which refer to international standards-setting bodies are discussed in other chapters of the book. Articles 3.4 and 3.5, which provide procedural obligations, fall entirely outside the scope of the discussion.

3.2 The concept of harmonization

There seems to be no uniform definition of harmonization in the literature. Leebron defines it broadly as 'making the regulatory requirements of governmental policies of different jurisdiction identical, or at least more similar'.[4] Sykes, on the other hand, uses a more narrow definition under which harmonization is understood as a process of reaching the same substantive requirements.[5] At the

[4] David W Leebron, 'Lying Down with Procrustes: An Analysis of Harmonization Claims' in Jagdish Bhagwati and Robert Hudec (eds), *Fair Trade and Harmonization: Prerequisites for Free Trade? Vol 1: Economic Analysis* (1996) 43.

[5] Alan O Sykes, 'The (Limited) Role of Regulatory Harmonization in International Goods and Services Markets' (1999) 2 Journal of International Economic Law 49, 50.

same time, Sykes identifies other methods of reducing the regulatory heterogeneity (or in other words making the regulatory requirements more similar). This particularly includes the establishment of specific constraints on national regulatory policies in order to decrease their adverse impact on international trade (eg transparency, non-discrimination, the least-restrictive alternative requirement) and recourse to the concept of mutual recognition (recognizing measures of other jurisdictions as equivalent to domestic requirements).[6]

This narrow conception seems to correspond closely with the understanding of harmonization under the SPS Agreement, which defines it as 'the establishment, recognition and application of common sanitary and phytosanitary measures by different Members'.[7] At the same time, it is worth noting that two other mechanisms for addressing regulatory heterogeneity that are identified by Sykes, also find their reflection in the SPS Agreement. Transparency, non-discrimination, and the least-trade-restrictive alternative requirements constitute the core of SPS disciplines while mutual recognition is recognized in Article 4 as the principle of equivalence. I will come back to these obligations in Chapter 6.

The academic debate over the advantages and drawbacks of international harmonization has a long history and appears to be unsettled. There is no place here to analyse in detail all the arguments that are advanced in the course of this discussion and I will briefly summarize only some of them. The case for harmonization can be made on different grounds. For instance, the fairness claim provides that harmonization is required in order to assure equal opportunities to all economic actors from the various jurisdictions.[8] Differences in national regulatory regimes affect their competitive situations and may give an unfair advantage to those producers which operate in the countries that impose less stringent requirements. Regulatory difference, the argument goes, is not a legitimate source of comparative advantage as opposed to technology, countries' endowments, location, or the market structure.[9] In consequence, international standards can ensure the establishment of a level playing field for all involved actors. From the economic perspective, harmonization is claimed to provide efficiency gains. Companies, instead of complying with different municipal requirements, may rely on one international standard, which under certain circumstances may decrease the cost of their operations. As explained by Sykes 'a firm may become subject to redundant regulatory requirements that increase their costs of doing business without being necessary to the attainment of regulatory objectives'.[10] Harmonization can also be justified on the basis of the economies of scale argument. If a company is forced to divide its operation into separate parts, each of them complying with

6 Ibid. 7 Annex A(2) of the SPS Agreement.

8 Leebron ('Lying Down with Procrustes') 60–62.

9 Graham Mayeda, 'Developing Disharmony? The SPS and TBT Agreements and the Impact of Harmonization on Developing Countries' (2004) 7(4) Journal of International Economic Law 737, 741.

10 Sykes ('The Limited Role') 54.

the regulatory requirements of a particular jurisdiction, this may considerably increase the unit cost and result in the loss of advantages of economies of scale.[11] There are also other additional costs that may be potentially borne by companies in the absence of internationally harmonized standards. This particularly includes information costs (ie costs of learning about regulatory requirements of different jurisdictions) and conformity assessment costs (ie costs of demonstrating compliance with different requirements of various jurisdictions).[12]

On the other hand, it remains unclear why regulatory differences cannot constitute a legitimate element of the comparative advantage that is enjoyed by a nation. On normative grounds, they seem to be similar to other factors that determine the competitive position of particular country.[13] It was also noted in the literature that in some cases, the higher standards may actually improve the competitive advantage in other sectors of domestic economy.[14] Others claim that regulatory competition between different jurisdictions is actually preferable as it allows experimentation with different alternatives and the determination of optimal choices.[15] Moreover, the heterogeneity as such may reflect genuine differences in tastes or income of citizens or may be the result of previous experience (so-called good heterogeneity in the nomenclature of Sykes). Indeed, it may be unreasonable to force the harmonization process in such a situation. A similar approach is taken by Leebron who also recognizes the existence of legitimate differences between states.[16]

The above suggests that harmonization may indeed provide a useful instrument to deal with certain types of regulatory heterogeneity. In particular, the harmonization efforts may be helpful, alongside other trade instruments, in addressing those differences (so-called bad heterogeneity) that result from the capture (eg protectionism) or inadequate information possessed by regulators (eg lack of information that a particular technology is able to achieve regulatory objectives).[17] At the same time, one has to be careful when overriding the legitimate differences between countries. As accurately summarized by Sykes 'when the cross-border effects of regulation are important, the purely regulatory competition of economic theory, and the complete regulatory harmonization that lies at the end of the continuum, are almost never sensible'.[18]

[11] Ibid, 55. [12] Ibid, 55–6.

[13] Michael Trebilcock and Robert Howse, 'Trade Liberalization and Regulatory Diversity: Reconciling Competitive Markets with Competitive Politics' (1998) 6(1) European Journal of Law & Economics 5, 14.

[14] Michael J Trebilcock and Robert Howse, *The Regulation of International Trade* (2005) 508 (noting that higher standards may 'confer a competitive advantage in some sectors, where those standards may create incentives for the development of... technologies that can then be exported to other countries').

[15] Alan O Sykes, 'Regulatory Competition or Regulatory Harmonization? A Silly Question?' (2000) 3 Journal of International Economic Law 257, 259.

[16] Leebron ('Lying Down with Procrustes') 68.

[17] Sykes ('The Limited Role') 57. [18] Sykes ('Regulatory Competition') 263.

3.3 Relevant international organizations

The SPS Agreement recognizes standards of certain international bodies as a reference point for the assessment of the legality of national SPS measures. The list of relevant international organizations includes the Codex responsible for food safety and the IOE for animal health and safety. In addition, standards developed under the auspices of the Secretariat of the IPPC in cooperation with regional organizations operating within its framework have the same status. The IPPC is responsible for the development of phytosanitary standards (life and health of plants).[19]

This reference to international standard-setting bodies had a significant impact on the operations of the above organizations. Prior to 1995, they received rather limited attention from the international community and served as specialized fora for scientific and technical discussion between experts.[20] Standards were voluntary without any legal sanctions for non-compliance.[21] The standard-setting process was based on consensus and extensive consultation with the aim of finding a common technical ground. Although the system was operating smoothly and controversies were rare, the consensual nature also meant that problematic areas were not harmonized. This changed with the adoption of the SPS Agreement, which linked the international standards with the world trade system. Suddenly, standard-setting bodies found themselves in the centre of trading controversies with serious economic consequences. Consensus, which was the common practice in the old days, is now sometimes replaced by majority voting.

In theory, the enumeration of international standards that are relevant under the SPS Agreement is open-ended. Annex A4(d) provides that for the matters not covered by the sister organizations,[22] relevant standards may also include documents promulgated by other international organizations. The SPS Agreement establishes two preconditions: an organization needs to be open for membership to all WTO Members, and it has to be identified as relevant by the SPS Committee. Note that this enigmatic formulation, rather than establishing a clear standard, raises a number of difficult legal questions. First, it is not

[19] Again for the sake of simplicity, when I refer to relevant international standard-setting organizations, I also mean the IPPC. Technically speaking the IPPC is a semi-institutionalized international convention and not a body or an international organization.

[20] Frode Veggeland and Svein Ole Borgen, 'Negotiating International Food Standards: The World Trade Organization's Impact on the Codex Alimentarius Commission' (2005) 18(4) Governance: An International Journal of Policy, Administration and Institutions 675, 684 (discussing the status of the Codex Alimentarius systems before 1995).

[21] David G Victor, 'The Sanitary and Phytosanitary Agreement of the World Trade Organization: An Assessment after Five Years' (1999–2000) 32 New York University Journal of International Law & Policy 890, 890.

[22] In the WTO jargon, all three standard-setting bodies (Codex, OIE and IPPC) are conventionally referred to as the sister organizations.

clear how the expression 'for matters not covered by the above organizations' should be understood. Does it refer to existing standards, the working agenda of the Codex, OIE, and IPPC, or rather the mandate of the sister organizations? A textual interpretation seems to exclude the narrow reading of Annex A4(d). The SPS Agreement refers to international organizations ('for matters not covered by the *above organizations*') and not to the existing standards. The second approach, although more compatible with the textual basis in the Agreement, has undesirable pre-emptive effects. The mere fact that one of the sister organizations works on a particular standard would automatically remove the subject matter of that future standard from the scope of Annex A4(d), without providing any guarantee that it will actually be adopted. The third approach is also over-expansive as in practice it prohibits the recognition of any other standard. Note that the mandate of the sister organizations is sufficiently broad to cover all possible issues relating to food safety and protection of animal/plant life and health, matters that are covered by the SPS Agreement. Consequently, an expansive interpretation would effectively deprive Annex A4(d) of any legal meaning. These objections would point to a narrow approach. Despite the fact that such an interpretation is not fully compatible with the textual basis in Annex A4(d), it provides a mechanism which is preferable from a pragmatic point of view. The SPS Committee may be guided in this process by different informal criteria such as transparency of the procedures used by the relevant organization in the standard-setting process, its impartially and openness, the consensus-based character of decision-making, or the level of the WTO Members' participation.[23]

The SPS Agreement does not define what may qualify as another international organization. To use the term 'international' arguably removes from the scope of Annex A(4)d all regional organizations.[24] Literal reading of the provision also suggests that subparagraph 4(d) is limited to those entities that are recognized as international organizations under public international law. This conclusion seems to be supported by the language of the TBT Agreement, which distinguishes between international bodies and systems.[25] Consequently, one may argue that the WTO generally differentiates between these two types of international cooperation with the SPS Agreement limited to one specific form (ie participation in international organization). However, a closer look at the relevant provisions of the SPS Agreement shows that the distinction between international organizations (bodies) and processes may be more apparent than

[23] Harm Schepel, *The Constitution of Private Governance. Product Standards in the Regulation of Integrating Markets* (2005) 188.

[24] This conclusion also seems to be supported by the fact that regional organizations are open only to WTO Members from a particular region. This textual interpretation was also confirmed by the SPS Committee; for additional discussion, see section 3.4 below.

[25] Annex 1(4) of the TBT Agreement (defining those two types of cooperation as either a body or a system whose membership is open to the relevant bodies of at least all WTO Members).

real. The SPS Agreement consistently refers to the Codex, OIE, and IPPC as international organizations,[26] despite the fact that the IPPC is merely an international convention (in the nomenclature of the TBT Agreement it would be labelled as a process). This would suggest that the form of cooperation is of secondary importance and that standards developed within the framework of international conventions or agreements may be equally relevant in the context of the SPS Agreement. Again, from a pragmatic point of view, it seems that there is no need to impose any excessively restrictive prerequisites here. After all, the SPS Committee retains a complete control over the process of identification.

It is also not clear whether the intergovernmental character of international cooperation (either in the form of organization or process) is one of the preconditions under Annex A(4)d. According to Scott, the language of the provision excludes non-intergovernmental bodies.[27] In support of her argument, Scott points out that while the SPS Agreement 'speaks of international organizations open for membership to all Members, the TBT Agreement refers to international bodies...whose membership is open to the relevant bodies of at least all Members'.[28] This difference in the language may indicate that the SPS Agreement establishes a higher threshold by actually requiring all relevant organizations to have intergovernmental character. Although such interpretation is textually correct, it provides unsatisfactory results by removing from the scope of the SPS Agreement a number of important actors that specialize in the production of standards. If one recognizes their expertise and experience in the SPS field, such an approach can hardly be seen as appropriate. Moreover, note that non-intergovernmental organizations such as the International Organization for Standardization (ISO) already participate as observers in the work of the SPS Committee. This practice may be regarded as a form of de facto recognition of their relevance in the SPS context.

One should also not overlook that the importance of subparagraph 4(d) may actually be limited. The SPS Agreement requires a prior identification of the relevant international organization by the SPS Committee. The consensual character of the Committee's decision-making[29] makes this task very difficult. WTO Members are frequently divided as to which standards (other than those promulgated by the sister organizations) should be relevant in the context of the SPS Agreement. This reluctance is, for example, reflected in the discussion that took

[26] See eg, 'the existence of eradication or control programmes, and appropriate criteria or guidelines which may be developed by the relevant international organizations' (Article 6.1) that also includes standards developed within the IPPC; 'Members should encourage and facilitate the active participation of developing country Members in the relevant international organizations' (Article 10.4).

[27] For example, International Organization for Standardization (ISO), which is composed of non-governmental national bodies responsible for standardization.

[28] Joanne Scott, *The WTO Agreement on Sanitary and Phytosanitary Measures. A Commentary* (2007) 245–6.

[29] Article 12.1 of the SPS Agreement.

place in the Committee when considering the request from the Secretariat of the Convention on Biological Diversity (CBD) for observer status in the Committee's meetings. As a consequence of the opposition from the United States and Canada, the SPS Committee has ultimately failed to grant such status and admits the CBD on a meeting-by-meeting basis.

3.3.1 Codex Alimentarius Commission

The Codex is an organization that was established in 1963 by the FAO and the WHO. Currently Codex has 181 member countries and one member organization (the EC).[30] As a general rule, membership is open to any country that is a member nation or an associate member of the FAO and/or WHO.[31]

The Codex Alimentarius Commission meets twice a year in plenary sessions; between sessions the Codex system is administered by the Executive Committee, which acts on behalf of the Commission. Technical elaboration of standards is done by the Codex committees and ad hoc task forces responsible for the scientific research on specific issues (eg Task Force on Food Derived from Biotechnology). A special place in the Codex standard-setting process is occupied by three independent scientific committees that are responsible for technical assessment: the Joint FAO/WHO Expert Committee on Food Additives (JECFA), the Joint FAO/WHO Meetings on Pesticide Residues (JMPR) and the Joint FAO/WHO Expert Meetings on Microbiological Risk Assessment (JEMRA).[32]

The primary goal of the organization is protecting consumers' health and ensuring fair practice in international food trade. This task is performed through the development of different documents, including standards, codes of practice, principles, guidelines, and recommendations (eg codes of conduct and recommended methods). Although the language used by the Codex is not entirely compatible with the SPS Agreement, the substantive analysis of the Codex norms allows for the identification of all three types of documents that are referred to by the SPS Agreement (ie standards, guidelines, and recommendations).[33]

Codex standards can be divided into four different groups: commodity standards, residue standards, standards on methods of analysis and sampling, and general standards. Commodity standards set the requirements on the essential attributes of a particular product (eg what can be considered as tuna) together with additional requirements such as the maximum levels for additives. In practice the food safety aspect is frequently addressed through reference to general

[30] The membership as of July 2009 <http://www.codexalimentarius.net/web/members.jsp?lang=EN>, last visited 17 July 2009.

[31] Article 2 of the Statutes of the Codex Alimentarius Commission.

[32] JECFA is responsible for assessment of safety of food additives, contaminants, naturally occurring toxicants, and residues of veterinary drugs in food, JMPR for pesticide residues in food and JEMRA for assessment and control of microbiological hazards in foods (eg salmonella).

[33] FAO/WHO, *Understanding the Codex Alimentarius* (3rd edn, 20 July 2009) 11.

or residue standards. The commodity standards are mostly of a non-scientific character and often rely on industrial practices.[34] Such standards fall into the ambit of the SPS Agreement only if they touch upon the food safety issue, otherwise they are covered by the TBT Agreement. The residue standards define a maximum level of residues of pesticides, veterinary drugs, or additives in foods and they are expressed in quantitative figures as maximum residue levels (MRLs). MRL is calculated on the basis of ADI, which is conceptualized as the maximum amount of a specific substance in food that can be ingested over a lifetime without an appreciable health risk. The process of establishing ADIs and MRLs is extremely scientific. As opposed to commodity standards, residue standards are important in the context of the SPS Agreement.[35] Standards on methods of analysis and sampling are designed as a tool for guaranteeing the quality and consistency of analytical methods, whereas general standards set broad requirements for whole categories of food (eg labelling of pre-packaged foods).

The Codex guidelines comprise two types of documents: general principles, which determine policies in selected key areas (eg Principles for the Risk Analysis of Foods Derived from Modern Biotechnology) and interpretative guidelines, which interpret the general principles or standards (eg conditions for production, marketing, and labelling of organic foods).[36] In some cases the general principles do not constitute a separate document but are simply included in a relevant standard or a code of practice (eg for food addictives).

All other documents may be labelled as different types of recommendations. This includes, for instance, codes of practices that define 'the production, processing, manufacturing, transport and storage practices for individual foods or groups of foods that are considered essential to ensure the safety and suitability of food for consumption'.[37]

Technically speaking, Codex standards (as well as other types of its documents) are adopted by simple majority voting.[38] In the case of regional standards, only member states belonging to the region or relevant group of countries have a voting right.[39] In practice, members tend to adopt decisions by consensus. The Rules of the Procedure of the Codex Alimentarius support this practice by requiring the Commission to 'make every effort to reach agreement on the adoption or amendment of standards by consensus'.[40] However, if consensus is not possible, Codex standards may be adopted by simple majority voting. The process of standard adoption itself is complex and consists of eight distinctive steps.[41] The

[34] Cf Victor ('Assessment') 887.

[35] The residue standards were central in *EC – Hormones* and *US/Canada – Continued Suspension*.

[36] FAO/WHO, *Understanding the Codex*, 11. [37] Ibid.

[38] Rule VIII.2 of the Rules of Procedures of the Codex Alimentarius Commission, in Codex, *Procedural Manual* (18th edn, 2008) 6–16.

[39] Ibid, Rule VIII.3. [40] Ibid, Rule XII.2.

[41] For details on the specific steps in the Codex procedure, see 'Procedures for the Elaboration of Codex Standards and Related Texts' in Codex (*Procedural Manual*) 29–37.

accelerated and simplified procedure is available if there is an immediate need for standard adoption or if there is a broad consensus between the members with respect to content of the standard. Note that both ordinary and accelerated procedures give numerous occasions for members to comment on a proposed standard. This, at least in theory, guarantees inclusiveness of a debate and improves the overall legitimacy of standards.

In contrast to the OIE and the IPPC, the Codex does not have any form of alternative dispute settlement system that would be available for its members.

3.3.2 World Organization for Animal Health

The OIE was created in 1924 as the International Office of Epizootics and currently has 174 members.[42] The OIE is responsible for the development of sanitary standards[43] but it also serves as a forum for dissemination of information among its members and a provider of technical assistance. The highest authority of the OIE is the International Committee composed of delegates from all members. Day-to-day operations of the organization are managed by the Central Bureau of the OIE headed by the Director General. The International Committee is responsible for formal adoption of standards, which are technically developed within specific scientific commissions. A special ad hoc commission may also be set up to examine technical or scientific issues. Besides, there are regional commissions responsible for regional cooperation, and permanent working groups that monitor and review developments in their area of competence.

The OIE develops two types of standards: (i) health standards on trade in animals and their products, which aim to ensure sanitary safety of international trade in terrestrial and aquatic animals and products derived from them, and (ii) biological standards on diagnostic techniques that provide rules on disease diagnosis. The first category of standards is included in the Terrestrial Animal Health Code (for mammals, birds, and bees) and the Aquatic Animal Health Code (for fish, molluscs, and crustaceans). The Diagnostic Manual for Aquatic Animal Diseases and the Manual of Standards for Diagnostic Tests and Vaccines are collections of diagnostic standards. The OIE does not distinguish, as a separate type of document, between guidelines or recommendations.

The development and adoption of standards has a much simpler and more informal character than the Codex. An initial draft is prepared by a small ad hoc group of experts and passed over to the relevant commission, which evaluates its scientific quality. A special report is subsequently distributed among delegates, who can submit their comments. Depending on these comments, either an

[42] The membership as of July 2009 (<http://www.oie.int/eng/OIE/PM/en_PM.htm?e1d1> last visited 20 July 2009).

[43] Information on the webpage of the OIE (<http://www.oie.int/eng/OIE/en_objectifs.htm#2> last visited 20 July 2009).

additional review is undertaken or a recommendation is sent to the International Committee. The Committee adopts a standard on the basis of consensus. If there is no consensus, a standard is still adopted provided that no more than 10 members are against.[44]

The OIE, contrary to the Codex, has is its own dispute settlement mechanism. The process may be initiated only upon mutual agreement of the parties and its outcome is legally binding only if both parties agree in advance to this effect. The dispute needs to be of a technical nature and is decided by a panel of experts. A report of a panel is confidential and is transmitted only to the OIE Director General and the parties. At least in theory, the alternative mechanism has a number of advantages compared to a traditional WTO dispute settlement system.[45] It is less expensive than the WTO counterpart and allows for the resolution of a dispute on a technical level, which may help to reduce political tension. The mechanism also seems to be the best option for the settlement of smaller disputes, which are not worth being taken to the WTO. An additional advantage lies in the fact that a panel may not only condemn a disputed measure but also propose available alternatives. This positive character of the dispute settlement mechanism should, at least in theory, facilitate reaching an agreement between the parties. Despite these advantages, the OIE dispute settlement system has been used only twice.[46]

3.3.3 International Plant Protection Convention

The IPPC is an international treaty adopted in 1951 which came into force on 3 April 1952. Currently, there are 170 states (including the EC as an international organization) that are parties to the Convention.[47] The Convention aims at securing 'common and effective action to prevent the spread and introduction of pests of plants and plant products'[48] while facilitating international trade. Despite the fact that the IPPC is a convention, it also has a quasi-institutional character. In particular, it has a permanent body, the Commission on Phytosanitary Measures (CPM), responsible for the administration and implementation of its objectives, with a membership that is open to all contracting states.[49] The CPM is served by the permanent IPPC Secretariat, which is in charge of logistics and technical support as well as coordination of information exchange. The IPPC also serves as an umbrella structure for regional plant protection organizations (RPPO). RPPOs

[44] FAO/WHO, *Report of the Evaluation of the Codex Alimentarius and Other FAO and WHO Food Standards Work* (2002) 22.

[45] See SPS Committee, OIE Dispute Mediation Process, Communication from the World Organization for Animal Health (OIE) of 12 October 2006, G/SPS/GEN/731.

[46] Such an unsuccessful record may be explained to some extent by the existence of a concurrent dispute settlement system within the WTO (including a consultation process) as well as informal processes, which are available within the SPS Committee.

[47] The number of the parties as of July 2009 (<http://www.fao.org/Legal/TREATIES/004s-e.htm> last visited 20 July 2009).

[48] Article I.1 of the IPPC. [49] Article XI.3 of the IPPC.

are intergovernmental organizations acting as coordinating bodies at a regional level, responsible for the development of regional phytosanitary standards.

The standard-setting process is even less complicated than in the case of the OIE, and is simply composed of three stages (draft, consultation, and approval). The central role in the process is played by the scientific Standards Committee, which reviews submitted drafts and makes recommendations on the further course of action. Both contracting parties and the RPPOs may comment on draft proposals during the consultation stage. Those comments are subsequently reviewed and assessed by the Standards Committee, and can be included in the proposal. The formal adoption of an international standard is done by the CPM and requires consensus. If this is not attainable a qualified majority of two thirds of votes suffices.[50] The IPPC develops only one type of document (standards), nevertheless, they appear under different names (eg Glossary of Phytosanitary Terms, Guidelines for Surveillance).[51]

The IPPC, similarly to the OIE, offers an alternative dispute settlement mechanism for all matters relating to the interpretation and application of the Convention.[52] The disputes are decided by a committee of experts consisting of five members including two that are proposed by the parties.[53] The report of the committee is non-binding and in principle is disclosed only to the parties to the dispute, but in some circumstances may be provided to the competent bodies of international organizations responsible for resolving trade dispute (ie a panel or the Appellate Body).[54] The IPPC system has similar advantages as the OIE counterpart. Nevertheless, it has not yet been used.[55] The adoption by the IPPC Subsidiary Body on Dispute Settlement of the special IPPC Dispute Settlement Manual was one of the efforts to popularize the recourse to this alternative mechanism. Unfortunately, up to now it has not yet produced the expected results.

3.3.4 International standard-setting bodies in practice

The overall assessment of the functioning of the international standard-setting bodies seems to be rather positive. As noted with respect to the Codex, international standards are seen by both developed and developing countries as a 'vital component of food control systems designed to protect consumer health and... international trade'.[56] Despite this common positive opinion, there are also some areas which draw a substantial amount of criticism. The major problem perhaps relates to under-inclusion of members from developing countries, and in particular the least developed countries, in the standard-setting process.

[50] Article XI.6 of the IPPC.
[51] As of 17 July 2009 there are 32 IPPC standards.
[52] Article XIII of the IPPC.
[53] SPS Committee, IPPC Dispute Settlement Manual, Submission by the Secretariat of the International Plant Protection Convention, 22 June 2007, G/SPS/GEN/787, 12.
[54] Ibid, 16.
[55] Ibid, 5.
[56] FAO/WHO (*Report of the Evaluation*) 6 and 30.

Although generally developing countries consider international standards as a helpful reference point for their SPS measures, there is also a feeling that standards sometimes reflect the interests of the developed countries (of both industry and consumer protection groups). For instance, some claim that scientific evaluation of risks is usually based on the nutrition habits of developed nations, without a proper consideration of diet patterns existing in the developing countries of Africa and Asia.[57] Similarly, the identification of priorities in the agenda is also heavily influenced by the members from developed countries. This affects the allocation of resources, which go in many cases to the research areas of particular interest to developed countries. The major reason which stands behind this limited role of most developing countries is of a financial nature as they do not always consider participation in the standard-setting process as a priority among many pressing demands for resources.[58] Equally important is a general lack of sufficient expertise that would allow meaningful participation in the work of international standard-setting bodies.

Another concern of developing countries relates to the continuous proliferation of national SPS standards, despite the harmonization efforts undertaken on the international level. This is particularly problematic because national standards frequently aim at a higher level of protection than their international counterparts. There are different reasons underlying this process. The most common motive is a difference in what is considered to be acceptable risk. Developed countries may simply regard international standards as insufficient to meet their very high levels of protection. Moreover, the production of national standards is normally faster compared to the process that takes place on the international level. Even if an international standard is finally approved, it is sometimes politically difficult for domestic regulators to change already operating measures.[59] Obviously there are also instances when a divergence is a result of the pressure from powerful national interest groups rather than of real health concerns.

Some commentators are worried about the use of majority voting in the standard-setting process, which may potentially undermine the legitimacy of international standards. The main criticism goes to the Codex, as an entity with the most liberal rules (simple majority). On the other hand, the OIE system is frequently referred to as a sensible compromise (the standard is adopted providing that no more than 10 members are against) between a pragmatic need for effective operation and sovereign rights of members to maintain measures that guarantee specific ALOP. There are also those who fear that the process of standard-setting is still dominated by industry without an appropriate input from other interest groups.[60] Consequently, standards, instead of reflecting general concerns of society, may actually promote the interest of specific groups.

[57] Ibid, 49. [58] Ibid, 39. [59] Ibid, 84.

[60] Victor ('Assessment') 886; as of 2002 out of 151 NGOs with observer status at the Codex, around 70% represented industry, 22% professional bodies, while only 8% consumer/public interest (FAO/WHO (*Report of the Evaluation*) 19). Unfortunately there is no more recent data.

Regrettably, the numerous initiatives that were undertaken during the last decade have produced only minor improvements. Different programmes seem to lack an overall and long-term perspective while the technical and financial assistance provided by developed countries constantly fails to meet demand. The situation is slightly better with respect to effectiveness in the operation of standardization bodies. The process of standard-setting has accelerated and expanded with a number of new issues included in the agenda (eg GMOs). There is also some moderate progress in enhancing the legitimacy of the standard-setting process by improving transparency (eg expanded web pages with access to different information) and assuring participation of different actors.[61] Independently, some national delegations also started to include representatives of non-industry NGOs. The involvement of consumer protection groups from developed countries, however, also generates its own problems as they are seen as an inspiration for the establishment of standards at levels that are not achievable by the industries of developing WTO Members.

3.4 Standards, guidelines, and recommendations

The SPS Agreement confers a quasi-legislative authority to international standards by recognizing them as a reference point for national regulatory actions. At the same time, the Agreement also confirms an institutional independence between the WTO and the sister organizations. The WTO dispute settlement bodies do not undertake any inquiry into the standard-setting process and are only concerned with its final outcome. Consequently, it is irrelevant from the perspective of the SPS Agreement whether a standard was accepted unanimously or by mere majority.[62] In the context of *EC – Hormones*, this meant that certain residue standards for hormones were regarded as relevant, despite the fact they were adopted by the Codex with only a slight majority of votes.[63] It is equally irrelevant when a standard was adopted. The pre-SPS Agreement standards have exactly the same status as those which were adopted after the Agreement came into force.[64] Similarly, it does not matter whether a WTO Member is a member of a particular standard-setting body which promulgated a standard.[65] The obligation of Article

[61] For example, the Codex has recently made some serious efforts to enhance the participation of NGOs as observers (including consumer protection groups). Despite those efforts, the significant majority of them are still industrial associations (Scott (*Commentary*) 247).

[62] Panel Report, *EC – Hormones (US)*, para 8.69.

[63] 33 Codex members for, 29 opposing and 7 abstaining from the vote, see Codex Alimentarius Commission, Report of the 21st Session, List of Standards and Related Texts Adopted by the 21st Session of the Codex Alimentarius Commission, ALINORM 95/37 (8 July 1995) para 45.

[64] Panel Report, *EC – Hormones (US)*, para 8.69.

[65] There are some countries which are WTO Members and not contracting parties to the IPPC (eg Gabon, Mongolia) or the OIE (eg Liechtenstein, Papua New Guinea).

3.1 also applies to those standards which were adopted by the relevant organization, even before a country became its member. Last but not least, contrary to the TBT Agreement, SPS standards need to be formally approved. A mere draft that is still in a regulatory pipeline is not considered to be a relevant standard.[66]

The SPS Agreement does not make any distinction between standards, guidelines, and recommendations as to their legal effect and they always appear in the text of the Agreement in a single reference.[67] As a consequence, their qualification within the relevant international standard-setting body is of secondary importance. This homogenous approach of the SPS Agreement to different norms produced by international standard-setting bodies was the subject of some criticism in the literature. As observed by Victor with respect to Codex documents, guidelines were normally adopted when it was not possible to agree on more concrete commodity or residue standards and were never meant to have the same effect as standards.[68] This was also reflected in the General Principles of Codex Alimentarius, which described guidelines and recommendations as having merely an advisory character.[69] Nowadays, this criticism has lost some of its force. The Codex, in an attempt to integrate its activities with the new legal framework established by the SPS Agreement, has progressively unified its approach to different types of documents, for example by requiring the same voting procedure. The text of the General Principles was amended in order to reflect the uniform nature of standards, guidelines, and recommendations. The IPPC has also introduced a homogenous system, which includes both specific rules and more general guidelines. What seems to be more important in this context is the substantive content of particular documents and not its classification. For instance, residue standards tend to be more precise, providing specific requirements in the form of exact numerical value, which leaves WTO Members a very limited discretion. Guidelines and recommendations are formulated in very general terms and provide greater leeway in the choice of implementing measures.[70]

Another problematic issue is the status of regional standards which are developed both within the IPPC[71] and the Codex.[72] The reference in the SPS Agreement to international standards suggests that regional norms are not

[66] Cf Article 2.4 of the TBT Agreement (providing that 'where technical regulations are required and relevant international standards exist or *their completion is imminent*, Members shall use them, or the relevant parts of them, as a basis for their technical regulations' (emphasis added)).

[67] For example, Article 3.1, 3.2 and 3.3 of the SPS Agreement.

[68] Victor ('Assessment') 886.

[69] Codex, *Procedural Manual* (16th edn, 2006) 30 ('The Codex Alimentarius also includes provisions of an advisory nature in the form of codes of practice, guidelines and other recommended measures intended to assist in achieving the purposes of the Codex Alimentarius').

[70] Cf SPS Committee, Clarification of References to Codex Text, Draft Response to the Codex Alimentarius Commission, Note by the Chairman, 13 March 1998, G/SPS/W/86/Rev 1, para 5.

[71] Article II.1 of the IPPC.

[72] The Codex regional standards include the Regional Standard for Canned Humus with Tehena, Regional Standard for Canned Foul Medames, Regional Standard for Tehena, Code of Hygienic Practice for the Preparation and Sale of Street Foods, and several Regional Guidelines for Codex Contact Points and National Codex Committees.

relevant for the purpose of Article 3.[73] This may actually have some advantages since under the regional standards 'it was easier for a coalition that favoured protectionism to control a process because like-minded industrial interests were often concentrated in particular regions'[74] while the process, which requires participation of members coming from different parts of the world, makes that more difficult. On the other hand, it is hard to find a rational for differentiation between regional and international standards in those disputes which involve countries from the same region. Note that regional standards are adopted in a similar procedure to international ones, except for the fact that only countries belonging to a particular region may vote. The adoption process is equally transparent and gives an opportunity for every participant to express its concerns and objections. At the same time, they may be best suited to address SPS particularities of the region. Again, it seems that the problem was partially resolved by adjustments that took place in international standard-setting organizations. The Codex has transformed the majority of its regional standards into international ones. Regional standards still play a role in the context of the IPPC framework (within the RPPOs). This is also a field where the sentiments for regional harmonization are particularly strong. As noted in the literature, plant safety may be an issue with respect to which a regional harmonization 'would be the most appropriate level for the convergence of regulations (since plant diseases tend to be region specific, and so does the response to them in terms of standards)'.[75] Although regional standards cannot be regarded as relevant under Article 3, underlying scientific material may be at least used for the purpose of Article 2.2.[76]

3.5 Substantive obligations

Article 3.1 of the SPS Agreement requires WTO Members to base their SPS measures on international standards. Those measures which conform to such standards are deemed to be necessary to protect human, animal or plant life or health, and presumed to be consistent with the relevant provisions of the SPS Agreement and GATT 1994 (Article 3.2). Pursuant to Article 3.3, a WTO Member aiming at the different level of SPS protection (in the majority of the cases higher) than would be achieved by a measure based on the relevant international standards may still introduce and maintain SPS measures, however, only if there is sufficient scientific justification.

Initially, Article 3 was described as a two-dimensional norm. The *Hormones* panel saw the obligation to base measures on international standards (Article 3.1)

[73] SPS Committee, Clarification of Reference to Codex Texts, para 6.

[74] Victor ('Assessment') 888–9.

[75] Doaa A Motaal, 'The "Multilateral Scientific Consensus" and the World Trade Organization' (2004) 38(5) Journal of World Trade 855, 861.

[76] Cf section 2 of this chapter.

as a general rule, with Article 3.2 merely explaining the legal consequences of the compliance. Consequently, in order to be based on an international standard a measure had to conform to that standard.[77] On the other hand, Article 3.3 was depicted as an exception providing WTO Members with a possibility of adopting measures that deviate from international standards. According to the panel, these two options exhausted the whole meaning of Article 3. The nature of the relationship between Article 3.1 and 3.3 also determined the allocation of the burden of proof between the parties. It was for a complainant to establish the violation of positive obligation of Article 3.1 by demonstrating that a measure did not conform to international standards. On the other hand, it was for a defendant to show that such a measure could nevertheless be saved under Article 3.3 as having a scientific justification.[78]

The Appellate Body disagreed and held that Article 3 provided three different options that are available for WTO Members: a measure may conform to international standards, it can be merely based on them, or it may deviate completely.[79] According to the Appellate Body, Article 3.1 'excludes from its scope of application the kinds of situations covered by Article 3.3 of that Agreement'.[80] In other words, these two provisions apply to different situations: failure to meet the conditions of Article 3.3 does not mean that Article 3.1 is violated. As support for its position, the Appellate Body pointed out the semantic difference between the expression 'to base' and 'to conform', with the first one denoting a weaker relationship than conformity. The Appellate Body also advanced the structural argument, noting that both terms were found in two different paragraphs of Article 3. It believed that this was a deliberate decision of the drafters, which implied different consequences for each provision. Finally, the Appellate Body referred to the purpose of Article 3. After identifying that its aim was to 'promote the harmonization of the SPS measures of Members on as wide a basis as possible',[81] it also observed that 'harmonization of SPS measures of Members on the basis of international standards is provided in the Agreement, as a *goal*, yet to be realized *in the future*'.[82] The interpretation, which equated two terms ('to base' and 'to conform'), would force the harmonization process and transform those standards into binding norms. This would, however, go against the object and aim of the SPS Agreement.[83] According to the Appellate Body, the SPS Agreement only encourages WTO Members to use international standards by establishing a presumption of consistency for conforming measures. In other words, Article 3.2 may be regarded only as an incentive for WTO Members to pursue the harmonization process.

[77] Panel Report, *EC – Hormones (US)*, para 8.72. [78] Ibid, para 8.86.

[79] Appellate Body, *EC – Hormones*, paras 170–72.

[80] Ibid, para 104. [81] Ibid, para 177.

[82] Ibid, para 165 (emphasis in the original).

[83] Ibid (noting 'we cannot lightly assume that sovereign states intended to impose upon themselves the more onerous, rather than the less burdensome, obligations by mandating conformity or compliance with such standards, guidelines or recommendations').

Such conceptualization also has important consequences for the allocation of the burden of proof between the parties to the dispute. If none of the provisions is an exception, it is always for a complainant to show a prima facie case of inconsistency with both subparagraphs. As noted by the Appellate Body, a Member is 'not penalized by exemption of a complaining Member from the normal burden of showing a *prima facie* case of inconsistency with Article 3.1 or any other relevant article of the SPS Agreement or of the GATT 1994'.[84] Thus, as rightly observed by Scott, 'in formal terms, the existence of international standards bears little on the allocation of burden of proof'.[85] It is always a complainant that needs to establish a prima facie case of inconsistency, irrespective of whether a measure conforms to or deviates from international standards.

3.5.1 Measures conforming to international standards

An SPS measure that conforms to international standards 'embodies... [such] standard completely and, for practical purposes, converts it into a municipal standard'.[86] This high level of compatibility also requires that the level of protection, which is reflected in a standard and a measure, is exactly the same. Note also that Article 3.2 does not predetermine the margin of discretion that is actually enjoyed by WTO Members. That depends more on the content of the particular international standard. For example, as already mentioned, guidelines are normally formulated in a very broad and general language and conformity with them may be achieved in different ways. In such a situation diverse SPS measures can take the advantage of the presumption provided by Article 3.2.

Measures conforming to international standards benefit from the presumption of consistency with WTO law. This simple rule is, however, not an example of clarity. Note that Article 3.2, when identifying the scope of the presumption, uses the expression 'relevant provisions of... [SPS] Agreement and of GATT 1994'. This may suggest that the presumption does not cover the entire agreements but is only limited to selected provisions. This is less problematic under the GATT 1994, which is composed of different obligations, some of them being simply irrelevant for assessment of SPS measures. However, even here it is difficult to identify all the relevant provisions. Intuitively one would include not only Article XX (the general exception) but also Article I (the national treatment principle), Article III (the most-favoured nation principle), and Article XI (the prohibition of quantitative restrictions). Such an interpretation would fall in line with the language of subparagraph 8 of the Preamble to the SPS Agreement, which speaks about 'the provisions of GATT 1994... [that] relate to the use of sanitary or phytosanitary measures, in particular the provisions of Article XX(b)'. The expression 'in particular' clearly suggests that there are other provisions of

[84] Ibid, para 171. [85] Scott (*Commentary*) 260.
[86] Appellate Body Report, *EC – Hormones*, para 170.

GATT 1994 which may be relevant for the assessment of SPS measures. Similar wording is included in Article 2.4 of the SPS Agreement, which again refers to 'the provisions of GATT 1994 which relate to the use of sanitary or phytosanitary measures, in particular the provisions of Article XX(b)'.

This, however, sits uneasily with the character of SPS measures, which makes difficult any analysis of the measures under the disciplines of Articles I, III, and XI of GATT 1994.[87] As will be discussed in Chapter 6, section 6.4.1(d), the SPS Agreement, in recognition of the special nature of the SPS Agreement, provides its own specific formulations of both the national treatment and the most-favoured nation principle (eg the SPS Agreement does not rely on the concept of like products, instead it prohibits discrimination between Members where identical or similar conditions prevail and between a Member's own territory and that of other WTO Members). The same is true for Article XI of GATT 1994. WTO Members sometimes need to introduce SPS quantitative restrictions (eg importation bans) in order to address a particular risk. What seems to be even more important is the fact that a single measure cannot be simultaneously compatible with the positive obligations of GATT 1994 and its general exception. Note that the recourse to Article XX(b) presupposes a prior violation of one of the provisions of GATT 1994 (ie Article I or III). This fact seems to support the position that in practice only Article XX(b) is covered by the presumption.

The scope of the presumption under the SPS Agreement is equally ambiguous. One interpretation would be to understand the expression 'relevant' as simply meaning 'applicable' and assume that the presumption covers all provisions of the SPS Agreement.[88] This would mean that a measure which conforms to an international standard is refutably immune from any attack under the SPS Agreement. This narrow approach would call for a more limited scope of the presumption, which would only apply to those provisions of the SPS Agreement that correspond to the content of a particular standard whereas other provisions will simply not be covered by the presumption. Article 5.5, which requires WTO Members to avoid arbitrary or unjustifiable distinctions in the levels of protection they consider to be appropriate, may serve as a good example here. Note that the majority of standards are simply concerned with a specific risk agent and do not say anything about consistent application of different national SPS measures. It is not difficult to imagine a situation where a measure which is compatible with the relevant international standard also fails to meet the requirement of

[87] But cf *EC – Biotech Products*, where Argentina requested the panel to find that product-specific measures were inconsistent with certain obligations of the SPS Agreement. Simultaneously, Argentina also claimed that those measures violated Article III:4 of GATT 1994. The panel, after finding that the first sentence of Annex C(1)(a) was violated, continued its analysis under GATT 1994, where it found that Argentina failed to establish a prima facie case of inconsistency (Panel Report, *EC – Biotech Products*, paras 7.2509–7.2516).

[88] Cf Scott (*Commentary*) 258; see also, Panel Report, *EC – Hormones (US)*, para 8.72 (noting that 'Article 3.2... introduces a presumption of consistency with both the SPS Agreement and GATT').

Article 5.5 (eg if another measure provides lower ALOP in a different but comparable situation). This fact could be an argument for a narrow interpretation. Despite this, there are good reasons for accepting a broader conceptualization of the presumption. First, it provides a clear and unambiguous rule, which is easy to apply. Consequently, there is no need to determine in each instance which obligations of the SPS Agreement find their reflection in a particular standard. Secondly, one also needs to keep in mind that the presumption of Article 3.2 is a rebuttable one. Therefore, it is always possible for a complainant to claim that a particular measure, despite being in conformity with an international standard, nonetheless violates certain obligations of the SPS Agreement (eg Article 5.5). Accepting the broader character of the presumption does not change anything in this respect. Thirdly, it also guarantees consistency of results under both agreements. A measure that is presumed to be compatible with the disciplines of GATT 1994 will enjoy the same status under the SPS Agreement.

The reference to GATT 1994 which appears in Article 3.2 is problematic in itself. Remember that pursuant to Article 2.4, measures that conform to the SPS Agreement are presumed to be consistent with the relevant provisions of GATT 1994. The second sentence of Article 3.2 establishes another presumption of consistency with the SPS Agreement for those measures which conform to international standards. If one combines these two norms, it is immediately apparent that the first sentence of Article 3.2 is an unnecessary repetition of what is already available under the SPS Agreement. One way of escaping from this conclusion is to conceptualize Article 2.4 as providing irrefutable presumption (ie a measure that conforms to the SPS Agreement is irrefutably presumed to be GATT-consistent), while Article 3.2 will provide rebuttable presumption (ie measures that conform to international standards are refutably presumed to be GATT- and SPS-consistent). This would differentiate the effect of those two provisions as under Article 2.4 no proof to the contrary would be possible. This approach, however, does not find sufficient support in the language of the SPS Agreement and also seems to be incompatible with the cumulative nature of the obligations provided by GATT 1994 and the SPS Agreement.[89]

The character of the presumption provided by Article 3.2 is equally unclear. Its language suggests that some distinction between 'deemed' and 'presumed' should be made. Remember that according to this provision, a measure conforming to international standard is *deemed* to be necessary to protect human, animal, or plant life or health and *presumed* to be consistent with the relevant provisions of the SPS Agreement and of GATT 1994. The ordinary meaning of the verb 'deem' means 'to regard as' or 'to consider',[90] whereas 'to presume' is defined as 'to take for granted as being true in the absence of proof to the contrary'.[91] The

[89] Cf Chapter 2, section 2.4.2.

[90] *The American Heritage, Dictionary of the English Language* (4th edn, 2004) <http://dictionary.reference.com/browse/deem> (last visited 15 July 2009).

[91] Ibid.

first expression is stronger and may imply different types of consequences. This also finds support in the general approach of the Appellate Body, which stressed that 'the different words are designed to convey different meanings' and '[a] treaty interpreter is not entitled to assume that such usage was merely inadvertent on the part of the Members who negotiated and wrote that Agreement'.[92] Such an interpretation is additionally encouraged by the structure of Article 3.2. Reading those two expressions as synonyms will deprive the first part of Article 3.2 of any legal meaning. A measure which is presumed to be compatible with the relevant provisions of the SPS Agreement and of GATT 1994 has to be regarded as necessary since the necessity test constitutes one of the requirements under both agreements.

So what are the available options? First, 'deemed' may be understood as providing a different and arguably stronger form of presumption. In practice, this could be translated into a higher threshold for any proof to the contrary that would apply to the necessity requirement of Article 2.2. Secondly, the expression may be interpreted as providing a form of non-rebuttable presumption. This would mean that SPS measures conforming to international standards are always considered to be necessary to protect human, animal, or plant life and health, while the conformity with other provisions of the SPS Agreement is subject to rebuttal. Finally, one may also simply disregard the language of the provision and accept that there is no real difference between those two expressions.

The problem with the first approach is that WTO law does not provide any standard which might be qualified as a stronger form of presumption. Further, there are no compelling reasons which could justify the existence of two different legal standards in this context. Why should one of the presumptions be stronger than the other? The same argument also applies to the second approach. Reading the first part of Article 3.2 as providing for an irrefutable presumption would result in a double standard: irrefutable presumption of necessity and rebuttable presumption with respect to the other requirements of the SPS Agreement. This interpretation also fails to address those situations where a standard turns out to be inappropriate (eg due to the latest scientific developments), however it has not yet been changed. In such a situation irrefutable presumption would produce absurd consequences: a measure could be regarded as necessary despite that fact that, for example, it lacks a sufficient scientific basis. Note also that the irrefutable presumption would transform international standards into binding norms, the interpretation which was clearly rejected by the Appellate Body. The third approach appears to be free from such drawbacks. Although it requires a rather liberal reading of Article 3.2 and reduces the conjunction to a single standard, it also seems to provide the best available solution. It guarantees consistency in the treatment of different obligations of the SPS Agreement and does not lead to undesirable results.

92 Appellate Body Report, *EC – Hormones*, para 164.

The SPS case law does not explicitly resolve that problem. Although the Appellate Body generally stated that the presumption of Article 3.2 is a rebuttable one,[93] it did not specify whether its statement referred to the whole article or just to the second part. The Appellate Body in *US – Continued Suspension* was probably slightly more precise when it read the conjunction as providing a single standard of rebuttable presumption.[94] However, since this observation was made on the margins of the main discussion, its value may be limited.

3.5.2 Measures based on international standards

The Appellate Body, when examining Article 3.1, observed that 'to be based means to be "founded" or "built" upon or "is supported by" '[95] and added that a measure that is based on international standards 'may adopt some, not necessarily all, of the elements of the international standard'.[96] As already mentioned, this type of relationship is weaker than a full conformity. Of course, in some cases it will be not possible to base a measure on an international standard without actually conforming to it. Codex MRL standards may serve as a good example. They provide a numerical value determining the maximum level of residues that is considered to be without any health consequences. This precise formulation leaves no discretion to WTO Members who may only accept or reject such a standard.

Regrettably, the above observations of the Appellate Body on the nature of the relationship, which is implied in the term 'to base on', was not further elaborated in SPS case law. In consequence, it remains unclear what level of compatibility it demands. The case law decided under Article 5.1 of the SPS Agreement and 2.4 of the TBT Agreement may be of some help here.[97] Note that Article 5.1 uses the same language in requiring WTO Members to base their measures on risk assessment. The expression 'to base on' was understood as indicating an objective relationship between two elements. Its examination consists of a comparison between the conclusions of a risk assessment (which under Article 3.1 would refer to an international standard) and SPS measure, in order to examine their compatibility.[98] The applicable standard was described rather vaguely as a form of rational relationship test. The Appellate Body was slightly more precise when it analysed the term 'as a basis for', which appears in Article 2.4 of the

[93] Ibid, para 170.

[94] Appellate Body, *US – Continued Suspension*, para 532 (noting that 'there is a rebuttable presumption that SPS measures that conform to international standards, guidelines or recommendations are necessary to protect human, animal or plant life or health, and... [are] consistent with the relevant provisions of this Agreement and of GATT 1994').

[95] Appellate Body Report, *EC – Hormones*, para 163.

[96] Ibid, para 171.

[97] Article 2.2 of the SPS Agreement also contains the very same language and provides that 'Members shall ensure that any sanitary or phytosanitary measure... is based on scientific principles'. However, this part of Article 2.2 has not yet been addressed by the case law.

[98] Cf Appellate Body Report, *EC – Hormones*, para 192.

TBT Agreement. However, even here, the Appellate Body failed to identify the required threshold precisely. It only held that 'there must be a very strong and very close relationship between two things in order to be able to say that one is "the basis for" the other'.[99] This implies a more demanding standard, which is not satisfied when the technical regulation is merely informed in its overall scope by the international standard.[100] Taking into account the functional and textual similarities between Article 2.4 of the TBT Agreement and Article 3.1,[101] one may assume that the term 'to base on' under the SPS Agreement also needs to be construed as requiring a rather strong and close relationship.[102] Irrespective of the approach that one prefers, what seems to be clear under Article 3.1 is that the 'based on' requirement calls for a substantive relationship between a measure and a standard. The mere fact that a WTO Member has considered a standard before the adoption of a measure is not sufficient to meet the required threshold.

The case law does not unambiguously settle whether a measure, in order to be considered as 'based on', needs to reflect the same level of protection as the one which is guaranteed by the particular standard. The panel in *EC – Hormones* observed that Article 3.3 opposes measures based on international standards with measures that result in a higher or different level of protection.[103] Thus, the level of protection is one of the factors which determines whether a measure can be considered as being based on an international standard.[104] The position of the Appellate Body on this issue is unclear. On the one hand, it criticized the panel for reading 'much more into Article 3.3 than can be reasonably supported by the actual text of Article 3.3'.[105] This may suggest a general disapproval of the reasoning of the panel, including its conclusions on the required level of protection. On the other hand, the Appellate Body also added that 'the correctness of the rest of the Panel's intricate interpretation and examination of the consequences of the Panel's litmus test, however, have to be left for another day and another case'.[106] This leaves a door open for finding in the future that the level of protection is a relevant factor for the purpose of Article 3.1.

A textual interpretation of Article 3 indeed supports the panel approach. Article 3 undeniably links the obligation to base a measure on international standards with the level of protection. Note that Article 3.3, which determines situations where a WTO Member may deviate from international standards, establishes a dichotomy between measures that are based on international standards and those that aim for higher or different level of protection. Its first sentence speaks about measures which result in a higher level of SPS protection than would be

[99] Appellate Body Report, *EC – Sardines*, para 245.

[100] Ibid, para 247.

[101] Note that both articles provide the obligation to base domestic measures on international standards.

[102] Cf Appellate Body Report, *EC – Sardines*, paras 242–4.

[103] Panel Report, *EC – Hormones (US)*, para 8.72.

[104] Ibid, para 8.73.

[105] Appellate Body Report, *EC – Hormones*, para 168.

[106] Ibid.

achieved by measures *based on* the relevant international standards. Similarly, the second sentence refers to measures which result in a level of SPS protection different from that which would be achieved by measures *based on* international standards. This language clearly suggests that under the SPS Agreement a measure may be considered as based on an international standard only if it achieves the same level of protection as the relevant standard. This conclusion will also indicate, in line with the position taken by the Appellate Body under Article 2.4 of the TBT Agreement, that the actual link between an SPS measure and an international standard needs to be very strong. Of course, one may imagine a situation where a measure is 'based on' a standard (in a colloquial sense) without reflecting the same level of protection. However, under the textual interpretation of Article 3.3, such a measure could not be qualified as the one which is based on an international standard.

One may also ask about the legal consequences of basing measures on international standards without, however, conforming to them. The case law is silent on this issue and any observations are necessarily speculative. Obviously, the presumption of Article 3.2 is not available here. On that basis, some commentators believe that there are no advantages to merely basing national SPS measures on an international standard.[107] Thus, the second option under Article 3, the argument goes, is actually without any legal meaning.[108] Others, however, believe that a measure which is based on an international standard, should be automatically considered as based on a risk assessment that was used by the relevant international organization in drafting its standard.[109] Consequently, no examination under Articles 2.2 (the second and third sentence) and 5.1 to 5.3 is necessary. A closer examination of the provisions of the SPS Agreement shows that actually both positions may tell part of the story.

Remember that the first sentence of Article 3.3 requires scientific justification only for those measures which result in a higher level of SPS protection than would be achieved by measures based on international standards. The second sentence, which prohibits inconsistencies with other provisions of the SPS Agreement, refers to the measures which result in a level of SPS protection that is

[107] Rüdiger Wolfrum, Peter-Tobias Stoll, and Anja Seibert-Fohr (eds), *WTO—Technical Barriers and SPS Measures, Max Planck Commentaries on World Trade Law, Vol 3* (2007) 421. Cf also, Joost Pauwelyn, 'The WTO Agreement on Sanitary and Phytosanitary (SPS) Measures as Applied in the First Three SPS Disputes' (1999) 2(4) Journal of International Economic Law 641, 656.

[108] Wolfrum at al (*Technical Barriers and SPS Measures*) 421; the commentators note that one meaning of the second option would be to allocate differently a burden of proof under Article 3.1 and 3.3. Consequently, it would be for a complainant to show that the measure which is based on international standards nevertheless violates some of the obligations of the SPS Agreement. In case of measures deviating from standards completely, the burden of proof would rest on a defendant. However, this approach, as the commentators admit, was explicitly rejected by the Appellate Body.

[109] Denise Prévost and Peter Van den Bossche, 'The Agreement on the Application of Sanitary and Phytosanitary Measures' in Patrick FJ Macrory, Arthur E Appleton, and Michael G Plummer (eds), *The World Trade Organization: Legal, Economic and Political Analysis* (2005) 274.

different from the one which would be achieved by international standards. This language clearly suggests that Article 3.3 is only applicable to measures which achieve a higher (different) level of protection as compared to international standards. If one subscribes to the interpretation proposed in this book, that a measure, in order to be based on international standards, needs to reflect the same level of protection, the logical conclusion is that Article 3.3 is simply inapplicable to such measures.[110] Since it is Article 3.3 that triggers the application of the other disciplines of the SPS Agreement, one may argue that actually none of the other provisions would be applicable to such measures. This, however, leads to absurd consequences. A measure which conforms to international standards will be presumed (refutably) to be consistent with the provisions of the SPS Agreement, whereas a measure which is merely based on an international standard will be completely and permanently immune as none of the other SPS provisions will apply. In practice, this will give an effect equivalent to irrefutable presumption.

The above suggests that in legal terms WTO Members do not gain anything by basing their measures on international standards. There are two exceptions to this rule. First, Article 5.8 provides a right of WTO Members to request an explanation of the reasons for a particular SPS measure only if a measure is not based on international standards. *A contrario* the obligation does not exist with respect to measures based on international standards. Secondly, one may also argue that the notification procedure provided by Annex B is inapplicable here. Note that pursuant to subparagraph 5, an SPS regulation needs to be notified only if its content is not *substantially* the same as an international standard. Therefore the exemption from the notification requirement goes beyond the measures that conform to international standards, and also covers those measures which are substantially the same. If one subscribes to the view presented here, that in order to qualify a measure as based on an international standard, there has to be a very strong and close relationship between such a measure and a standard (reflected in the same level of protection); Annex B(5) would become inapplicable to this category of measures.

There are also some practical advantages in basing measures on international standards. Inapplicability of Article 5.8 and Annex B(5) will probably increase difficulties that are faced by a complainant when making a prima facie case of inconsistency. As noted by the Appellate Body, a right to receive information under Article 5.8 provides a useful mechanism, which helps to collect data necessary to meet the burden of proving a prima facie case.[111] A defendant itself may also benefit from the fact that the measure is based on an international standard. In particular, it may rely on evidence and scientific data, which were used in the standard-setting process, to show that its measure is supported by sufficient

[110] Cf Appellate Body Report, *EC – Hormones*, para 104.
[111] Ibid, para 102. This was also confirmed by the panel in *Japan – Apples*, para 8.41.

scientific evidence. Note, however, that this is merely an advantage in a fact-finding process and not a form of legal presumption.

3.5.3 Measures deviating from international standards

Article 3.3 provides that a WTO Member may adopt SPS measures which result in a higher level of SPS protection than that provided in international standards, if there is scientific justification or 'as a consequence of the level of sanitary or phytosanitary protection a Member determines to be appropriate in accordance with the relevant provisions of paragraphs 1 through 8 of Article 5'.[112] According to the Appellate Body, this is the third option that is available to WTO Members under Article 3. Literal reading of the provision would indicate that the SPS Agreement introduces a logical disjunction ('or'), which envisages two separate modes of compliance. Such an approach to Article 3.3 would have important consequences. As the first part of the alterative does not explicitly refer to Article 5 of the SPS Agreement, it may be argued that scientific justification is possible without any formal risk assessment as prescribed by Articles 5.1 to 5.3.[113] Thus, a WTO Member will be able to defend its SPS measure by simply relying on scientific principles and evidence in accordance with Article 2.2.

The case law, however, went in a different direction. The Appellate Body, after textual analysis of the provision, held that the disjunction provided by Article 3.3 was more apparent than real.[114] This conclusion was based on three arguments. First, the Appellate Body found that the second sentence of Article 3.3 required measures which resulted in a level of SPS protection different from that which would be achieved by measures based on international standards to comply with all other provisions of the SPS Agreement. According to the Appellate Body, this also included Articles 5.1 to 5.3.[115] The second argument was based on the definition of scientific justification as provided by the footnote to Article 3.3.[116] In particular, the Appellate Body observed that the examination and evaluation of available scientific information might have 'appear[ed] to partake of the nature of the risk assessment required in Article 5.1 and defined in paragraph 4 of Annex

[112] Article 3.3 of the SPS Agreement.

[113] Cf the appellant's (EC) submission in *EC – Hormones*, para 88.

[114] Appellate Body Report, *EC – Hormones*, para 176.

[115] Note that the interpretation proposed by the Appellate Body is only one of the possible interpretations. The reference to 'any other provision of this Agreement' may actually mean all other provisions which are not mentioned in the first sentence ('notwithstanding the above').

[116] The footnote to Article 3.3 of the SPS Agreement provides as follows: '[f]or the purposes of paragraph 3 of Article 3, there is a scientific justification if, on the basis of an examination and evaluation of available scientific information in conformity with the relevant provisions of this Agreement, a Member determines that the relevant international standards, guidelines or recommendations are not sufficient to achieve its appropriate level of sanitary or phytosanitary protection'.

A of the *SPS Agreement*'.[117] Thirdly, the Appellate Body advanced a functional argument and looked to the object and the purpose of Article 3 and the SPS Agreement as a whole. On the one hand, the SPS Agreement aims at promoting harmonization of SPS measures on as wide a basis as possible. On the other hand, it also recognizes and safeguards the right and duty of WTO Members to guarantee a level of protection which they deem to be appropriate. The Appellate Body believed that 'the requirements of a risk assessment under Article 5.1 . . . are essential for the maintenance of the delicate and carefully negotiated balance in the *SPS Agreement* between the shared, but sometimes competing, interests of promoting international trade and of protecting the life and health of human beings'.[118] This led the Appellate Body to conclude that under both parts of the disjunction, a WTO Member is obliged to possess an adequate risk assessment that reasonably supports a relevant SPS measure.

The approach of the Appellate Body may be criticized from the perspective of international customary rules of treaty interpretation. On a textual level, it ignores the explicit language of Article 3.3, which establishes a disjunction rather than a conjunction. A reading of Article 3.3 that equates the two alternative parts, effectively turns the word 'or' into 'and'.[119] The result also remains questionable in the light of the principle of effective treaty interpretation. As already mentioned in Chapter 2, the principle was recognized as one of the relevant interpretative guidelines in the context of WTO law and which requires an interpreter to read different provisions so as to give meaning to each of them. An interpretation that transforms two separate obligations into a single requirement can hardly be seen as reaching that standard.

At the same time, one also needs to admit that the text of Article 3.3 by itself is not an example of clarity[120] and the interpretation proposed by the Appellate Body is defendable on textual grounds. Article 3.3 may indeed be read as requiring WTO Members, under both alternative parts, to comply with all other obligations of the SPS Agreement, including its risk assessment disciplines. The approach of the Appellate Body is probably also correct from a pragmatic point of view. Accepting justification of a measure in the form of a mere collection of scientific evidence will obviously reduce the importance of the regime provided by Article 5.1 to 5.3. Why should a WTO Member perform a risk assessment if it is sufficient to show that a measure is based on scientific evidence? This conclusion appears to be counter-intuitive, as risk assessment obligations occupy a central position in the legal framework established by the SPS Agreement. Thus, despite some objections which may be raised against the interpretation proposed by the Appellate Body, the final outcome seems to be acceptable.

[117] Appellate Body Report, *EC – Hormones*, para 175.
[118] Ibid, para 176.
[119] Wolfrum et al (*Technical Barriers and SPS Measures*) 424.
[120] Cf Appellate Body Report, *EC – Hormones*, para 175.

3.5.4 Measures that are less demanding than international standards

A specific type of measure deviating from international standards deserves a separate analysis. There seems to be some disagreement in the literature as to the permissibility under the SPS Agreement of measures which are less demanding than international standards (ie aiming at the lower level of protection than that provided by such standards). In other words, the question is whether the Agreement establishes any minimum harmonization requirement. The problem is not purely academic and may arise as a real WTO controversy. As noted by Victor, '[in case of] manufactured goods, such as processed foods, there is often a substantial premium in efficiency for producers that can export to a market governed by a single standard'.[121] The existence of local standards that are lower than international counterparts may also give an additional advantage to domestic producers as compared to foreign operators, which have to comply with higher, and presumably more costly, international standards. Although a lower standard is in principle available to both domestic and foreign entities, the latter may be obliged to observe higher standards on the basis of its own national law or due to requirements imposed on other exporting markets.

Part of the literature submits that Article 3 may actually require tightening 'national regulations against the countries that fall below the minimum of the international standard'.[122] Note that the first sentence of Article 3.3 only permits the deviation from relevant international standards in those instances where a measure results in a higher level of SPS protection than that provided by such standards. Textual interpretation would therefore suggest that WTO Members are precluded from adopting measures which aim at the lower level of protection.[123] This interpretation may be additionally supported by the language of the Preamble to the SPS Agreement. In particular, its second preambular statement identifies the improvement of the sanitary and phytosanitary situation in all WTO Members as one of the goals of the Agreement.[124] Requiring WTO Members to align with minimal international standards may be seen as a way of reaching that objective. The case law may also suggest that WTO Members can only adopt and maintain measures which are stricter than international standards. Except from the *Hormones* panel, the WTO jurisprudence, when addressing the obligations of Article 3, has consistently referred to the possibility of

[121] Victor ('Assessment') 884.

[122] Scott (*Commentary*) 261, citing Robert Howse, 'A New Device for Creating International Legal Normativity: The WTO Technical Barriers to Trade Agreement and "International Standards"' (unpublished manuscript) 16.

[123] Joost Pauwelyn, 'Non-Traditional Patterns of Global Regulation: Is the WTO "Missing the Boat"' in Christian Joerges and Ernst-Ulrich Petersmann (eds), *Constitutionalism, Multilevel Trade Governance and Social Regulation* (2006).

[124] It specifically provides: '[d]esiring to improve the human health, animal health and phytosanitary situation in all Members'.

adopting measures that provide higher, and not different, levels of protection.[125] For instance, the Appellate Body in one of its recent decisions observed that 'while use of international standards is encouraged, the *SPS Agreement* recognizes the right of WTO Members to introduce or maintain an SPS measure, which results in a higher level of protection than would be achieved by measures based on such international standards'.[126]

This approach to Article 3.3, although textually defensible, does not appear to be persuasive. Note that the language of the provision is very ambiguous and equally permits quite different interpretations. Remember that the second sentence of Article 3.3 refers broadly to measures which result in *different* levels of protection. This expression clearly covers situations where a WTO Member departs from the international standard in both directions—upwards as well as downwards. One may rationally assume that the use of such language is not accidental and reflects the general acceptance of the SPS Agreement for measures that differ from international standards, including those which result in a lower level of protection.[127] Article 3.1 is another provision which seems to support this broader interpretation. It points to subparagraph 3 as only one of the possible instances where a Member may deviate from international standards ('except as otherwise provided for in this Agreement, and *in particular* in paragraph 3'). Although none of the other provisions explicitly addresses such a situation, this language may again suggest a general recognition by the SPS Agreement of the possibility of adopting measures that differ from international standards.[128] Last but not least, the Preamble to the SPS Agreement reaffirms that no Member should be prevented from adopting or enforcing SPS measures and qualifies this recognition only with the need to guarantee that these measures are not applied in a manner which would constitute a means of arbitrary or unjustifiable discrimination or a disguised restriction on international trade.[129] Excluding a priori measures that aim at the lower level of protection would be incompatible with this declaration.[130]

There are other important reasons which speak against the conceptualization of Article 3 as providing a minimum threshold of SPS protection. First, such

[125] For example, Appellate Body Report, *Japan – Agricultural Products*, para 79; Panel Report, *EC – Biotech Products*, para 7.2981.

[126] Appellate Body Report, *US – Continued Suspension*, para 532.

[127] Cf Panel Report, *EC – Hormones (US)*, para 8.72 (observing that '[Article 3.3] applies more specifically to measures which result in a *higher level* of sanitary... protection than would be achieved by measures based on the relevant international standards' or measures 'which result in a *level* of sanitary... protection *different* from that which would be achieved by measures based on international standards') (emphasis in original).

[128] Scott (*Commentary*) 264.

[129] Recital no 1, Preamble to the SPS Agreement.

[130] Scott also points to recital no 6, which emphasises that the SPS Agreement does not generally require Members to change their appropriate level of protection of human, animal, or plant life or health and notes that 'there is nothing to suggest that this bites only at one end of the protective spectrum' (Scott (*Commentary*) 254).

interpretation would go against the idea of the promotion of international free trade, which is an overall objective of the WTO system.[131] Under the 'minimum requirement' approach, WTO Members would be required to restrict trade more than they deem appropriate.[132] Secondly, this may also produce a disequilibrium between the rights and obligations of different WTO Members. Note that in the majority of cases, an option to adopt and maintain measures which are less stringent than international standards will be used by Members from developing countries. On the other hand, developed countries will tend to take measures that aim for a higher level of protection. Consequently, a reading of Article 3.3 may result in a double standard that will affect developing countries more than developed ones.[133] Thirdly, the narrow interpretation appears to be incompatible with other parts of the WTO jurisprudence. Remember the finding of the Appellate Body that 'harmonization of SPS measures of Members on the basis of international standards is projected in the Agreement, as a *goal*, yet to be realized *in the future*'.[134] Reading the SPS Agreement as establishing a minimum standard requirement will transform those standards into binding norms, the quality that was explicitly rejected by the Appellate Body.[135] Fourthly, a tendency of the SPS case law to describe Article 3 as providing a possibility to adopt measures with higher, rather than different, level of protection may be explained by the fact that all disputes decided until now were concerned with the upward deviation. In other words, there was simply no need for the WTO dispute settlement bodies to articulate any rules that are applicable to downward departure from international standards. At the same time, it is a routine approach of case law to address only specific circumstances of the dispute rather than to pronounce general criteria.

The above strongly supports the interpretation of Article 3, which preserves a right of WTO Members to depart from international standards in both directions: upwards when adopting measures that result in a higher level of protection than that provided by international standards, but equally downwards when a measure establishes less stringent requirements than those which are reflected in international standards.

3.6 Conclusions

Article 3 is probably one of the most obscure provisions in the whole SPS Agreement. It does not set any clear normative standards and is open to different competing interpretations. This is probably a result of the disagreement that existed between the drafters, with one group of countries supporting the

[131] Prévost and Van den Bossche ('Agreement') 275.

[132] Cf Scott (*Commentary*) 262.

[133] Ibid, 265.

[134] Appellate Body Report, *EC – Hormones*, para 165 (emphasis in original).

[135] Scott (*Commentary*) 263.

establishment of strict harmonization disciplines and others seeking guarantees for rather unrestrained national regulatory freedom in the area of SPS risk. The compromise which was born from these competing interests has not achieved necessary clarity. If one considers this lack of precision, the approach taken by the Appellate Body appears to be fully understandable. In accordance with the principle of *in dubio mitius*, and in the absence of the proof to the contrary, one may rationally assume that WTO Members impose on themselves less burdensome obligations rather than more onerous.[136] Consequently, Article 3.1 and 3.3 apply to different situations with Article 3.2 operating as a kind of reward for complying with international standards. Conceptualizing Article 3.1 and 3.3 as a rule and an exception would strengthen the harmonization objective of the SPS Agreement, without a clear indication of such an intent in the text of the relevant provision, and would further constrain the ability of WTO Members to set their own level of protection (including a situation where there is no standard). In such a case it would always be the defendant who needs to justify its measure. The approach of the Appellate Body also seems to find support in the more general literature on harmonization. Although the harmonization may provide a useful instrument to address certain types of regulatory heterogeneity, it seems to be the undesirable option when heterogeneity between various jurisdictions results from genuine and legitimate differences. The level of risk that a particular country is prepared to tolerate is arguably one of such legitimate concerns. At the same time, the SPS Agreement provides a number of other instruments which help to detect and eradicate instances of what is called by Sykes 'bad heterogeneity'.

The case law holds that Article 3 provides three options that are available to WTO Members. According to the Appellate Body, a WTO Member may conform to international standards, only base its measures on such standards, or deviate from them completely. Each of these situations brings its own specific legal consequences. In the first instance, a measure is immune to the challenge under both the SPS Agreement and GATT 1994. In this context, the chapter submits that the presumption is broad in its character and covers all requirements of the SPS Agreement. Although under GATT 1994 the presumption seems to be limited to Article XX(b), in practice this guarantees the overall compliance with this Agreement. The chapter also proposes to disregard the ambiguous language of Article 3.2 ('deemed' and 'presumed') and apply in its context a uniform standard of rebuttable presumption.

The legal effects resulting from basing a measure on an international standard seem to be limited to disciplines of Article 5.8 and Annex B(5). At the same time, the chapter recognizes that practical advantages may be substantial. The inapplicability of the above provisions increases difficulties that are faced by a complainant when making a prima facie case of inconsistency. On the other hand, a defendant is able to rely on evidence and scientific data that were used as

[136] Appellate Body Report, *EC – Hormones*, para 165.

a basis for a standard, which makes its evidentiary obligations much easier. The chapter also submits that a measure, in order to be considered as being based on international standards, has to reflect the same level of protection as is provided by such a standard. The support for this conclusion is found in the language of Article 3.3, which establishes a dichotomy between measures that are based on international standards and those which aim for a different level of protection.

Under the third option, a WTO Member needs to guarantee that its measure has sufficient scientific justification. As clarified by the case law, such justification has a form of formal risk assessment. Although this interpretation disregards the textual basis of Article 3.3, it nevertheless seems to be correct as accepting the possibility of justifying a measure without having a risk assessment would considerably reduce the importance of Article 5.1 to 5.3, the provision that is central to the whole SPS Agreement. In the context of the third option, this chapter also submits that the SPS Agreement does not establish any minimum SPS standards. While the text of the Agreement may support a different interpretation, a number of important considerations strongly speak against such an approach. It is not only incompatible with the overall goal of WTO law of promoting rather than restricting trade, but it also produces disequilibrium between rights and obligations of WTO Members. Under the 'minimum requirement' obligation WTO Members would be allowed to deviate from standards upwards, but such move in the opposite direction will be prohibited. This in turn will put a greater burden on Members from developing countries who are the most frequent users of this option.

4

Science, Risk Assessment, and the SPS Agreement

4.1 Introduction

The SPS Agreement ascribes a special role to science. As a general rule, the SPS Agreement requires scientific justification for all measures that do not conform to international standards. This obligation is provided by Article 2.2 which asks WTO Members to base SPS measures on scientific principles and not to maintain them without sufficient scientific evidence. Article 5.1 translates these requirements into the obligation of a formal risk assessment. The risk assessment should, among other things, take into account 'available scientific evidence', providing risk managers with scientific evaluation of potential risks posed by a regulated product. If such evidence is insufficient to perform a risk assessment, a WTO Member may take a provisional SPS measure on the basis of 'available pertinent information', seeking at the same time additional data necessary for a more objective assessment of risk.

So what is the exact role of science under the SPS Agreement? Whose science could be relied on as a legitimate basis for an SPS measure? If minority scientific opinions are relevant, do they need to meet any minimal epistemic threshold? Is there any place for other considerations, such as cultural, economic, and political factors, in the process of risk assessment? What is the level of required scrutiny when examining scientific backup for national measures? As the SPS Agreement is silent on these issues, it has been primarily the role of WTO jurisprudence to provide relevant answers. Although some problems have already been addressed by the case law, others still await authoritative interpretation.

This chapter starts with a discussion of science-based obligations provided by Article 2.2 (requirement to base a measure on scientific principles and not to maintain it without sufficient scientific evidence). The second part concentrates on the risk assessment disciplines of Article 5.1 to 5.3 and is divided into two separate sections, one concerning the requirements imposed on the substantive content of risk assessment, and the second analysing rules that pertain to the implementation of its results. The third part addresses the problem of the applicable standard of review in science-related trade disputes, while the last part offers some observations on the role that is played by science in the context of the

SPS Agreement. Although Article 5.7 is an important element in the science–law puzzle, it is nevertheless discussed in a separate chapter. This approach appears to be justified when one considers the complexities that arise in the context of this article. This, in turn, asks for a separate and independent examination.

4.2 Scientific principles and sufficient scientific evidence (Article 2.2)

The central science-based provision of the SPS Agreement is Article 2.2, which requires WTO Members to base their SPS measures on scientific principles and prohibit their maintenance without sufficient scientific evidence (the second and third sentences). Article 2.2 employs the conjunction ('and'), meaning that both obligations need to be met simultaneously. The case law has not yet defined the term 'scientific principles', tending to evaluate both elements (scientific principles and sufficient scientific evidence) in one analysis that actually concentrates only on the second aspect.[1] The ordinary meaning of the word 'principle' denotes a general rule or law which shows how a particular theory is put into practice.[2] This language suggests that the obligation to base a measure on scientific principles requires a certain methodological soundness and rigour in the process through which the evidence is collected and assessed. The elements which are normally associated with the rigour of scientific methods include objectivity, peer review, falsifiability (hypotheses can be tested and previous results verified or refuted), and impartiality.[3] Note also that the requirement to base a measure on scientific principles is relevant not only for the collection and evaluation of scientific evidence (Article 2.2) but also for the risk assessment (Article 5.1)[4] as well as the adoption of provisional measures taken in the circumstances of insufficiency of scientific evidence (Article 5.7).[5] The methodological quality may also play an important role when assessing minority scientific opinions.

The expression 'sufficient scientific evidence' has received far more attention in the case law. The word 'scientific' was defined by the Appellate Body as 'having or

[1] For example, Panel Report, *EC – Biotech Products*, para 7.1427.

[2] Cf Dictionary.com, Random House, Inc (<http://dictionary.reference.com/browse/principle> last visited 20 July 2009), which defines the term 'principle' as 'a fundamental, primary, or general law or truth from which others are derived'.

[3] Cf First Submission of the US in *EC – Hormones*, 18 August 1997, WT/DS26/R/USA, 21 (identifying such elements as: the observation of phenomena in nature or under controlled conditions, the systematic classification of empirical data, the measurement of empirical quantities and calculation of probable errors and significant deviations, as well as the formation of a hypothesis).

[4] Cf the observation of the compliance panel in *Australia – Salmon* that the 'level of objectivity to be achieved in a risk assessment must be such that one can have reasonable confidence in the evaluation made, in particular in the levels of risk assigned' (Panel Report, *Australia – Salmon (Article 21.5 – Canada)*, para 7.47).

[5] For details see Chapter 5, section 5.4.2.

appearing to have an exact, objective, factual, systematic or methodological basis and relating to, or exhibiting the methods or principles of science'.[6] In another case, the panel used tautology to define scientific evidence as 'evidence gathered through scientific methods, excluding by the same token information not acquired through a scientific method'.[7] Interestingly, this indicates that the concept of 'scientific evidence' also presupposes the existence of certain methodological rigour. Although such interpretation blurs the distinction between two parts of the conjunction, and may be criticized on this ground, it seems to be unavoidable. As correctly noted by the EC in *EC – Biotech Products*, 'there [seems] to be a degree of overlap, insofar as if it were demonstrated that a particular piece of evidence were not "scientific", it would presumably also be possible to argue that it was not based on scientific principles'.[8] It is also worth noting that from an epistemological point of view there is no specific moment when mere information transforms into scientific evidence. The scientific community uses different standards to assess the value of evidence. These particularly include reproducibility of observations, consistency with relevant theories, clarity of data, ability to explain the observations with plausible hypothesis, inability to disprove the underlying hypothesis with experiments, reinforcement of observations through several independent kinds of evidence, and ideas that are interconnected.[9] It is a concentration of those elements that determines whether a specific piece of information may be regarded as scientific.

Sufficiency was conceptualized in the SPS jurisprudence as a relational concept which requires an 'adequate relationship between two elements,... between the SPS measures and scientific evidence'.[10] The adequate relationship was understood as a 'rational or objective one'[11] and is determined on a case-by-case basis. Consequently, the approach or methodology used for the examination of sufficiency ultimately depends on the specific circumstances of the case.[12] Although the case law has failed to establish any clear test for the evaluation of a rational relationship, it has at least identified some factors such as 'the characteristic of the measure at issue, quality and quantity of scientific evidence' that are relevant in this process.[13] The applicable test is substantive and requires a panel to

[6] Appellate Body Report, *EC – Hormones*, fn 172 (referring to the ordinary meaning of the word 'scientific').

[7] Panel Report, *Japan – Apples*, para 8.92.

[8] Panel Report, *EC – Biotech Products, Annex F 'Replies by the European Communities to Questions Posed by the Panel in the Context of the Second Substantive Meeting 7 March 2005'*, para 98.

[9] Daniel M Byrd III and C Richard Cotheren, *Introduction to Risk Analysis: A Systemic Approach to Science-based Decision-making* (2005) 27; William Bechtel, *Scientific Evidence: Creating and Evaluating Experimental Instruments and Research Techniques, PSA: Proceedings of the Biennial Meeting of the Philosophy of Science Association, Vol 1* (1990) 561.

[10] Appellate Body Report, *Japan – Agricultural Products*, para 73.

[11] Ibid. [12] Appellate Body Report, *Japan – Apples*, para 164.

[13] Appellate Body Report, *Japan – Agricultural Products*, para 84.

make an enquiry into the substance of evidence.[14] Interestingly, WTO dispute settlement bodies, when examining sufficiency under Article 2.2, also frequently relied on the standards developed in the context of Article 5.1. Accordingly, minority scientific opinions coming from qualified and respected sources may satisfy the requirement of sufficiency.[15] The distinction between the probability and possibility thresholds, which is reflected in the definitions of quarantine and food-borne risk assessment, also has to be considered.[16] This means that scientific evidence pertaining to quarantine risks will be found to be sufficient only if it relates to the examination of the probability of adverse effects.[17] On the other hand, in the case of food-borne risk, mere evaluation of possibility will suffice. The evidence as such has to be specific and relates to particular risk (eg the entry, establishment, or spread of the bacteria that causes fireblight disease) and particular requirements capable of reducing or eliminating that risk (eg the effectiveness of chlorine treatment in eliminating the bacteria).[18] The concept of sufficiency also implies a certain margin of precaution or deference on the side of the national government (at least in all cases where risk is irreversible). As noted by the Appellate Body 'a panel charged with determination... whether "sufficient scientific evidence" exists to warrant the maintenance by a Member of a particular SPS measure, may, of course, and should, bear in mind that responsible, representative governments commonly act from the perspective of prudence and precaution where risks are irreversible'.[19]

On the other hand, the Appellate Body upheld the finding of the *Apples* panel that the disproportion between the risk and the SPS measure implies a lack of rational or objective relationship.[20] This would suggest that the examination of rationality actually involves a kind of strict proportionality test under which a panel balances the restrictiveness of a measure with relevant risk. For the *Apples* panel this meant that the complex phytosanitary measure against fireblight which consisted of various product and process requirements, was over-restrictive as compared to identified risks. In the context of its analysis, the panel also introduced the notion of 'negligible' risks. The panel determined that it was unlikely that infected cargo crates could serve as a pathway for fireblight. The same conclusion was reached with respect to subsequent transmission (once apples reach the territory of Japan) from imported apples to host plants. On that basis, the panel concluded that risk was 'negligible' and held that if the risk is negligible, while the SPS measure is strict, no rational relationship would be found.[21] At the same time, it is not clear whether a concept of proportionality

[14] For example, Panel Report, *Japan – Apples*, paras 8.123–8; Catherine Button, *The Power to Protect. Trade, Health and Uncertainty in the WTO* (2004) 179.

[15] Appellate Body Report, *Japan – Agricultural Products*, para 77.

[16] Ibid, para 78.

[17] For details see Chapter 4, section 4.4.1.

[18] Panel Report, *Japan – Apples*, para 8.92.

[19] Appellate Body Report, *EC – Hormones*, para 124.

[20] Appellate Body Report, *Japan – Apples*, para 164.

[21] See also Chapter 6, section 6.4.1.

and negligible risk was seen by the panel as a generally applicable standard. Note that the *Japan – Apple* case was concerned with a typical quarantine risk. It might be the case that such a standard would not be introduced in a dispute concerning measures that address human health risks.

The Appellate Body indirectly touched upon the issue of negligible risk in *US – Continued Suspension*, when it discussed the standard of review applicable to the assessment of scientific evidence. The *Suspension* panel concluded, on the basis of the opinions provided by its experts, that there was no appreciable risk of cancer from residues of oestradiol-17β in meat in meat products. The Appellate Body disagreed and criticized the panel for over-stepping its mandate. According to the Appellate Body, 'it was not the Panel's task... to determine whether there is an appreciable risk of cancer arising from the consumption of meat from cattle treated with oestradiol-17β'.[22] The above passage may suggest that once a risk is ascertainable (it is more than just a theoretical uncertainty), a panel is not permitted to conduct any further examination as to the extent of risk. This would preclude the application of the concept of negligible risk as well as any strict proportionality analysis as introduced by the *Apples* panel. As discussed in Chapter 6, this is an important development as both concepts, strict proportionality and negligible risk, may interfere with the sovereign right of WTO Members to establish ALOP.

4.2.1 Relationship between Articles 2.2 and 5.1

Articles 2.2 and 5.1 are closely related and 'should (be) constantly read together. Article 2.2 informs Article 5.1: the elements that define the basic obligations set out in Article 2.2 impart meaning to Article 5.1'.[23] Article 5.1 was also described as a specific application of the basic obligations provided in Article 2.2,[24] in the sense that it 'may be seen to be marking out and elaborating a particular route leading to the same destination set out in Article 2.2'.[25]

This close interrelation between two articles was translated in the case law into a negative presumption applicable to SPS measures that do not conform to risk assessment. As explained by the Appellate Body, a measure that violates Article 5.1 should be 'presumed... not to be based on scientific principles or to be maintained without sufficient scientific evidence'.[26] It may be disputed whether this conclusion is indeed correct. As will be discussed in the subsequent sections, the requirements of risk assessment go well beyond the evaluation of scientific evidence (eg quarantine risk assessment requires assessing efficiency of measures that might be applied). Thus, not every violation of Article 5.1 indicates that a

[22] Appellate Body Report, *US – Continued Suspension*, para 614.
[23] Appellate Body Report, *EC – Hormones*, para 180.
[24] Appellate Body Report, *Japan – Agricultural Products*, para 82.
[25] Panel Report, *Australia – Salmon*, para 8.52.
[26] Appellate Body Report, *Australia – Salmon*, para 137.

measure is not based on scientific principles or maintained without sufficient scientific evidence (eg quarantine risk assessment may simply fail to evaluate other risk mitigating alternatives). Moreover, one may also imagine a situation in which a WTO Member does not possess formal risk assessment because the existence of risk is so evident that it constitutes a 'common sense'. However, since the presumption is well established in the case law, a moderate call would be to conceptualize it as rebuttable.

Given the more general character of Article 2.2 the positive presumption is not available here, as Article 5.1 does not exhaust the whole meaning of Article 2.2. As noted by the Appellate Body, an additional examination of the measure under Article 2.2 is not excluded, even if such a measure is found to be consistent with Article 5.1.[27] Assuming that the relationship between two provisions may be described in terms of a general and specific rule, this conclusion is correct. The specific obligation cannot exhaust the whole normative content of the general rule. However, in practice it is difficult to imagine a possible violation of Article 2.2 if a measure is compatible with Article 5.1. Although the language of Article 2.2 (to maintain) may suggest that the provision is more concerned with the dynamic aspect of scientific justification, while Article 5.1 appears to concentrate on the static facet (to base), the case law has interpreted Article 5.1 as actually reflecting both aspects.[28]

The above is also true for those instances when a measure is found to violate Article 2.2. The logic of the general–specific rule also means that not every violation of this provision will be covered by Article 5.1.[29] Nevertheless, it seems unlikely that an SPS measure may be considered to be based on a proper risk assessment if it is not supported by sufficient scientific evidence, or if it is not firmly based on scientific principles. The last situation, when a measure is incompatible with Article 2.2, is a rather easy one. As mentioned above, Article 5.1 contains a number of specific requirements which appear to exceed the obligations provided by Article 2.2. Consequently, the additional assessment of a measure under Article 5.1 cannot be excluded.

4.3 Scientific justification (Article 3.3)

Article 3.3 allows WTO Members to adopt SPS measures which result in a higher (or different) level of SPS protection than that provided in the international standards, guidelines, or recommendations, if there is a scientific justification or as a consequence of the level of national ALOP as determined in accordance with the specific provisions of Article 5. As already discussed in Chapter 3,

[27] Appellate Body Report, *EC – Hormones*, para 250.

[28] Panel Report, *EC – Biotech Products*, para 7.3034; see Chapter 6, section 6.4.4.

[29] Panel Report, *Australia – Salmon*, para 8.52.

section 3.5.3, the SPS case law equates the two alternatives and requires a risk assessment for all measures that deviate from international standards.

4.4 Risk assessment (Article 5.1–5.3)

The analysis of this section is divided into two parts. The first concentrates on the content of the risk assessment, particularly its nature and required components, while the second part analyses the relationship between risk assessment and SPS measures. However, before going into details of the obligations of the SPS Agreement, it is worth briefly reviewing how risk assessment is conceptualized by the sister organizations. This should provide a useful starting point for the subsequent discussion on disciplines of the SPS Agreement.

The Codex defines risk assessment as a scientifically based process that is composed of four distinctive steps: (i) hazard identification, (ii) hazard characterization, (iii) exposure assessment, and (iv) risk characterization.[30] Hazard identification is understood as a detection of biological, chemical, and physical agents (eg toxins) which may be present in food, capable of causing adverse health effects. Hazard characterization involves the quantitative/qualitative evaluation of the nature of identified adverse effects caused by such an agent.[31] Exposure assessment consists in quantitative/qualitative evaluation of exposure of relevant population groups to the agent under review (eg exposure via food). Risk characterization is a final quantitative/qualitative estimation, including attendant uncertainties, of the probability of occurrence and severity of known or potential adverse health effects in a given population based on the three preceding steps.[32]

The OIE conceptualizes risk assessment slightly differently.[33] According to the Terrestrial Animal Health Code, hazard identification, which is defined similarly to the Codex,[34] precedes risk assessment and constitutes a triggering factor of the whole process. The OIE considers this step as being separate from both risk assessment and management. As far as risk assessment is concerned, the OIE distinguishes four interrelated steps: (i) release assessment, (ii) exposure assessment, (iii) consequence assessment, and (iv) risk estimation. The first step is understood as the evaluation of potential biological pathway(s) necessary for an importation

[30] Codex, *Procedural Manual* (18th edn, 2008) 74.

[31] The *Procedural Manual* also provides that 'for chemical agents, a dose response assessment should be performed, while for biological or physical agents, a dose-response assessment should be performed if the data are obtainable' (ibid).

[32] Ibid.

[33] Risk assessment is defined in the Terrestrial Animal Health Code as 'evaluation of the likelihood and the biological and economic consequences of entry, establishment and spread of a *hazard* within the territory of an *importing country*' (emphasis in original) (OIE, *Terrestrial Animal Health Code*, Glossary).

[34] Hazard identification is understood as a process of identifying the pathogenic agents which could potentially be introduced in the commodity considered for importation (ibid).

activity to introduce pathogenic agents into a particular environment together with the estimation of probability. The exposure assessment consists of describing the biological pathway(s) necessary for exposure of animals and humans in the importing country to the pathogenic agents released from a given risk source, and estimating the probability of such exposure. The third step analyses the relationship between specified exposures to a pathogenic agent and the consequences of such exposures together with the probability. During the risk estimation phase, one is expected to integrate the results of preceding steps to produce an overall measurement of risk associated with the hazard identified at the beginning.[35]

The IPPC distinguishes between two types of risk assessment, depending on its subject. The first one (for quarantine pests) is defined as the evaluation of the probability of the introduction and spread of a pest and the magnitude of the associated potential economic consequences.[36] The second one (for regulated non-quarantine pests or RNQP[37]) is conceptualized as the evaluation of the probability that a pest in plants for planting affects the intended use of those plants with an economically unacceptable impact.[38] The IPPC takes a similar line as the OIE, and recognizes identification of the pest (whether an organism is a pest at all) and a pathway (ie commencement of importation of new commodities) as a distinctive step from pest risk assessment. This preliminary action serves as a triggering factor for successive risk assessment. The assessment as such can be divided into four phases: (i) pest categorization, (ii) assessment of introduction and spread, (iii) assessment of economic impact, and (iv) conclusions. The first phase consists of the determination of whether a pest is a quarantine pest or RNQP. Depending on the categorization, the subsequent steps of the risk assessment differ. The second phase comprises the identification of the endangered area and assessment of the probability of the introduction and spread of the pest (quarantine pests) or assessment of whether the plants for planting are, or will be, the main source of pest infestation, in comparison to other sources of infestation of the area (RNQP). The third step is an evaluation of the potential economic consequences of the introduction and spread of the pest. According to the IPPC, this also includes assessment of environmental impacts. The conclusions are based on the previous findings and provide the estimates of overall pest risk.[39]

Although the Codex and OIE methodologies technically differ, their approach is substantially the same. Both prescribe hazard identification, its subsequent characterization, assessment of potential exposures, and final estimation of risks.

[35] Ibid, Article 2.2.4.

[36] A quarantine pest is understood as a pest of potential economic importance to the endangered area, which is either not present there, or present but not widely distributed while being officially controlled (ISPM No 05, Glossary of Phytosanitary Terms (2007) 15).

[37] RNQP is defined as a pest whose presence in plants for planting affects the intended use of those plants with an economically unacceptable impact and which is therefore regulated within the territory of the importing country (ibid); the main difference between quarantine pest and RNQP is the fact that the latter is generally present in the territory of the importing country.

[38] Ibid, 13.

[39] ISPM No 2, Framework for Pest Risk Analysis (2007) 12.

In both cases, risk can be expressed qualitatively or quantitatively. The IPPC model of risk assessment expresses its own particularities as it includes the evaluation of economic consequences resulting from the entry or spread of a pest. Nevertheless, the underlying idea in IPPC risk assessment is the same. The process aims at the assessment of risk associated with a particular hazard on the basis of likelihood and potential adverse effect.

4.4.1 The substantive content of risk assessment

(a) What constitutes risk assessment under the SPS Agreement?

The SPS Agreement distinguishes between assessment of risks to life and health of humans, animals, and plants attributable to pest and disease (pest/disease risks or quarantine risk assessment) and risks to life and health of humans and animals arising from the presence of certain substances in food, beverages, and feedstuffs (food-borne risk assessment). The first type of assessment is defined as an 'evaluation of the likelihood of entry, establishment or spread of a pest or disease within the territory of an importing Member according to the sanitary or phytosanitary measures which might be applied, and of the associated potential biological and economic consequences'.[40] With respect to food-borne risks, the definition provides that risk assessment is an 'evaluation of the potential for adverse effects on human or animal health arising from the presence of additives, contaminants, toxins or disease-causing organisms in food, beverages or feedstuffs'.[41] The subsequent discussion will analyse in more detail the importance of this distinction.

The structure of the risk assessment was conceptualized separately for each type of risk. In the first case, a risk assessment was structured as a three-step analysis. Initially, a risk assessor needs to identify two sets of data: 'the diseases [or pests] whose entry, establishment or spread a Member wants to prevent within its territory as well as potential biological and economic consequences associated with the entry, establishment or spread of these diseases [or pests]'.[42] The second step requires an assessment of likelihood with respect to each set of data. During the third phase risk assessors are expected to evaluate the likelihood of entry, establishment or spread of an identified disease or pest according to the SPS measures that might be applied.[43] As clarified by the case law, the last element calls for the evaluation of SPS alternatives,[44] which arguably makes it more difficult for WTO Members to adopt an *ex post* justification. Interestingly, there is no equivalent requirement under the second type of risk assessment. One may only speculate about the reasons for such divergence. One possible explanation would

[40] Annex A (4) of the SPS Agreement. [41] Ibid.

[42] Appellate Body Report, *Australia – Salmon*, para 121. [43] Ibid.

[44] Panel Report, *Japan – Apples*, para 8.283; that finding was subsequently upheld by the Appellate Body.

be that food-born risk assessment deals primarily with human health risks. As will be discussed in section 4.4.1(b) below, although quarantine risk assessment may also evaluate this type of risk, its main concern is plant and animal life and health. Consequently, a less demanding standard for food-borne risk assessment may be seen as a conscious decision of the drafters who wanted to grant WTO Members with the additional discretion when regulating human health risks. I will come back to this issue in the subsequent section.

The assessment of food-borne risks is conceptualized as a two-step analysis. The first step consists of the identification of adverse effects to human or animal health and life arising from the presence of certain substances such as additives and toxins in food, feedstuffs, and beverages. This step can be further divided into two distinctive phases: (i) identification of the additives, contaminants, toxins, or disease-causing organisms in particular food, beverages, or feedstuffs and (ii) identification of any possible adverse effects on human or animal health.[45] If such adverse effects are found, the second step of the analysis requires the evaluation of the potential of occurrence of these effects.[46]

(b) Likelihood v Potential—Two different concepts?

As mentioned in section 4.4.1(a), the SPS Agreement makes a distinction between two types of risk assessment. According to the Appellate Body, each type requires its own level of 'likelihood'. Assessment of quarantine risks demands the evaluation of the likelihood of entry, establishment, or spread of a pest or disease. In the case of food-borne risks, the SPS Agreement speaks only about the potential for adverse effect. The Appellate Body, referring to the ordinary meaning of those terms, equated likelihood with probability,[47] whereas potential was understood as a mere possibility.[48] In other words, the first category requires a higher level of 'probability' compared to the second one and arguably is more difficult to establish.[49] Consequently, when assessing pest and disease risks 'it is not sufficient that a risk assessment conclude that there is a possibility of entry, establishment or spread of diseases',[50] a panel should rather look for the probability of entry, establishment, or spread of diseases.[51]

[45] Panel Report, *US – Continued Suspension*, para 7.507.

[46] Panel Report, *EC – Hormones (US)*, para 8.98; although the Appellate Body said that the 'utility of a two-step analysis may be debated' it also admitted that 'it does not appear . . . to be substantially wrong' (Appellate Body Report, *EC – Hormones*, para 184). The subsequent case law followed this approach.

[47] Appellate Body Report, *Australia – Salmon*, fn 70; the Appellate Body specifically said that '[i]n view of the very different language used in Annex A(4) for the two types of risk assessment, we do not believe that it is correct to diminish the substantial differences between these two types of risk assessments'.

[48] The Appellate Body said in a different case that the ordinary meaning of 'potential' relates to 'possibility', see Appellate Body Report, *EC – Hormones*, para 184.

[49] See also, Panel Report, *EC – Biotech Products*, para 7.3184.

[50] Appellate Body Report, *Australia – Salmon*, para 123.

[51] Ibid.

It is not clear whether the drafters of the SPS Agreement deliberately introduced the distinction between possibility and probability.[52] One may argue that food-borne risk assessment is concerned with human and animal health risks, while quarantine risk assessment is limited to threats for animals and plants. Less demanding requirements with respect to food-borne risks giving WTO Members greater room for manoeuvre, would be justified by the importance of the values that are protected by an SPS measure. However, this interpretation does not fit with the language of the SPS Agreement. Note that Annex A(1)(c) explicitly recognizes the existence of human health risks arising from diseases carried by animals, plants, or products thereof, or from the entry, establishment, or spread of pests. The outbreak of BSE—a disease that can be transmitted from animals to humans causing a new variant Creutzfeldt-Jakob disease—serves as a good real life example. Since both types of risk assessment may relate to human health, the distinction established by the SPS Agreement does not appear to be convincing. What is the rationale for requiring different thresholds from two types of risk assessments if both of them may relate to potential risks for human health? Is it a fact that the majority of regulated SPS risks for human health are concerned with food products? Although this may look like a pragmatic explanation, on the normative level the answer seems to be unsatisfactory. As a matter of principle, the same values (eg human health) require identical protection irrespective of the source of the risk. One way out from this problem is to simply disregard the distinction made by the Appellate Body between 'potential' and 'likelihood' and apply the uniform standard of 'potential' to both types of risk assessment if they concern human health. Such an approach will ensure consistency in the treatment of such risks.

The distinction between probability and possibility is also problematic on a technical level. The case law has never clearly articulated what is meant under the SPS Agreement by the evaluation of probability as opposed to assessment of mere possibility, assuming perhaps that they are self-explanatory scientific terms. Regrettably, this is not the case. As noted by one of the experts in *Australia – Salmon* 'the requirement of a [quarantine] risk assessment is to evaluate the *probability* of risk... [as] one can *always* construct a *possible* infection transmission scenario, however improbable, and therefore demonstration of the possibility of successful transmission'.[53] The same opinion was expressed by another expert in *US – Continued Suspension* who observed in the context of genotoxicity of residues of oestradiol-17β that 'I have a problem with that word potential in my responses. So it is almost like saying is it possible, yes almost anything is possible, but in a real situation—is it likely to occur at a significant level?—no.'[54] In

[52] Cf Joost Pauwelyn, 'The WTO Agreement on Sanitary and Phytosanitary (SPS) Measures as Applied in the First Three Disputes' (1999) 4 Journal of International Economic Law 641, 647.

[53] Panel Report, *Australia – Salmon*, para 8.79 (emphasis in the original).

[54] Panel Report, *US – Continued Suspension, Annex G, 'Transcript of the Panel's Joint Meeting with Scientific Experts on 27–28 September 2008'*, para 716.

the same case, the Appellate Body appeared to subscribe to this broad reading of possibility when it observed, although without going into details, that 'to examine the "potential" for adverse effects is to ask whether those adverse effects could ever occur'.[55] If one agrees with such conceptualization of possibility, the required threshold in food-borne risk assessment will be so low that it will simply turn a test under the definition in the second sentence of Annex A(4) into pure formality. One may rationally assume that the applicable standard is higher than that described by both Dr Wooldridge and Dr Guttenplan (the experts assisting the panels) as well as briefly mentioned by the Appellate Body.

So what is really meant by 'possibility' under the SPS Agreement? One approach is to assume that probability is a quanitative estimation whereas possibility is a purely qualitative concept. If the distinction between these two is to be meaningful, the probability has to be conceptualized as a quantitative term. This position would also find support in a part of scientific literature that equates quantitative calculations with the assessment of probability of occurrence of adverse effect, while the qualitative measurement is seen as the evaluation of the possibility of causal link without indicating its likelihood.[56] Consequently, a merely qualitative evaluation would not be sufficient to meet a definition of quarantine risk assessment as provided by the SPS Agreement. This would also suggest that finding a quarantine risk that is not reducible to a quantitative dimension would not allow a WTO Member to adopt an SPS measure. A danger of such an approach is that it may interfere with the right of WTO Members to establish their own ALOP for those quarantine risks which cannot be expressed in quantitative terms. This interpretation, of course, sits uneasily with the other findings of the SPS case law which confirmed the possibility of performing a full qualitative evaluation with respect to both categories of risk assessment, including quarantine risk assessment where one is expected to evaluate probability.[57] For example, the compliance panel in *Australia – Salmon* was satisfied with the full qualitative assessment of quarantine risks. What seems to be uncontroversial is that even if the quarantine risk assessment requires a quantitative calculation of probabilities, qualitative elements may still be present in such an assessment. This would go in line with the observation of the Appellate Body that both types of assessment are not limited to the matters 'that are susceptible of quantitative analysis by the empirical and experimental laboratory methods'.[58]

[55] Appellate Body Report, *US – Continued Suspension*, para 572.

[56] Vern Walker, 'The Myth of Science as Neutral Arbiter for Triggering Precaution' (2003) 26 Boston College International & Comparative Law Review 197, 200. Cf also Douglas Crawford-Brown, Joost Pauwelyn, and Kelly Smith, 'Environmental Risk, Precaution and Scientific Rationality in the context of WTO/NAFTA Trade Rules' (2004) 24(2) Risk Analysis 461, 466 (noting that 'the term "likelihood", which has been equated to "probability", seems to call for a more formal calculation of probabilities, consistent with the idea of probability as used in quantitative risk assessment').

[57] Appellate Body Report, *Australia – Salmon*, para 124 (noting that likelihood may be expressed either quantitatively or qualitatively).

[58] Appellate Body Report, *EC – Hormones*, para 187.

Another approach would be to regard possibility as merely a different name for qualitative probability. Consequently, probability could be expressed either qualitatively or quantitatively but its qualitative expression will constitute at the same time an instance of possibility. Note that this interpretation would go in the same direction as the Codex approach. Codex does not use the notion of possibility and consistently relies on the concept of probability, recognizing at the same time, that it can be expressed quantitatively and qualitatively.[59] As a matter of consistency, one would expect that a risk assessment which is performed for food-borne risks within the relevant international standard-setting body uses the same threshold as required from domestic risk assessors. However, even this approach appears to be incompatible with the existing case law. The Appellate Body has consistently emphasized the importance of the distinction between possibility and probability. If possibility is defined as qualitative probability this distinction will be lost. Moreover, the *Biotech* panel, elaborating on this distinction, struck down a number of national safeguard measures relating to quarantine risks by finding that the scientific reports which were relied on by particular EC Member States, failed to evaluate probability and were limited to mere examination of possibility.[60] This clearly indicates that qualitative probability is a different concept from possibility.

It is perhaps safe to conclude that the assessment of possibility entails some less rigorous standard than qualitative evaluation of probability. At the same time, the required threshold is arguably higher that the standard described by Drs Wooldridge and Guttenplan. However, this also means that the border between two concepts will remain obscure with possibility having ambiguous and undefined meaning.

(c) Minimum magnitude of risk

The Appellate Body held that no minimum magnitude of risk needs to be demonstrated in a risk assessment (eg 1:1,000,000).[61] This lack of a *de minimis* threshold indicates that any ascertainable risk may be a subject of domestic regulation. The concept of 'ascertainable' is crucial in this context. As observed by the Appellate Body, risk that a WTO Member intends to regulate needs to be ascertainable, in the sense of having some indication of potentiality.[62] This also meant for the Appellate Body that 'theoretical uncertainty is not the kind of risk which, under Article 5.1, is to be assessed'.[63] The last concept was defined as a form of uncertainty that is 'inherent in the scientific method and which stems

[59] For example, Codex, *Procedural Manual*, 73 (defining risk characterization as qualitative and/or quantitative estimation of the probability of occurrence).

[60] For example, Panel Report, *EC – Biotech Products*, para 7.3044.

[61] Appellate Body Report, *EC – Hormones*, para 186; Appellate Body Report, *Australia – Salmon*, para 124.

[62] Appellate Body Report, *US – Continued Suspension*, para 569.

[63] Appellate Body Report, *EC – Hormones*, para 186.

from the intrinsic limits of experiments, methodologies, or instruments deployed by scientists to explain a given phenomenon'.[64] Accordingly, ascertainability of risk serves as a bottom line which delimits the scope of 'legitimate risks' under the SPS Agreement.

Having said that, it still remains unclear where to draw the line between theoretical uncertainty and minimally ascertainable risk. For example, one of the experts in *EC – Hormones* described risk from eating meat treated with oestrogens for growth promotion purposes as ranging between zero to one person in million.[65] The panel described this estimate as not a scientifically identified risk. This comes close to the notion of theoretical uncertainty which always remains in scientific research. The Appellate Body, however, disagreed and criticized the panel for introducing a threshold requirement.[66] According to the Appellate Body, risks that are characterized as '0 to 1 in a million' may be legitimately regulated by WTO Members. Yet in another case, the panel, when faced with the same risk (same in the sense of its probability and not possible consequence), labelled it as negligible and struck down the SPS measure. The panel probably borrowed the concept of negligible risk from one of its experts who identified the standard scientific definition of 'negligible' as a likelihood of between zero and one in a million.[67] This time the Appellate Body upheld the panel's finding. Although the panel did not refer in its reasoning to theoretical uncertainty, it seems that negligible risk may actually represent the same quality. Negligible risks are risks which cannot be excluded (the upper limit of the range) but may actually not exist (the lower limit of the range). This residual uncertainty is very similar to a theoretical one.[68]

(d) As appropriate to the circumstances

The expression 'as appropriate to the circumstances' has been interpreted in the case law as giving to risk assessors a certain level of flexibility when performing risk assessments.[69] According to the *Salmon* panel, Article 5.1 gives an opportunity 'to assess the risk, on a case-by-case basis, in terms of product, origin and destination, including, in particular, country specific situations'.[70] In *Japan – Apples* this meant that phytosanitary status of Japan as a fireblight-free country

[64] Appellate Body Report, *Japan – Apples*, para 241.
[65] Panel Report, *EC – Hormones (US)*, fn 331.
[66] Appellate Body Report, *EC – Hormones*, para 186.
[67] Panel Report, *Japan – Apples*, para 8.149.
[68] Cf also the observation of Dr Boisseau who said that 'if we cannot say that the risk is zero at low doses, the fact is that we simply don't know, we cannot say that it exists, nor can we say that it does not exist' (Panel Report, *US – Continued Suspension, Annex G, 'Transcript of the Panel's Joint Meeting with Scientific Experts on 27–28 September 2008'*, para 721).
[69] Appellate Body Report, *EC – Hormones*, para 129.
[70] Ibid, para 8.71. This flexibility is also recognized by the standard-setting organizations. For example, the OIE provides that risk assessment should be 'flexible to deal with the complexity of real life situations' (OIE, Terrestrial Animal Health Code, Article 2.1.3).

combined with the specific climatic conditions which made Japan a potentially favourable environment for the spread of disease, had to be considered as relevant factors when deciding whether its risk assessment met the requirements of Article 5.1.[71] Consequently, the phytosanitary vulnerability of Japan might have justified a more conservative risk assessment than would otherwise be possible. Yet another panel saw the subject of the risk (ie whether risk is to human, animal, or plant life or health) as a relevant factor in determining whether the risk assessment is 'as appropriate to the circumstances'.[72] Again, the report does not explain what exactly the panel had in mind. One reading would be to assume that the WTO dispute settlement bodies are ready to grant WTO Members a larger margin of discretion when a measure is concerned with human health risks.

The panel in *EC – Biotech Products*, following the line of previous case law, repeated that the term 'as appropriate to the circumstances' purports a certain degree of flexibility and noted that when there is little available scientific evidence, this phrase 'may provide a measure of flexibility in terms of how (but not whether) the applicable elements of the Annex A(4) definition, including the likelihood evaluation, are satisfied'.[73] This flexibility was also recognized in the recent ruling in *US/Canada – Continued Suspension* where the Appellate Body held that methodological difficulties posed by the nature and character of the particular substance (oestradiol-17β) and risk which is evaluated may affect what qualifies as a proper risk assessment. On the other hand, it is still not clear whether the phrase 'as appropriate to circumstances' may be used to address problems that are faced by developing countries when performing risk assessments.[74] The text of the Preamble to the SPS Agreement which mandates recognizing difficulties that may be encountered by developing Members in the formulation and application of SPS measures, appears to support such an interpretation.[75] In practice, this could be translated into a less rigorous specificity test or a lower epistemic benchmark for minority scientific opinions that constitute a basis for SPS measures.

At the same time, the SPS case law consistently holds that flexibility which is implied in the phrase 'as appropriate to the circumstances' cannot 'annul or supersede the substantive obligation . . . to base the sanitary measure in dispute (irrespective of the products that measures may cover) on a risk assessment'.[76] In the same vein, the Appellate Body noted that such flexibility 'does not excuse the risk assessor from evaluating whether there is a connection between the particular substance being evaluated and the possibility that adverse effects may arise'.[77]

[71] Panel Report, *Japan – Apples*, para 8.240.

[72] Panel Report, *Australia – Salmon*, para 8.71.

[73] Panel Report, *EC – Biotech Products*, para 7.3053.

[74] Joanne Scott, *The WTO Agreement on Sanitary and Phytosanitary Measures. A Commentary* (2007) 97.

[75] Recall that the Preamble explicitly recognizes that developing Members may encounter special difficulties in the formulation and application of SPS measures in their own territories.

[76] Panel Report, *Australia – Salmon*, para 8.70.

[77] Appellate Body Report, *US – Continued Suspension*, para 562.

The *Biotech* panel also derived from the language of 'as appropriate to the circumstances' a specific temporal requirement. After noting that the phrase was not qualified by any temporal limitation such as reference to the circumstances existing at the time of adoption of an SPS measure, the panel held that 'at any given time, SPS measures must be based on an assessment of risks which is appropriate to the circumstances existing at that time'.[78] That finding has important implications. A measure is expected to have scientific backup in the form of risk assessment during its whole regulatory life. Thus, WTO Members are obliged to follow changes in scientific knowledge in order to guarantee that their risk assessments take into account the latest developments.[79] The *Suspension* panel, in a finding not reviewed by the Appellate Body, went in the same direction and connected the phrase 'as appropriate to the circumstances' to the evolution of risk assessment over time.[80]

(e) *Specificity of risk assessment*

Risk assessment needs to be specific. According to the Appellate Body, this means evaluating the specific potential (or likelihood) of harm arising from a particular SPS risk. Thus, in *EC – Hormones*, the WTO dispute settlement bodies expected to be provided with the evaluation of the carcinogenic or genotoxic potential of residues of hormones used for growth promotion purposes that were present in meat and meat products. In *Japan – Apples*, the evaluation of 'entry, establishment or spread of fire blight through [US] apple fruit as a separate and distinct vector'[81] was necessary. Similarly, in *US/Canada – Continued Suspension*, the EC was required to assess the possibility that specific adverse effects resulted from the presence of residues of oestradiol-17β in meat or meat products as a consequence of cattle being treated with the hormones for growth promotion purposes. As explained by the panel, this meant that the EC had to evaluate the existence of causal connection between the consumption of such meat and the possibility of adverse health effects.[82] A general discussion on a particular SPS risk, such as the evaluation of the entire category of hormones or collection of various host plants, does not satisfy the specificity requirement.[83]

The strict specificity requirement, as developed by the SPS case law, was met with some criticism in the literature. For instance, Sykes notes that insistence of the Appellate Body to produce highly particularized studies may result in the practical impossibility of regulating low-level risks (ie where a particular chemical or biological agent is responsible for certain adverse health effects, but

[78] Panel Report, *EC – Biotech Products*, para 7.3031.

[79] This does not exclude the possibility of relying on minority scientific opinion if such opinion is not falsified by new scientific developments.

[80] Panel Report, *US – Continued Suspension*, para 7.605.

[81] Appellate Body Report, *Japan – Apples*, para 200.

[82] Panel Report, *US – Continued Suspension*, para 7.513.

[83] Appellate Body Report, *EC – Hormones*, para 200.

only with respect to a small part of the population or over a long period of time). Those types of risks are frequently difficult to measure, either quantitatively or qualitatively, and easily escape traditional assessment methods. How does one provide the specific risk assessment which will evaluate the risk connected with the presence of a particular substance in a particular product, if the presumed effect is, for example, one in a million? As elegantly explained by Sykes in the context of risks from hormone residues in meat products 'it is likely impossible to know with any degree of statistical confidence whether these small residues, when added to the diets of people who are exposed to the same hormones from many other sources (not to mention numerous other carcinogens), do or do not cause a few more cases of cancer at the margin'.[84] This in turn may undermine the right of WTO Members to establish the level of protection they deem appropriate. Remember that the SPS case law on several occasions has confirmed that WTO Members are free to regulate any ascertainable risk and adopt any ALOP, including zero-risk level. However, if the general studies showing general risk from a particular substance do not satisfy the specificity requirement, it may appear that in low-risk situations the right to decide on the ALOP will remain illusory.

This problem is well exemplified by the discussion in *US/Canada – Continued Suspension* on risks resulting from cumulative exposure to different hormones. As noted by the EC, when discussing risks resulting from the consumption of meat from cattle treated with oestradiol-17β, humans are exposed to multiple sources of hormones and hormone residues, some of them being endogenous and others artificial. According to the EC, it is simply impossible to assess accurately 'all cumulative and synergistic effects that may arise from all potential exposure patterns, including simultaneous exposure to several of these hormones'.[85] At the same time, it might be the case that there is no risk from consumption of hormone-treated beef alone but the accumulation of different sources can lead to overall risk. The same concern was expressed in the opinion of the EC Scientific Committee on Veterinary Measures Relating to Public Health (SCVPH)[86] and by one of the experts advising the panel. It is worth citing here a relevant paragraph in some length. Dr Boobis particularly observed that:

> [T]he risk from . . . low levels of exposure . . . would be such, that it would be necessary to study extremely large populations to detect any increase in cancer incidence. . . . Hence, in the risk assessment of the hormones used as growth promoters, it is questionable whether an increase in risk, even if it existed, could be detected in exposed populations.[87]

[84] Alan O Sykes, 'Domestic Regulation, Sovereignty and Scientific Evidence Requirements: A Pessimistic View' (2002) 3 Chicago Journal of International Law 353, 365.

[85] Panel Report, *US – Continued Suspension*, para 7.496.

[86] Ibid, para 7.534 (noting 'the difficulty of evaluating health effects at low dose is here compounded by the fact that the data on exposure of human population are exceedingly limited').

[87] Ibid, para 7.692.

Since it was not sufficient for the EC to establish (or at least to authenticate) the existence of the risk on the basis of general studies which showed carcinogenic effects of hormones (albeit in the higher concentrations), the EC failed to show that its risk assessment specifically evaluated the possibility that the adverse effects come into being, originate, or result from the consumption of meat or meat products which contain veterinary residues of oestradiol-17β.[88]

The report of the Appellate Body in the same case suggests, however, that the WTO dispute settlement bodies may adopt in the future a less demanding standard which will be able to guarantee that the right to establish ALOP for complex and low-level risks is truly meaningful. The Appellate Body particularly found that the specificity requirement, being a part of risk assessment disciplines, is qualified by the expression 'as appropriate to the circumstances'. According to the Appellate Body, this meant that methodological difficulties posed by the nature and characteristics of the particular substance (eg cumulative character of risk with only a small fraction that can be potentially attributed to residues of oestradiol-17β) may affect the level of required specificity. In the particular circumstances of the case, the EC would not be required to establish in its risk assessment a direct causal link between the possibility of adverse health effects and specific hormone residues. It would be sufficient to 'demonstrate that the additional human exposure to residues of oestradiol-17β in meat from treated cattle is one of the factors contributing to the possible adverse health effects'.[89] At the same time, according to the Appellate Body, it was also unnecessary to isolate a contribution that might be attributed to oestradiol-17β but rather to assess the overall risks resulting from the exposure to different sources of hormones, including naturally occurring hormones. Such a low threshold makes the whole assessment process much easier compared to a standard that was applied by the panel and in practice may remove the specificity requirement for risk of cumulative character. As far as other situations are concerned, it also gives a promise of a more attentive approach that takes into account difficulties that generally arise in low risk situations by allowing the consideration of methodological obstacles posed by the nature and characteristic of such risks.

(f) Risk assessment techniques of international organizations

Article 5.1 requires WTO Members to take into account risk assessment techniques developed by the relevant international organizations. Although the SPS Agreement does not explicitly specify which entity may be considered as a relevant organization and uses slightly different language than Annex A,[90] this

[88] Ibid, para 7.537.

[89] Appellate Body Report, *US – Continued Suspension*, para 562.

[90] Article 5.1 speaks about international organizations while other provisions of the SPS Agreement, including Annex A(3), refers to international standard-setting organizations.

undoubtedly includes all sister organizations.[91] The *Suspension* panel also held that risk assessment techniques developed by the JECFA, which is formally independent from the Codex, were relevant in the context of Article 5.1. In support of this conclusion, the panel referred to the functional connection that exists between the JECFA and the Codex.[92] The approach of the *Suspension* panel may also suggest that the concept of relevant international organizations actually includes other organizations, bodies or processes that are not directly linked to the sister organizations. In particular, the panel asked its experts to identify any relevant international guidance documents without imposing any predetermined limits as to the identity of such a body. In their response, experts referred to different documents issued by the WHO and the FAO as well as specialized scientific bodies such as the International Program on Chemical Safety (IPCS). The panel's interpretation, although it may be questioned on textual grounds (strictly speaking neither the JECFA nor the IPCS are international organizations), seems to be correct. Since this part of the obligations of Article 5.1 provides a rather soft requirement ('taking into account'), its broad reading does not interfere with the rights of the WTO Members.

In principle, there are two types of 'risk assessment techniques'. The first type, which can be labelled as general risk assessment techniques or meta-rules, establishes the disciplines for assessment of entire categories of risks. This includes, for example, a methodological distinction between different steps of risk assessment such as hazard identification, hazard characterisation, exposure assessment and risk characterization or rules on the role of non-scientific factors in a risk assessment. The second type relates to the assessment of specific risks, such as risks posed by residues of veterinary drugs in food and so on.

In line with the language of Article 5.1, the case law recognizes that taking into account risk assessment techniques means only that these techniques should be considered relevant, but 'a failure to respect each and every aspect of them would not necessarily, *per se*, signal that the risk assessment is not in conformity with the requirements of Article 5.1'.[93] According to the *Suspension* panel 'compliance with... [relevant] risk assessment techniques is not required by the SPS Agreement'[94] as WTO Members simply need to take them into account. The Appellate Body added in *Japan – Apples* that the SPS Agreement does not require WTO Members to follow any kind of special methodology of risk assessment.[95] In consequence, the obligation seems to be rather easy to satisfy. For example, in *US/Canada – Continued Suspension*, the mere fact that the EC was aware of the Codex and JECFA techniques and considered them when conducting its assess-

[91] Panel Report, *US – Continued Suspension*, para 7.446 (taking a contextual interpretative approach and finding that the context of Article 5.1 includes the Preamble, Articles 3.2, 3.4 and Annex A(3).

[92] Ibid, para 7.447.

[93] Panel Report, *Japan – Apples*, para 8.241.

[94] Panel Report, *US – Continued Suspension*, para 7.468.

[95] Appellate Body Report, *Japan – Apples*, para 205.

ment, was sufficient to meet the requirement. It was irrelevant that the methodology of the EC risk assessment differed from the Codex practice.

However, the above does not mean that risk assessment techniques developed by relevant international bodies are without any significance. As observed by one of the panels, a departure by a WTO Member from such risk assessment techniques may have some implications as 'reference to these risk assessment techniques can provide very useful guidance as to whether the risk assessment at issue constitutes a proper risk assessment within the meaning of Article 5.1'.[96]

(g) Science and other elements relevant for risk assessment

Risk assessment is a process in which science plays a central role. As already discussed in section 4.4.1(a), risk assessment is conceptualized under the SPS Agreement as either a two- or three-step analysis. In both cases, the individual steps are predominantly concerned with evaluation of scientific data. At the same time, it would be premature to conclude that science and scientific evidence exhaust the whole list of factors that are relevant in the context of risk assessment. The SPS Agreement as well as the corresponding case law recognizes that risk assessment is a much more nuanced and complicated process and cannot be reduced to simple evaluation of scientific information. However, when it comes to precise identification of other factors, the opinions in the literature diverge. There are those who believe that cultural preferences, societal values and subjective considerations relating to risk perception may play a role in the risk assessment.[97] There are, however, others who take a more conservative stance. Those scholars see rather limited roles for cultural factors in risk assessment,[98] claiming that their acceptance would interfere with the objectivity and integrity of the process. Consequently, the additional factors that may be taken into account in the assessment of risk are limited to control and enforcement issues (ie actual enforcement of SPS measures), SPS conditions in the exporting and importing country, and so on. This disagreement in the literature is paralleled by the lack of clarity in the case law which failed to identify what may qualify for, and what is excluded from, the risk assessment process. However, before going into details,

[96] Panel Report, *Japan – Apples*, para 8.241 (the panel also added 'In particular, it can shed useful light, in this dispute, on the US argument that Japan has failed to evaluate the likelihood of entry because it failed to consider all the steps in the pathway that would lead to apple fruit being a vector for the entry and transmission of the disease').

[97] Regine Neugebauer, 'Fine-Tuning WTO Jurisprudence and the SPS Agreement: Lessons from the Beef Hormone Case' (2000) 31 Law & Policy in International Business 1255, 1267; M Gregg Bloche, 'WTO Deference to National Health Policy: Toward an Interpretative Principle' (2002) 5(4) Journal of International Economic Law 825, 836; David Winickoff, Sheila Jasanoff, Lawrence Busch, Robin Grove-White, and Brian Wynne, 'Adjudicating the GM Food Wars: Science, Risk, and Democracy in World Trade Law' (2006) 30 Yale Journal of International Law 81, 97–100.

[98] Eg Warren H Maruyama, 'A New Pillar of the WTO: Sound Science' (1998) 32 International Lawyer 651, 673.

let us look at the treaty provision which identifies elements that are relevant in the context of risk assessment.

According to Article 5.2, WTO Members, when assessing risks, are required to take into account, as well as scientific evidence, other factors such as relevant processes and production methods, relevant inspection, sampling and testing methods, prevalence of specific diseases or pests, existence of pest- or disease-free areas, relevant ecological and environmental conditions, and quarantine or other treatments. Although the Appellate Body observed that the list is not exhaustive and other elements may be also relevant, it did not specify what kind of other factors may be taken into account.[99]

The function of Article 5.2 was initially seen as 'shed[ding] light on the elements that are of relevance in the assessment of risks'.[100] This finding was developed further in a more recent decision where the panel held that the purpose of Article 5.2 is more than just informative as it actually requires WTO Members to consider the enumerated factors.[101] Thus, Article 5.2 performs a dual function: descriptive and prescriptive. It describes—very imprecisely as the enumeration is open-ended—the substantive elements that should be included in the risk assessment.[102] It also imposes a procedural requirement to take into account those factors when performing a risk assessment. The panel saw those functions as two separate issues, each calling for an independent examination. Consequently, the question of whether conclusions of the EC risk assessment were supported by the scientific evidence (one of the factors enumerated in Article 5.2) was a separate issue from the question of whether the EC took the available scientific evidence into account when performing its assessment.[103] The same approach was employed by the panel with respect to the examination of potential abuses and misuses of hormonal implants.[104] As far as the available scientific evidence is concerned, the panel also explained that Article 5.2 mandates a WTO Member to consider a wide range of scientific information in order to ensure that its SPS measure is based on sufficient scientific data and is supported by scientific principles.[105] Although the panel was not clear on this point, it probably required a review of all types of relevant scientific data, including also those which do not support the conclusion of a particular risk assessment.

The Appellate Body report in *EC – Hormones* was indeed a promise to those who wanted to see risk assessment as incorporating a broad spectrum of factors, including cultural preferences and subjective considerations relating to risk

[99] Appellate Body Report, *EC – Hormones*, para 187.

[100] Panel Report, *Japan – Apples*, para 8.230.

[101] Panel Report, *US – Continued Suspension*, para 7.441

[102] Of course not all elements enumerated in Article 5.2 have to be included in a particular risk assessment; risk assessment is context-dependent and some of the factors may be simply irrelevant in assessing specific risk (eg process and production methods will be irrelevant when assessing risks related to import of unprocessed agriculture products).

[103] Panel Report, *US – Continued Suspension*, para 7.481.

[104] Ibid, para 7.483.

[105] Ibid, para 7.480.

perception and so on. In particular the Appellate Body criticized the panel for excluding from the scope of risk assessment 'all matters not susceptible of quantitative analysis by the empirical or experimental laboratory methods commonly associated with the physical sciences'.[106] According to the Appellate Body, this was an unjustified approach, as Article 5.2 enumerates different elements, some of them being not necessarily or wholly susceptible to laboratory investigation. According to its famous statement:

> risk that is to be evaluated in a risk assessment under Article 5.1 is not only risk ascertainable in a science laboratory operating under strictly controlled conditions, but also risk in human societies as they actually exist, in other words, the actual potential for adverse effects on human health in the real world where people live and work and die.[107]

The subsequent case law understood the above passage as principally referring to control and enforcement problems rather than as providing a possibility for broad inclusion to risk assessment of non-scientific factors such as risk perception. This seems to be compatible with the textual basis in the Agreement. Article 2.2 assigns a special role to science by requiring WTO Members to base their SPS measures on scientific principles, and not maintain them without sufficient scientific evidence. That role is subsequently highlighted in Article 3.3 which speaks about scientific justification as a condition for deviating from international standards. The definition of risk assessment, as provided in Annex A to the SPS Agreement, puts great emphasis on scientific evaluation of potential risks. This conclusion also appears to be supported by the position held by the Codex which consistently sees a role for non-scientific factors in risk management rather than in its assessment.[108]

Of course, it is not possible to establish a strict distinction between risk assessment and risk management as both phases overlap to some extent.[109] This means that some subjective elements, which are not of a purely scientific nature, will be present in a risk assessment. This was recognized in the SPS case law as early as *Australia – Salmon (Article 21.5 – Canada)*, where the panel admitted that risk assessment, and in particular its qualitative version, inevitably includes subjective elements relating to the weighing of scientific evidence or ascribing a value to risk (ie characterizing risk as low, moderate, or high).[110] Obviously, the existence of such elements is not prohibited by the SPS Agreement. In the more recent case, the Appellate Body also recognized that ALOP may be relevant in the context of risk assessment by influencing the scope and methods used in the assessment (eg how the potential risk is framed). In particular, a chosen ALOP may affect

[106] Appellate Body Report, *EC – Hormones*, para 187. [107] Ibid.

[108] Codex Alimentarius Commission, 'Criteria for the Consideration of the Other Factors Referred to in the Second Statement of Principle' (2001).

[109] See, Chapter 1, section 1.4.3.

[110] Panel Report, *Australia – Salmon (Article 21.5 – Canada)*, para 7.47.

a decision as to what kind of research needs to be performed for specific type of risks as well as the parameters of such research.[111]

The above observations do not, however, change the overall character of each phase of the risk analysis model. Risk assessment remains basically a scientific task where scientific elements dominate the process. Risk management is a value-laden task where the regulator decides whether and how to address a particular risk.[112] Acknowledging the existence of subjective factors in risk assessment does not undermine its scientific character. In fact, the opposite is true. Such factors are recognized precisely to guarantee the integrity of the scientific advice which is provided by risk assessors, since their identification allows consideration of the potential impact that they may have on the final outcome of risk assessment. This seems to be reflected in another passage from the *US – Continued Suspension* report, where the Appellate Body observed that the ALOP cannot 'affect the rigour or objective nature of the risk assessment, which must remain, in its essence, a process in which possible adverse effects are evaluated using scientific methods'.[113] Consequently, ALOP cannot predetermine the result of the risk assessment.

Two specific non-scientific elements that should be included in the risk assessment, pursuant to Article 5.2, require further elaboration. The subsections that follow will address them in more detail.

(h) Good veterinary practice

As noted in the preceding subsection, the observance of good veterinary practice (GVP) is one of the elements that should be considered in a risk assessment. In the SPS practice, GVP is defined as a use of veterinary medicines, including feed additives, according to the authorized use, in terms of dosages, applications, and withholding periods, in order to guarantee the safety of animal products.[114] The Appellate Body confirmed on two occasions that the SPS Agreement permits WTO Members to take into account in their assessment, risks arising from failures of exporting Members to comply with GVP and from difficulties of control, inspection, and enforcement of GVP requirements.[115] The Appellate Body also stressed that such evaluation constitutes an option for WTO Members and cannot be conceptualized as an obligation.[116] It is always the decision of a particular WTO Member as to whether to supplement its risk assessment with

[111] Appellate Body, *US – Continued Suspension*, para 534.

[112] For further discussion on the distinction between risk assessment and risk management in the context of the SPS Agreement, see Chapter 6, section 6.3.

[113] Ibid.

[114] The Codex uses terms 'good veterinary practice', 'good veterinary and husbandry practice', and 'good practice in the use of veterinary drugs' interchangeably (see Panel Report, *EC – Hormones (US)*, para 2.19).

[115] Appellate Body Report, *EC – Hormones*, para 205, Appellate Body Report, *US – Continued Suspension*, para 545.

[116] Appellate Body Report, *EC – Hormones*, para 206.

considerations relating to risks resulting from the failures to observe GVP (ie abuse or misuse of veterinary drugs). In practice, this can be done either as a separate assessment or as an element of the original risk assessment.[117]

The above standard was borne with some difficulties. For instance, the *Hormones* panel saw failures in GVP as a part of risk management rather than risk assessment and refused to consider them under Article 5.1.[118] The *Suspension* panel, although recognizing that abuses or misuses in the administration of hormones were elements of risk assessment, believed that it was not necessary to consider them before establishing that the EC risk assessment met the specificity requirement.[119] The panel reasoning implied that failures to observe GVP would became relevant only if the EC risk assessment was able to show risk from the hormones concentration levels found in meat that would occur when GVP was correctly applied. The Appellate Body rightly disagreed with this methodology, finding its conclusion to be premature.[120] Since the evaluation of risks resulting from the misuse or abuse in the administration of growth promotion hormones was included in the EC risk assessment, logically it was not possible to determine whether such an assessment was sufficiently specific without considering these data.

In this context, it is also interesting to note that international standard-setting bodies, when performing risk assessment, assume that GVP is followed.[121] This may look strange at first but when one considers the nature of international standards, such an approach is entirely logical. Risk assessment that is conducted within international standard-setting bodies, in order to be applicable to different circumstances has to be de-contextualized. This allows them to maintain the utility and universality of risk estimates (it would be impossible to assess the safety of a risk agent in 'normal' circumstances). On the other hand, the assessment that takes place on the national level is frequently context-specific and evaluates the safety of particular products coming from the predefined country with specific veterinary practice. This obviously requires consideration of GVP in such assessment.

(i) Economic factors in risk assessment

Economic factors are yet another element which sometimes have to be considered in risk assessment. Remember that one of the constitutive elements of quarantine risk assessment, as defined by the first sentence of Annex A(4), is the evaluation of potential economic consequences associated with the entry, establishment, or

[117] Appellate Body, *US – Continued Suspension*, para 546 (referring to the 1999 Opinion, which clarified that observance of GVP is an aspect of the exposure assessment).

[118] For details of the panel analysis, see Chapter 6, section 6.3.

[119] Panel Report, *US – Continued Suspension*, para 6.164.

[120] Appellate Body Report, *US – Continued Suspension*, para 547.

[121] Panel Report, *US – Continued Suspension, Annex D, 'Replies of the Scientific Experts to Questions Posed by the Panel'*, para 368.

spread of a pest or disease within the territory of an importing Member. The reference to economic considerations also appears in Article 5.3 which mandates, when assessing risks to animal and plant life and health, to take into account relevant economic factors. This includes the potential damage in terms of loss of production or sales in the event of the entry, establishment, or spread of a pest or disease, the costs of control or eradication in the territory of the importing Member, and the relative cost-effectiveness of alternative approaches to limiting risks.

The relationship between the first sentence of Annex A(4) and Article 5.3 is not entirely clear. Note that the scope of these two provisions is not exactly the same. While the first one applies to all quarantine risks, including human health risks arising from diseases carried by animals and plants or from the establishment or spread of pests, the second is only concerned with the assessment of risks to animal or plant life and health. Consequently, human health risks, irrespective of their nature, fall outside the scope of Article 5.3. On the other hand, the definition of the first sentence of Annex A(4) does not cover risks to animal life and health which arise from additives, contaminants, toxins, or disease-causing organisms in feedstuffs.

As correctly observed by Scott, the relationship between two provisions may be construed either as cumulative or autonomous. Under the first approach, Article 5.3 will be applicable to those situations which fall into the scope of the definition in the first sentence of Annex A(4) and 'within that category, only in cases concerning the protection of animal and plant life and health'.[122] In other words, the specific economic factors enumerated in Article 5.3 will be relevant for those measures which address quarantine risks, but with sole impact on animal and plant life and health. According to Scott, who seems to prefer this approach, the role of Article 5.3 would be to 'specify which economic consideration must be taken into account, as regards animal or plant life or health, when a prior obligation to have regards to economic consequences anyway exists'.[123] On the other hand, the autonomous approach would call for the independent application of both provisions as two separate sets of obligations which are concerned with different (although overlapping) situations. This would mean that economic considerations will constitute an element of every quarantine risk assessment. Simultaneously, a WTO Member would be obliged to take into account specific economic factors when assessing risks to animal and plant life and health. The potential overlap between the two disciplines would relate to quarantine risks to animal and plant life and health.

The textual interpretation seems to point to the autonomous approach. Note that neither Annex A(4) nor Article 5.3 contains any language that would indicate any interconnection. Furthermore, the nature of the obligation required by each provision is entirely different. The first sentence of Annex A(4) requires the

[122] Scott (*Commentary*) 103. [123] Ibid.

inclusion of the evaluation of the economic consequences of a risk assessment. Article 5.3, on the other hand, only mandates taking some specific economic factors into account. As noted elsewhere, 'taking into account' merely requires the consideration of relevant elements without a need to reflect them in an SPS measure (risk assessment).[124] Hence, the cumulative reading may actually reduce the importance of the first sentence of Annex A(4) as WTO Members would be able to satisfy the obligation to evaluate economic consequences by claiming that they took specific economic factors into account. Having said that, one also needs to recognize that the cumulative approach allows the conceptualization of the obligations of the SPS Agreement in a more harmonious way. In particular, it helps to avoid overlaps in the application of two provisions and guarantees a consistent approach to the same categories of risk. Under the cumulative approach, assessment of human health risks would not require the evaluation of economic consequences at all.[125] This would also prevent a double standard that is difficult to justify on normative grounds. Moreover, as noted above, the nature of the obligations established by each provision is different. One of the consequences resulting from this divergence is that a softer standard ('taking into account') would be applied to the assessment of risks to animal and plant life and health, while the quarantine risk assessment, pertaining to human health risk, would provide a more demanding requirement (ie, actual evaluation of economic factors in the risk assessment). Again, it is difficult to find a rationale for such an interpretative outcome. Last but not least, the autonomous approach may potentially interfere with the right of WTO Members to establish a level of protection that they deem appropriate. As discussed below, the case law requires the existence of an objective relationship between risk assessment and an SPS measure, meaning that conclusions of the risk assessment must reasonably support the SPS measure. If the evaluation of economic consequences is one of the elements of quarantine risk assessment (or quarantine risk assessment which is concerned with risks to animal life and health), this may imply that a final measure needs to be rationally related to such economic considerations. This obviously invites a form of proportionality test that calls for balancing the restrictiveness of a measure with its economic costs. A possible way out from this conclusion is provided by the cumulative approach. If the definition in the first sentence of Annex A(4) is conceptualized not as a separate substantive obligation but rather as a background provision which delimits the scope of the application of Article 5.3, WTO Members would only be required to take economic factors into account, and there would be no danger of compromising their ALOP. All these considerations strongly suggest

[124] For example, Panel Report, *Japan – Apples*, para 8.241.

[125] Note that even this approach does not introduce a consistent method with respect to animal health risks. There would be no obligation to consider economic factors for risks to animal life and health which arise from additives, contaminants, toxins, or disease-causing organisms in feedstuffs.

accepting the cumulative approach. Yet it is still to be seen which route will be taken by the SPS case law.

4.4.2 Implementation of risk assessment results

An SPS measure needs to be based on risk assessment. As discussed below in more detail, this obligation has a substantive character and requires an objective relationship between two elements, that is, between a risk assessment and a measure. The case law also made it clear that under Article 5.1 there is no minimum procedural requirement to take into account a risk assessment when enacting or maintaining an SPS measure.[126] This understanding of Article 5.1 was established after the initial recognition of the dual nature of this obligation, having both a procedural and a substantive dimension.[127] The lack of a procedural aspect specifically means that risk assessment can be performed by a Member *ex post*,[128] even at the time of the panel's proceeding.[129] Moreover, a WTO Member is not obliged to conduct its own risk assessment and such assessment may be carried out by another country or international organization and simply 'borrowed' by a Member.[130]

The above standard was criticized by some scholars. For instance, as observed by Hurst 'the Panel's test promotes harmonization by encouraging Members to evaluate the scientific basis for *all* their measures, not just those which are later disputed—and this evaluation will occur at the time of enactment, rather than at the advent of litigation'.[131] This, in turn, encourages gambling over the SPS measures, as most of them will never be challenged. One may also add that the obligation to consider scientific data before adoption of a measure would also improve the quality of a domestic risk regulatory process, without preventing WTO Members from determining the level of protection they deem to be appropriate. At the same time, the language used in Article 5.1 seems to be sufficiently broad to include both procedural and substantive obligations.

On the other hand, one also needs to recognize that reading Article 5.1 as a partially procedural requirement may be problematic. What would be the consequence of finding the violation of procedural requirement without substantive infringement of Article 5.1? How would the compliance look in such

[126] Appellate Body Report, *EC – Hormones*, para 189.

[127] Panel Report, *EC – Hormones (US)*, para 8.113 (noting that 'the Member imposing a sanitary measure needs to submit evidence that at least it actually took into account a risk assessment when it enacted or maintained its sanitary measure in order for that measure to be considered as based on a risk assessment').

[128] Panel Report, *EC – Biotech Products*, para 7.3030.

[129] Pauwelyn ('WTO Agreement SPS Measures') 649 (to be precise, both risk assessment and evidence need to be available at the date of fixing the panel's terms of reference).

[130] Appellate Body Report, *EC – Hormones*, para 190.

[131] David R Hurst, 'Hormones: European Communities—Measures Affecting Meat and Meat Products' (1998) 1(9) European Journal of International Law 14 at <http://207.57.19.226/journal/Vol9/No1/srlg.html> (last visited 25 July 2009).

a case? Would a WTO Member be required to adopt the very same measure, however, this time with prior consideration of risk assessment? This seems to lead to unacceptable consequences. If a measure is compatible with the substantive part of the obligation of Article 5.1, the procedural deficiency appears to be somehow unimportant. The opposite is also true. Finding a substantive violation makes the procedural deficiencies obsolete. Does it really matter whether a WTO Member took into account a risk assessment prior to the enactment of a measure if such a measure is not based on the assessment anyway? At the same time, it is worth noting that the absence of a procedural requirement under Article 5.1 does not mean that the SPS Agreement is only concerned with substantive compliance. According to Annex B, drafts of SPS measures having a potentially significant impact on the trade of other Members, and which substantially differ from international standards, are subject to a special notification procedure.[132] The notification should include, among other things, information on the rational of the proposed regulation which will obviously require a reference to scientific data. Furthermore, a WTO Member is obliged to engage in a discussion with other Members on the proposed measure.[133] It is difficult to imagine such a discussion without at least some exchange of scientific arguments. These obligations are supplemented by Article 5.8 which, although it does not prevent WTO Members from making *ex post* rationale for their measures, implicitly requires WTO Members to have some scientific backup for existing measures, even before any litigation takes place.[134] When all those obligations are taken into consideration, the rejection of the procedural character of Article 5.1 does not appear to be as serious as one might first expect. This observation combined with the potential problems that may arise in the compliance phase seem to provide a support for the approach taken by the Appellate Body.

(a) Objective relationship between the risk assessment and the SPS measure

The relationship between risk assessment and an SPS measure is described in the case law as an 'objective relationship between two elements', or 'an objective situation that persists and is observable between an SPS measure and a risk assessment'.[135] In practice, the examination of an objective relationship consists of a comparison between scientific conclusions reached in the risk assessment with the conclusions embedded in the SPS measure, in order to examine their similarity.[136] What, then, should be the level of compatibility between

[132] For details, see Chapter 2, section 2.2.3.
[133] Annex B(5)(b) of the SPS Agreement.
[134] Article 5.8 requires WTO Members to provide other Members in certain predefined circumstance with an explanation of the reasons for their SPS measures.
[135] Appellate Body Report, *EC – Hormones*, para 189.
[136] Ibid, para 192; see also Panel Report, *Australia – Salmon (21.5 – Canada)*, paras 7.79 – 7.83.

the results of the risk assessment and the SPS measure? The SPS case law has not developed any clear standard and prefers a case-by-case approach, in which 'account is taken of all considerations rationally bearing upon the issue of the potential adverse health effects'.[137] What seems to be the established practice is the absence of the requirement of full conformity. According to the Appellate Body, scientific conclusions of the risk assessment must only reasonably support the SPS measure under examination. This flexible standard is also reflected in the recognition that the existence of uncertainties and research constraints that have an impact on the level of scientists' confidence in the results of a risk assessment may influence the objective relationship. Consequently, a margin of discretion, which is implied anyway in the concept of an objective relationship, is even broader for those assessments that are conducted in circumstances of uncertainty. In practice, one assessment, depending on the approach of particular Members (less or more cautious), may provide a rationale for a wide variety of measures.[138]

Lack of the requirement for full conformity, of course, does not mean that a WTO Member may adopt an SPS measure which is completely unrelated to the outcome of a risk assessment.[139] The objective relationship has a substantive character, in the sense that risk assessment is expected to sufficiently warrant or support a domestic measure. The case law has identified two rather uncontroversial cases in which a risk assessment cannot be regarded as a rational basis for an SPS measure. This particularly includes a situation where a risk assessment does not evaluate relevant risk as well as modes of its mitigation.[140] Equally, a measure that introduces a total ban is not rationally (objectively) related to a risk assessment which found no evidence of any specific risk.[141]

In contrast to the standard that was developed under Article 2.2 (ie sufficiency of scientific evidence understood as a rational relationship), the test of the objective relationship is less controversial.[142] Whereas the first one involves an extensive examination of scientific evidence and subsequent conclusions (at least until the Appellate Body report in *US – Continued Suspension*), the test of an objective relationship appears to be more limited. The quality of scientific evidence under Article 5.1 is normally examined as a part of risk assessment analysis which comes before evaluation of the objective relationship. In consequence, panels somehow by implication identify lack of objective relationship once a risk assessment is found to be deficient.[143]

137 Appellate Body Report, *EC – Hormones*, para 194.
138 Ibid, para 7.3065.
139 Panel Report, *EC – Biotech Products*, para 7.3607.
140 Panel Report, *Australia – Salmon*, para 8.98.
141 Panel Report, *EC – Biotech Products*, para 7.3607.
142 For a more detailed discussion, see section 4.2.
143 For example, Appellate Body Report, *EC – Hormones*, para 208, Panel Report, *US – Continued Suspension*, para 7.578; but see Panel Report, *Australia – Salmon*, paras 8.93–8.99.

(b) Minority scientific opinions

The risk assessment may set out both the mainstream scientific opinion, as well as the opinions of scientists taking divergent views.[144] The Appellate Body explained that:

> In most cases, responsible and representative governments tend to base their legislative and administrative measures on 'mainstream' scientific opinion. In other cases, equally responsible and representative governments may act in good faith on the basis of what, at a given time, may be a divergent opinion coming from qualified and respected sources.[145]

Initially, there was some disagreement in the literature as to the existence of any epistemic threshold that might have been deduced from the above passage. Some argued that reference to the notion of 'qualified and respected sources' had to be seen as requiring from divergent opinions a certain level of reliability and scientific quality.[146] Others, however, held that any conclusion in this regard was premature.[147] In particular, it was submitted that the Appellate Body did not define what was meant by 'qualified and respected sources', while the divergent opinion expressed by one of the experts advising the panel in *EC – Hormones* was rejected not on the basis of its epistemic deficiencies but rather due to its generality.[148]

The subsequent case law, at least formally, has followed the first position. The panels and the Appellate Body consistently held that minority scientific opinions need to come from respected and qualified sources.[149] In one of the recent rulings, the Appellate Body explained that such opinion 'must... have the necessary scientific and methodological rigour to be considered reputable science'.[150] This meant that even though correctness of the view does not need to be accepted by the mainstream scientific community, a divergent opinion must be recognized as a defendable scientific claim.[151] This approach seems to be plausible. As noted by Sykes, unlimited reliance on scientific minority views would transform

[144] Appellate Body Report, *EC – Hormones*, para 194.

[145] Ibid.

[146] For example, Dale E McNiel, 'The First Case Under the WTO's Sanitary and Phytosanitary Agreement: The European Union's Hormone Ban' (1998) 39 Virginia Journal of International Law 89, 119; Hurst ('Hormones') 12.

[147] Steve Charnovitz, 'The Supervision of Health and Biosafety Regulation by World Trade Rules' (2000) 13 Tulane Environmental Law Journal 271, 279.

[148] Appellate Body Report, *EC – Hormones*, para 198.

[149] See eg, Panel Report, *US – Continued Suspension*, para 7.419; see also, Appellate Body Report, *EC – Asbestos*, para 178.

[150] Appellate Body Report, *US – Continued Suspension*, para 591.

[151] Note also that this approach closely corresponds with the position held by the European Court of Justice and the US Supreme Court, eg Case T-13/99 *Pfizer Animal Health v Council* [2002] ECR II-330; *Daubert v Merrell Dow Pharm Inc*, 43 F.3d 1311, 1319 (9th Cir 1995) (finding science from 'a recognized minority of scientists in their field' as an acceptable basis for regulatory decisions).

the requirement of risk assessment into 'minimal procedural hurdles',[152] as it is always possible to find an expert with a dissenting scientific opinion. The introduction of a certain epistemic threshold allows filtering out of 'junk science' and guarantees the operability of the SPS Agreement.

On the other hand, the SPS case law has failed to establish any clear criteria which could be used in the assessment of the credibility of minority scientific opinions. In the scientific practice, the plausibility of a claim is determined through different mechanisms, such as assessment of compatibility of such a claim with more general theories, logical coherence, or the ability to explain convincingly particular phenomena.[153] At the same time, it seems impossible to predetermine any threshold that would allow identification when a merely speculative claim is transformed into a scientific one. Any assessment will be context-dependent and will vary from situation to situation. Due to the high degree of specialization of scientific research, such assessment is probably possible only within a particular scientific community operating in a specific field. This means that the panel will need to rely heavily on the opinions of its experts when deciding whether a minority opinion holds a minimal epistemic status.

This formal acceptance of minority scientific views as a basis for an SPS measure should also be confronted with the practice of the case law. Some of the panels, despite recognizing the relevance of such opinions, conducted an analysis which sought to determine the best science in the field. This meant that opinions of dissenting experts were disregarded or underestimated. One example of such an approach may be found in the *Biotech* panel report in the part relating to the assessment of insufficiency of scientific evidence under Article 5.7. The panel ignored the minority opinions and decided a number of issues on the basis of what may be labelled as the best science approach.[154] The panel's analysis of the genotoxicity of oestradiol-17β in *US – Continued Suspension* is yet another example. As discussed in more detail in the subsequent section, the structure of the fact-finding process (eg the formulation of questions to experts) and the arbitral choices between conflicting and contradictory opinions led to the practical exclusion of minority scientific views.[155]

It was also submitted in the literature that the epistemic threshold which is required by WTO case law for minority scientific opinion, may operate to the exclusion of opinions that are inconsistent with the currently dominant Kuhnian paradigm.[156] Consequently, WTO Members will be barred from regulating risks that may be identified within other paradigms. This criticism seems to be misplaced. Note that Kuhn's theory describes paradigms as very general sets of

[152] Sykes ('Domestic Regulation') 366.

[153] Crawford-Brown et al ('Environmental Risk') 465.

[154] For details, see Chapter 5, section 5.4.2.

[155] Cf Appellate Body Report, *US – Continued Suspension*, para 610.

[156] Jeffery Atik, 'Science and International Regulatory Convergence' (1996–97) 17 Northwestern Journal of International Law & Business 736, 752.

practices that define a scientific discipline during a particular period of time. The scientific paradigms delimit what is to be observed and scrutinized, the kind of questions that are supposed to be asked and probed for answers as well as how these questions are to be structured, and how the results of scientific investigations should be interpreted.[157] Consequently, the paradigm defines what may be regarded as scientific claims and what may not. The paradigm shifts, which completely reorganize the way a particular section of science is conducted, are very rare. Between those periods of scientific revolutions, scientific research is done within boundaries delimited by the specific paradigms. Since the paradigm determines what is to be considered as scientific, any claim which falls outside will be merely a speculative assertion or superstition. In other words, one cannot speak about any scientific value of a claim which is incompatible with the paradigm. Note also that in practice this problem will be marginal since the paradigms are so general that they can easily accumulate very different scientific positions. Moreover, it seems that even if there are two competing paradigms, neither of them having an established position, a panel will nevertheless accept as valid scientific claims coming from either of them.

Since scientific minority opinions by definition belong to the same paradigm as mainstream science (otherwise they would not be regarded as scientific in the first place), discussing the status of claims formulated outside the dominant paradigm seems to be counter-productive. A better framework for understanding the nature of such opinions is provided by the concept of the core and the research frontier. The core is a set of theories, analytic techniques, and facts that are considered to be indisputable in the particular scientific community. The core is 'the starting point, the knowledge which people take as a given from which new knowledge will be produced'.[158] The research frontier, on the other hand, consists of 'all the work currently being produced by all active researchers in a given discipline'.[159] A particular component which belongs to the research frontier moves to the core when there is a substantial consensus between scientists on its plausibility. The mainstream science, as understood under the SPS Agreement, includes both the core and those parts of the research frontier that exhibit a higher level of consensus among active scientists (albeit insufficiently high so as to qualify as the core). Minority scientific opinions will consist of those claims that belong to the research frontier but also remain in contradiction to the mainstream part of this section, or those that are less conclusive or insufficiently substantiated.

What seems to be more problematic for scientific minority opinions is the specificity condition that is required by the SPS case law from risk assessment. As correctly noted by Peel, this requirement may result in the practical exclusion of many divergent opinions, as they are usually 'based in the kind of suggestive but

157 See generally, Thomas S Kuhn, *The Structure of Scientific Revolutions* (1996).
158 Stephen Cole, *Making Science: Between Nature and Society* (1992) 15.
159 Ibid.

not definite scientific evidence'.[160] One of the immanent features of minority scientific opinions is their lower level of conclusiveness and limited empirical basis (otherwise they will move either to the core or to the mainstream research frontier). Sometimes they establish little more than the theoretical possibility of risk rather than a concrete causal relationship between a risk agent and an adverse effect. This is a place where a panel should be particularly careful. Rejecting such scientific minority opinions only because they do not meet a specificity requirement may indeed interfere with the right of WTO Members to establish their ALOP. This problem is particularly important if a particular area of scientific research is new (eg genetic manipulations). As will be discussed in the next section, a proper construction of applicable standard of review may provide a feasible solution to this problem.

4.5 Standard of review applicable to scientific determinations

The applicable standard of review is an element which, alongside the substantive treaty rules, influences the vertical allocation of power between the WTO dispute settlement bodies and national governments.[161] In the context of the SPS Agreement, as in case of other treaties, *de novo* standards ultimately transfer that power to the international level as the panels are able to review all factual and technical findings made on a national level and may substitute them with their own. On the other hand, full deference preserves the state sovereignty as the review on the WTO level is limited only to the formal or procedural aspects of domestic risk determinations. This function was explicitly recognized by the Appellate Body when it stated that appropriate standard of review 'must reflect the balance established in [the SPS Agreement] between the jurisdictional competences conceded by the Members to the WTO and the jurisdictional competences retained by the Members for themselves'.[162]

The problem of relevant standard of review was one of the central issues in *EC – Hormones*. The EC argued in its appeal that the panel was obliged to apply a so-called 'deferential reasonableness standard'. The EC saw such a standard as provided by Article 17.6(i) of the Anti-Dumping Agreement[163] and claimed

[160] Jacqueline Peel, 'Risk Regulation under the WTO SPS Agreement: Science as an International Normative Yardstick?', Jean Monnet Working Paper No 2002/04, 71.

[161] Matthias Oesch, *Standards of Review in WTO Dispute Resolution* (2003) 28; Mari Presley, 'Sovereignty and Delegation Issues Regarding U.S. Commitment to the World Trade Organization's Dispute Settlement System' (1998) 8 Journal of Transnational Law & Policy 173, 184.

[162] Appellate Body, *EC – Hormones*, para 115.

[163] Article 17.6(i) of the Anti-Dumping Agreement specifically provides that 'in its assessment of the facts of the matter, the panel shall determine whether the authorities' establishment of the facts was proper and whether their evaluation of those facts was unbiased and objective. If the establishment of the facts was proper and the evaluation was unbiased and objective, even though the panel might have reached a different conclusion, the evaluation shall not be overturned'.

that it was also applicable to 'all highly complex factual situations, including the assessment of the risks to human health arising from toxins and contaminants'.[164] The Appellate Body disagreed with the EC, finding no textual basis. In particular, it observed that there was no indication that WTO Members intended to adopt the same standard as set out in Article 17.6(i) in the context of the SPS Agreement. Instead, the Appellate Body identified Article 11 of the DSU as providing the applicable standard of review, both with respect to factual and legal determinations.[165] The objective standard of review was described vaguely as neither a *de novo* review as such, nor total deference. According to the Appellate Body, neither of these two extremes provided a feasible option.[166] At the same time, the Appellate Body did not specify where exactly the applicable standard was to be located. Not surprisingly, this approach was criticised in the literature as lacking clarity and relying on the provision, which actually does not determine any standard of review. In particular it was noted that Article 11 of the DSU rather provides due process guarantees within the WTO dispute settlement system.[167] This aspect of Article 11 was in fact recognized in another passage of the report, where the Appellate Body observed that:

> the duty to make an objective assessment of the facts is... an obligation to consider the evidence presented to a panel and to make factual findings on the basis of that evidence. The deliberate disregard of, or refusal to consider, the evidence submitted to a panel is incompatible with a panel's duty to make an objective assessment of the facts.[168]

The Appellate Body was less ambiguous, and perhaps more generous, with respect to the standard of review that is applicable to factual findings made by panels. Accordingly, the intervention of the Appellate Body would only be possible in situations where the evidence had been deliberately disregarded or where there had been a wilful distortion or misrepresentation of the evidence. The threshold cannot be met by a simple mistake in the appreciation of evidence but requires an egregious error that calls into question the good faith of a panel.[169]

Despite this lack of clarity and the ambiguity as to the appropriate legal basis, the relevant passage from the *EC – Hormones* report became a point of reference for the subsequent case law.[170] The dispute settlement bodies consistently emphasised that panels cannot engage in a *de novo* review of scientific determinations made by Members. A panel was not expected to conduct its own risk assessment

164 Appellate Body, *EC – Hormones*, para 112.

165 Ibid, para 116. Note a certain inconsistency in the reasoning of the Appellate Body, as it held elsewhere that the DSU does not prescribe any particular standard of review (para 114).

166 Ibid, para 117 (noting that panels are poorly suited to engage in *de novo* review whereas a total deference could not ensure an objective assessment as foreseen by Article 11 of the DSU).

167 For example, Michael Trebilcock and Robert Howse, *The Regulation of International Trade* (1999) 70; Button (*Power to Protect*) 172–3.

168 Appellate Body Report, *EC – Hormones*, para 133.

169 Ibid.

170 For example, Panel Report, *US – Continued Suspension*, para 7.414.

but rather it should have assessed the quality of national assessment.[171] This, however, did not mean that a panel was required to 'accord to factual evidence of the parties the same meaning and weight as do the parties'.[172] Similarly, there was no obligation to grant a WTO Member with any degree of discretion in the manner in which it chooses, weighs, and evaluates scientific evidence.[173] A panel remained free in its decisions as to which evidence to choose in order to make factual findings and what weight to ascribe to such evidence. Consequently, under the margin of discretion enjoyed by a panel as a trier of facts, no precedence was given to the importing Member's evaluation of scientific evidence.[174] This broad mandate was accurately summarized in the *Suspension* report, where the panel explained its approach to scientific evidence in the following words:

> [W]hile, on some occasions, we followed the majority of experts expressing concurrent views, in some others the divergence of views were such that we could not follow that approach and decided to accept the position(s) which appeared, in our view, to be the most specific in relation to the question at issue and to be best supported by arguments and evidence.[175]

The above passage shows that the applicable standard of review came close to the *de novo* extreme. The panel could pick and choose between different scientific opinions and decide which of them was the most persuasive and specific to a particular problem. This broad discretion of the panel, implying a rather intrusive enquiry into scientific justification of a measure, was, however, not an invention of the *Suspension* panel. As discussed above, a similar approach was adopted in the *Apples* case when examining the sufficiency of scientific evidence under Article 2.2. The analysis performed by the *Biotech* panel under Article 5.7 may serve as yet another example.[176] At the same time, one has to admit that such a strict standard of review is less visible in the early SPS case law (eg *EC – Hormones*). That is also reflected in the literature, which labelled the initial approach of the Appellate Body as a deferential standard that left a great margin of discretion to national governments.[177]

The intrusive standard that was developed in time by SPS case law could be questioned on different grounds.[178] As noted in the literature, WTO panels

[171] Panel Report, *Japan – Agricultural Products*, para 8.32.
[172] Appellate Body Report, *Australia – Salmon*, para 267.
[173] Appellate Body Report, *Japan – Apples*, paras 162–3.
[174] Ibid, para 166.
[175] Panel Report, *US – Continued Suspension*, para 7.420.
[176] For details see Chapter 5, section 5.4.2(b).
[177] Ryan D Thomas, 'Where's the Beef? Mad Cows and the Blight of the SPS Agreement' (1999) 32 Vanderbilt Journal of Transnational Law 487, 507; Hurst ('Hormones') 16; Bloche (WTO Deference) (2002) 837.
[178] See eg, Natalie McNelis, 'The Role of the Judge in the EU and WTO: Lessons from the BSE and Hormones Cases' (2001) 4(1) Journal of International Economic Law 189, 205–6, making her argument for a deferential approach by reference to the concept of sovereignty; the problem with

are ill equipped to make complex scientific determinations.[179] Panelists do not possess sufficient knowledge and expertise to assess and evaluate the substance of scientific evidence underlying a particular national SPS measure.[180] Even if one recognizes the fact that panellists are supported by the scientific experts, they still need to decide how to interpret their responses, how to assess and choose between the conflicting views, what kind of interferences to make on the basis of such opinions, or what weight to attach to particular observations and so on. This obviously leaves a considerable margin of discretion to the panel and increases the possibility of a mistake. If one adds that the panel operates under tight time constraints and within the institutional infrastructure that lacks the capacity to 'undertake its own inquiry into the science',[181] it is hardly advisable to opt for a *de novo* review. This also means that in most cases it will be the defendant that is in a superior position to make scientific determinations. The above lack of expertise in scientific matters is anecdotally exemplified in the questions posed by the *Suspension* panel during one of its meetings with the experts. In the context of a discussion on the very complicated issue of the genotoxic potential of oestradiol-17β, the panel asked a number of elementary questions concerning the basis of toxicology.[182] One may only wonder how it was ultimately possible for the panel to make a complex factual judgment on, for instance, the existence of threshold in a dose–response relationship of oestradiol-17β, without possessing adequate literacy in scientific matters.

Lack of expertise is, however, not the only problem that is faced by the WTO dispute settlement bodies when evaluating a science-based measure. Assessment of risk is not a purely scientific enterprise and depends to a certain extent on the specific socio-cultural conditions of a particular country. As will be discussed in more detail in Chapter 5, the assessment inevitably involves subjective judgments of the assessors that reflect their attitudes toward particular risks, values of the relevant community in which the experts are acting, or other normative elements such as the required level of protection or treatment of uncertainties. Consequently, the same set of scientific data may produce entirely different risk estimates in various jurisdictions. Again, the panel seems to be worse placed to make such judgements compared to a WTO Member maintaining the measure.[183] If a panel ends up imposing on the WTO Members its own vision of science, these normative and context-dependent elements will be lost while the

this type of claims lays in the fact that every international treaty requires compromising national sovereignty.

[179] Button (*Power to Protect*) 181.

[180] Oesch (*Standards of Review*) 57.

[181] Andrew T Guzman, 'Dispute Resolution in SPS Cases' in Dan Horovitz, Daniel Moulis, and Debra Steger (eds), *Ten Years of WTO Dispute Settlement* (2007) 229.

[182] Panel Report, *US – Continued Suspension, Annex G Transcript of the Panel's Joint Meeting with Scientific Experts on 27–28 September 2006*, para 95 (particularly asking the experts to explain the meaning of *in vivo* and *in vitro* studies).

[183] Button (*Power to Protect*) 181.

ultimate determination will most probably fail to produce a correct estimation. It is also submitted in the literature that the costs of the potential mistake of the WTO dispute settlement bodies are considerably higher in SPS disputes than a case of other trade controversies.[184] These costs are not only concerned with the health or environmental effects resulting from the removal of a contested measure but may also relate to non-compliance of the losing party (which is assumed to be a more frequent situation compared to traditional trade disputes). This particularly includes costs of suspension of concession and costs to the credibility of the WTO as an institution.[185]

The above concerns caused a number of scholars to argue that WTO panels 'should not function as adjudicatory bodies reviewing the substantive scientific details underlying the parties' risk assessments'[186] but rather as administrative bodies that only supervise transparency and procedural aspects of the regulatory process. Such an approach would require the WTO dispute settlement bodies to refrain completely from reviewing the substance of scientific determinations and instead concentrate on transparency, arbitrary, and unjustifiable discrimination, or the least-trade-restrictive alternative disciplines. Science-based provisions should be applied only as procedural requirements (eg a panel would review whether a Member took into account scientific evidence).[187] On the other hand, there are others who believe that such total deference is inappropriate in the context of the SPS Agreement as it would 'disappoint hopes that the scope for health-based non-tariff barriers would be much reduced by the WTO Agreement'.[188] This concern indeed seems to be legitimate. Note that deferential standards of review which concentrate only on procedural compliance will make potential abuses of the SPS Agreement relatively easy. Moreover, if one resigns entirely from the scientific disciplines of the Agreement (in the substantive and not procedural sense), the actual regime will not be very different from the one provided by GATT 1994. However, the SPS Agreement was precisely introduced to remedy the shortcoming of the old GATT system which was widely seen as inadequate to deal with the SPS measures.[189] Consequently, although I agree that panels should adopt a rather deferential approach when examining scientific determinations made on the national level, the position that calls for a purely procedural approach appears to be equally unsatisfactory as *de novo* review.

The shift in the SPS jurisprudence that came with the Appellate Body report in *US/Canada – Continued Suspension* was truly unexpected and has real seismic

[184] For example, Alessandra Arcuri and Giuseppe Dari-Mattiacci, *Centralization versus Decentralization as a Risk-Return Trade-off*, Amsterdam Center for Law & Economics Working Paper No 2007–06 (arguing on the basis of portfolio theories for decentralized risk governance, which implies the existence of deferential standard of review).

[185] Guzman (Dispute Resolution) 230–1.

[186] Winickoff et al (Adjudicating the GM Food Wars) 109.

[187] Guzman (Dispute Resolution) 230–1.

[188] For example, Button (*Power to Protect*) 170.

[189] Cf Chapter 2, section 2.1.

effect. The Appellate Body reversed the panel on the majority of issues that were appealed, including the applicable standard of review. Interestingly, while the Appellate Body opted for a rather deferential approach to the examination of evidence produced on a national level, it also employed a more demanding standard of review with regard to the panel's legal determinations. In particular, the Appellate Body did not refer to the threshold of deliberate disregard and wilful distortion or misrepresentation of evidence and simply reversed the relevant panel's findings.

The Appellate Body, when addressing the standard of the review applicable to risk assessment, noted that the task of the panel under Article 5.1 was restricted to the determination as to whether such assessment was supported by coherent reasoning and respectable scientific evidence. This limited duty was opposed with the erroneous standard of review that would require a panel to assess correctness of a risk assessment.[190] Elaborating on the task of the panel, the Appellate Body identified four steps that should have been taken by the panel when assessing whether an SPS measure was based on risk assessment. Three of them related to a risk assessment as such, whereas the fourth one concentrated on the relationship between conclusions of a risk assessment and an SPS measure. According to the Appellate Body, the panel must:

- identify the scientific basis upon which the SPS measure was adopted; that basis may be provided by both minority and mainstream scientific opinion;
- verify whether the scientific basis comes from respected and qualified sources; arguably this requirement will be a pure formality in the case of mainstream science, whereas minority scientific opinions may require a more detailed examination in order to assess whether they meet a minimum epistemic threshold (but nothing more than that);
- assess whether the reasoning articulated on the basis of the scientific evidence is objective and coherent; this meant to the Appellate Body that the panel should review whether a conclusion reached by a WTO Member finds sufficient support in the scientific evidence relied upon; and
- determine whether the results of the risk assessment sufficiently warrant the SPS measure; again these results may come from both minority and mainstream scientific circles.[191]

This methodology, although not without some ambiguity, provides the most precise formulation of the applicable standard of review in the SPS case law. It clearly asks a panel to limit the scope of its examination to reasonableness of national determinations. A panel is not entitled to ascertain whether such findings are correct (eg in the light of the mainstream or best science). Once the scientific claim is defensible on methodological grounds (it cannot be fully falsified

[190] Appellate Body Report, *US – Continued Suspension*, para 590.
[191] Ibid, para 591.

by other evidence) it may constitute a basis for a measure even if, in the panel's opinion, the particular scientific problem is better explained by other evidence. The standard of review articulated by the Appellate Body has an obvious impact on the whole structure of the panel's fact-finding process. It redefines the role of the experts and determines the extent of information that may be sought by the panel. In particular, a panel may seek the assistance of experts in order to verify whether the individual steps identified above are met. Consequently, it will possible for a panel to rely on experts in the identification of scientific basis, the determination of a minimal epistemic threshold with respect to scientific views reflected in a risk assessment, and the evaluation of a relationship between scientific evidence and conclusions of a risk assessment. Experts may also be consulted with regard to the fourth element (the requirement of objective relationship). At the same time, the Appellate Body warned that this task cannot be transformed into a *de novo* review where the panel asks whether 'the experts would have done a risk assessment in the same way and would have reached the same conclusions as the risk assessors'.[192]

It might be interesting to look at how this new standard of review is to be applied in practice. The Appellate Body's review of the panel analysis on the genotoxicity of oestradiol-17β, which was one of the central issues in the EC assessment, may serve as a useful example. The risk assessment, referring to different scientific studies, claimed that some metabolites of oestradiol-17β were genotoxic and could act as a tumour initiator and promoter. The EC risk assessment also stated that certain population groups, such as prepubertal children, were specifically vulnerable to risk whereas the genotoxic mechanism of oestradiol was such that it was not possible to establish a safety threshold.[193] The panel, instead of analysing whether such conclusions were supported by coherent reasoning and respectable scientific evidence, engaged in a general enquiry on the current state of science with respect to genotoxicity of oestradiol-17β.[194] On some occasions, this review turned out to be a search of what was 'widely accepted by the broader scientific community, rather than a discussion on the evidence relied upon by the European Communities' risk assessment'.[195] In other words, the panel aimed to establish the best science and measure the EC risk assessment against such benchmark. This approach is particularly visible in the treatment of the scientific opinions that were expressed by the experts advising the panel. The panel tended to disregard some of them if they were not shared by the majority. For example, one of the experts, Dr Cogliano, noted that the conclusion that no threshold could be identified for oestradiol-17β reflected a certain view on the genotoxic mechanism which was equally scientifically unverified as the

[192] Ibid, para 592. [193] Ibid, para 599.

[194] For example, Panel Report, *US – Continued Suspension*, paras 7.562–3.

[195] Appellate Body Report, *US – Continued Suspension*, para 602.

opposed view that such a threshold in fact existed.[196] The panel disregarded this remark and continued its analysis on the basis of the opinions of the other three experts. Yet in many other instances, the panel chose some views in a rather arbitral manner without explaining why competing statements of other experts were not considered to be relevant.[197] According to the new four-step methodology pronounced by the Appellate Body, the proper analysis would require a panel to find first the scientific basis in the EC risk assessment for the conclusion of genotoxicity of oestradiol-17β. As a second step, the panel should verify whether such a scientific basis comes from respected and qualified sources (ie by establishing a minimum epistemic threshold). The third step would require the assessment of the reasoning articulated on the basis of the scientific evidence as to its objectiveness and coherence. The last step will consist of the determination of the objective relationship between risk assessment as a whole, including the evaluation of genotoxicity, and the SPS measure (ie a permanent ban on the import of beef from cattle treated with oestradiol).

Although the above standard is much closer to the deferential end of the spectrum, it would be too far to conclude that it is a fully procedural one. The verification of a scientific basis requires substantive inquiry in order to assess the existence of a minimum epistemic threshold. A similar substantive element is also included in the evaluation of objectivity and coherence in the reasoning made on the basis of the scientific evidence and in the determination of the relationship between risk assessment and an SPS measure (whether it sufficiently warrants a measure). The review of the panel's finding on the genetoxicity of oestradiol-17β that was offered by the Appellate Body shows that the panel is not expected to subscribe to any scientific view presented by the defendant and to concentrate its analysis only on the procedural compliance. Its proper task is to determine whether the evidence, interferences, and connection between risk assessment and measure are reasonable. At the same time, the Appellate Body was clear this does not involve the assessment of correctness. This standard, although not fully deferential, leaves WTO Members with a great degree of discretion as to how to assess scientific data and what kind of interferences to make on this basis. It also greatly reduces the need of the panel to engage in a detailed examination of scientific evidence and deciding which scientific view is better. Once it is established that a particular claim is at least scientifically defensible, the task of the panel ends. Of course, the final appraisal of the new approach announced by the Appellate Body will only be possible with the new cases coming to the WTO dispute settlement system. Nevertheless, on its face, it looks very close to the standard of reasonableness which is provided by Article 17.6(i) of the Anti-dumping Agreement.

[196] Panel Report, *US – Continued Suspension, Annex D, 'Replies of the Scientific Experts to Questions Posed by the Panel'*, para 186.

[197] Ibid, para 6.113 (disregarding the opinion of Dr Guttenplan and concluding that there was no evidence that oestradiol-17β is genotoxic *in vivo*).

4.6 Science as a benchmark in international trade/health disputes

Against the above analysis, it is necessary to ask what is the precise role that is performed by science under the SPS Agreement. Theoretically, at least three different functions may be distinguished. First, science can be seen as a method of reducing regulatory heterogeneity between WTO Members, alongside harmonization or non-discriminatory rules. Science appears to be one of the most universal discourses available to human kind. This is particularly true with respect to disciplines such as mathematics and physics, but also biology, chemistry, and toxicology. Consequently, the reliance on science will arguably result in a greater convergence between the different national regulatory measures than would be possible under any other criteria. Secondly, science may be regarded as a sophisticated instrument for detecting and sorting out those measures which have a protectionist rather than a health protective character. In other words, science may provide a form of a test that would help to indirectly determine the real purpose of a measure. As noted by Sykes, the requirement of scientific evidence 'aids in motive review, and helps to sort regulations between those that are protectionists and those that seek to promote some legitimate, non-protectionist regulatory objective'.[198] Under this approach, the function of science would be limited to determining whether an SPS measure is a genuine health regulation (principally in terms of technical paradigm of risk) or a form of disguised protectionism.[199] Science would therefore circumscribe 'the regulatory activity of national governments so as to limit the abuse of putatively scientific claims for protectionist purposes'.[200] Thirdly, science may provide a mechanism for the general elimination of 'irrational' barriers to international trade. This goal goes beyond the pure non-protectionism principle as it aims at the removal of all restrictions that do not find scientific justification.[201] This would mean that the SPS Agreement intends to prevent 'unjustified regulation per se, whether or not such a regulation creates a competitive disadvantage for foreign goods *vis-à-vis* domestic goods'.[202] The SPS Agreement with its science-based requirements would go beyond the traditional WTO disciplines of combating protectionism by imposing a new standard of rationality.

[198] Sykes ('Domestic Regulation') 355.

[199] John J Barceló III, 'Product Standards to Protect the Local Environment—the GATT and the Uruguay Round Sanitary and Phytosanitary Agreement' (1994) 27 Cornell International Law Journal 755, 768.

[200] David A Wirth, 'The Role of Science in the Uruguay Round and NAFTA Trade Disciplines' (1994) 27 Cornell International Law Journal 817, 818.

[201] Jan Bohanes, 'Risk Regulation in WTO Law: A Procedure-Based Approach to the Precautionary Principle' (2002) 40 Columbia Journal of Transnational Law 323, 326.

[202] Robert Hudec, 'Science and Post-discriminatory WTO Law' (2003) 26 Boston College International & Comparative Law Review 185, 187.

It seems that to a different degree each of the above functions underlies the SPS disciplines. Probably the first task is the most limited one. The introduction of a common benchmark, being to some extent an objective category, indeed contributes to the creation of a common playground between the different regulatory systems. It delimits plausible (in the sense of being acceptable by other trading partners) arguments for adoption and maintenance of an SPS measure and sets the frames of legitimate discourse. As noted in the literature, 'science brings constraints on valid lines of argument in the face of data and methods used to estimate risk'.[203] A particular measure would be considered as justifiable only if it finds a support in scientific findings. At the same time, one has to recognize that different standards that were developed in the case law indicate that this function may be of limited importance. As discussed above, the SPS jurisprudence accepts minority scientific opinions as legitimate grounds for domestic regulations. In the same vein, the SPS Agreement only requires a rational relationship between a measure and the conclusions of risk assessment. This standard gives WTO Members an additional margin of discretion as to the choice of a measure than would be possible under the full conformity threshold. In addition, WTO Members, when performing risk assessments and deciding on the appropriate regulatory response, may take a precautionary approach that may result in the adoption of measures which are more stringent compared to other jurisdictions. Obviously this contributes rather to the proliferation of measures than to their uniformity. Moreover, the universality of science and scientific methods should also not be overstated. Science itself is to a certain degree a product of culture and may differ between countries.[204] As noted by Atik, the reliance on science in international trade disputes may in some cases lead to fragmentation rather than convergence due to the high economic stakes involved. He explains that scientific research concentrates in so-called 'scientific centres of authority', which operate in the particular cultural settings, including different scientific traditions, different priorities in their agendas and so on.[205] This in turn may produce multiple, mutually exclusive scientific views.

As far as the second function is concerned, science indeed has the potential to detect protectionism. In the absence of scientific evidence confirming the existence of risk, in theory one can assume that a measure was adopted due to other, presumably protectionist purposes rather than as a response to health or environmental risks.[206] Such interpretation would also be compatible with the traditional approach of the WTO and GATT 1947 laws which seek in the first place the elimination of discriminatory measures. At the same time, it is necessary to recognize that science is not a perfect proxy for protectionism and may be

[203] Crawford-Brown et al ('Environmental Risk') 462.

[204] Atik ('Science') 740. [205] Ibid, 755.

[206] Donald H Regan, 'What are Trade Agreements for? Two Conflicting Stories Told by Economists, with a Lesson for Lawyers' (2006) 9(4) Journal of International Economic Law 951, 968.

both over- and under-inclusive. A measure that lacks a scientific basis does not necessarily have to be protectionist. There may equally be other reasons standing behind its adoption which are not reducible to the protection of national producers against competition from foreigners.[207] Note that the SPS Agreement as such does not provide any possibility of justifying non-protectionist measures that are not supported by scientific evidence in the form of risk assessment. The opposite situation is also possible. A measure, despite being based on science, can still be motivated by protectionism with health considerations serving only as a useful excuse. As noted in the literature 'there is still rational flexibility in scientific arguments, allowing both the development and application of risk estimates to be captured and used strategically to foster trade barriers'.[208]

The above indicates that that role of science under the SPS Agreement is closest to the third function, which is the elimination of irrational, from the scientific perspective, national barriers to international trade. Thus, scientific requirements go beyond the detection of protectionism and aim at the removal of all unjustified measures that may impede trade. This obviously differs from the traditional GATT disciplines and presumably constitutes a step further in the process of international integration.[209] For some authors this change represented a real paradigm shift in international economic law from negative to positive integration. Under the new standard of the SPS Agreement (and also the TBT Agreement), a measure 'could constitute a violation even if it were not discriminatory, that is even if it were not such as to afford protection to domestic production'.[210]

It is not clear whether this function of science was indeed contemplated by the drafters. One may only speculate that the introduction of the harmonization disciplines by the SPS Agreement could indicate that the idea of the positive integration was not alien. The same is suggested by one of its preambular statements which pronounces the desire to establish a multilateral framework of rules and disciplines to guide the development, adoption, and enforcement of SPS measures in order to minimize their negative effects on trade. This formulation, which only refers to the consequences of measures without indicating protectionism as its main concern, may imply that the disciplines of the SPS Agreement are more far-reaching than traditional GATT rules. Another part of the Preamble goes in the same direction and identifies elimination of both discriminatory measures and measures that constitute a disguised restriction as the main preoccupation of the agreement.[211]

[207] Cf Appellate Body, *EC – Hormones*, para 245 (where the Appellate Body identified a number of objectives pursued by the EC measure, none of them having protectionist character).

[208] Crawford-Brown et al ('Environmental Risk') 462.

[209] Cf Alberto Alemanno, *Trade in Foods. Regulatory and Judicial Approaches in the EC and the WTO* (2007) 245.

[210] Henrik Horn and Joseph HH Weiler, 'European Communities—Trade Description of Sardines: Textualism and its Discontent' in Henrik Horn and Petros C Mavroidis (eds), *The WTO Case Law of 2002* (2005) 251.

[211] Recital No 1, Preamble to the SPS Agreement.

The role that has been assigned to science under the SPS Agreement is subject to both intense criticism and genuine appraisal. Some critics argue that the introduction of science-based criteria may undermine the sovereign right of WTO Members to set their own level of protection.[212] Indeed, as discussed above, some of the science-based standards developed by the case law seem to interfere with the right of a Member to establish its ALOP.[213] Another line of criticism emphasises that the principal aim of the WTO system is the promotion of socio-economic welfare. Such welfare consists of both material and non-material elements, including a popular perception of safety. Reliance on purely scientific criteria cannot guarantee the maximization of such social welfare[214] as popular fear in itself is a social cost. Thus, the domestic risk regulations, which are based on purely popular perceptions of risk, may sometimes be reasonable, even if they are scientifically unfounded. Howse makes a similar argument referring to the concept of utility. He particularly notes that 'if citizens believe they need a certain regulation, however "deluded" such belief is, their utility will be reduced if they do not get it, in the sense they will believe themselves exposed to a risk they believe to be significant'.[215] Consequently, strict scientific criterion may actually lead to the reduction of the overall socio-economic welfare/utility rather than its increase. Other authors criticize the use of science on the grounds of democratic values. For instance, Bohanes submits, on the basis of subjective welfarism approach,[216] that authorities should not override, in the name of technical rationality, democratically selected preferences and beliefs of their own citizens. Therefore, the ultimate decisions on how to react to potential risks (irrespective of whether they are substantiated by scientific evidence) have to respect the general preferences of the citizens.[217] According to Howse, democracy requires that the ultimate decision rests on citizens and not to honour their 'choice is in fact to favour an artificial and cryptically elitist conception of democratic deliberation'.[218] From a pragmatic point of view, it is argued that the application of the SPS Agreement, which prefers scientific findings over the popular perception of the risk, possibly 'endangers public support for the [international] trade regime'.[219] Moreover, since the compliance with the rulings which go against the public perception of risk is politically extremely difficult, 'the credibility of the WTO is also harmed

[212] Joanne Scott, 'On Kith and Kine (and Crustaceans): Trade and Environment in the EU and WTO' in Joseph HH Weiler (ed), *The EU, the WTO and the NAFTA—Towards a Common Law of International Trade* (2000) 157 (particularly pointing out the exclusion of theoretical risk as a legitimate subject of regulation).

[213] Cf Chapter 6, section 6.4.1.

[214] Bohanes ('Risk Regulation') 355.

[215] Robert Howse, 'Democracy, Science and Free Trade—Risk Regulation on Trial at the WTO' (1999–2000) 98 Michigan Law Review 2329, 2337.

[216] The subjective welfarism approach claims that 'government, even, or perhaps especially in a democracy, should attend exclusively to conception of welfare as subjectively held by its citizens' (Bohanes ('Risk Regulation') 360, citing Cass Sunstein, *Free Markets and Social Justice*).

[217] Ibid, 361. [218] Howse ('Democracy') 2337.

[219] Charnovitz ('Supervision of Health') 302.

in the complaining state, which observes a legal victory before a panel or the AB [Appellate Body], but does not get the benefits of that victory'.[220] Another group of critics questions the objectivity of science and scientific risk assessment. Those authors particularly stress that science is not a neutral criterion able to operate as an objective benchmark in SPS disputes.[221] Science, and in particular technical risk assessment, is infiltrated with non-scientific assumptions, policy considerations, and value judgements. In particular 'choices regarding risk identification, methodological design, sampling, and extrapolation assumptions can involve substantial policy decisions that must be made prior to, and throughout, the risk-assessment process'.[222] In consequence, requiring allegiance to a one-dimensional vision of science (mainstream science or the best science) will be inevitably reductionist. As explained by a group of eminent social scientists 'a significant body of social science comparing the treatment of risk-based decision-making across national political systems demonstrates how differences in issue framing and science policy can lead to systematic transnational variations in the assessment of health, safety, and environmental risks'.[223] There are also those who believe that a proper role of WTO law, including the SPS Agreement, should be limited to the detection of protectionist measures. The SPS Agreement, with its strict scientific disciplines that condemn non-discriminatory measures which lack technical rationality, clearly goes beyond that task.[224] According to Regan 'domestic political/regulatory process, for all its possible failings, will generally do a better job than an international tribunal of identifying, and balancing, and finding the best way to promote domestic interests'.[225]

On the other hand, the advocates of science-based standards submit that the introduction of this benchmark may actually help national governments to maintain their autonomy in the area of human and environmental health risk regulations since the scientific requirement can serve as a shield against the pressure from other states to weaken national regulations. In effect, the SPS Agreement does not undermine national sovereignty but rather plays a supportive role.[226] The criticism based on the concept of socio-economic welfare and social costs may be tried with four counter-arguments. First, it does not take into account the external costs of the regulations (ie costs borne by the foreign traders or producers). It also underestimates the ancillary costs borne by the nationals (ie costs of consumers due to increase of prices or limitations in possible choices) as well as long term costs (ie *ex post* costs of enforcing the regulation).[227] If one approaches

220 Guzman ('Dispute Resolution') 231.

221 See generally Vern R Walker, 'Keeping the WTO from Becoming the "World Trans-Science Organization": Scientific Uncertainty, Science Policy, and Factfinding in the Growth Hormones Dispute' (1998) 31 Cornell International Law Journal 251; cf Chapter 1, section 1.4.3.

222 Winickoff et al ('Adjudicating the GM Food Wars') 95.

223 Ibid, 97.

224 Ibid, 86; Regan ('What are Trade Agreements for?') 987.

225 Regan ('What are Trade Agreements for?') 986.

226 Atik ('Science') 741.

227 Howard F Chang, 'Risk Regulation, Endogenous Public Concerns, and the Hormones Dispute: Nothing to Fear but Fear Itself' (2004) 77 Southern California Law Review 743, 762.

the concept of welfare/utility maximization from a more holistic perspective that takes into account different costs borne by different social groups, including foreigners, the outcome of the final calculation is less than certain. Secondly, regulatory responses to risks which are perceived rather that scientifically identified, may create an incentive for different interests groups to promote new fears. Those new fears will, of course, entail additional costs which should also be taken into account.[228] Thirdly, it seems that in some cases fears of the general public can be reduced through less stringent means than regulations that impose specific requirements on the regulated product (eg an importation ban). This particularly includes dissemination of information on potential risks, current scientific knowledge with regard to a specific risk issue, impact of uncertainties on final risk estimates, and so on. Consequently, if one wishes to rely on the concept of socio-economic welfare or utility, it is necessary to ensure that the general public is well informed on different aspects of the risk situation. As a result of this information exchange, it may appear that in some cases no regulation is really necessary. Fourthly, the reliance on the science may actually increase the total welfare (utility) as it allows the concentration of limited resources on those risks which find support in scientific findings as opposed to risks which are merely perceived. If the resources are used for the last category, some other risks may simply be left out, generating in the future its own costs. In consequence, 'deciding whether regulation would minimize social costs would entail a complex calculation'[229] and cannot be assumed on an a priori basis.

Criticism that refers to the democratic values may be addressed on three different levels. First, it is submitted that formally democratic decisions are not always the outcome of genuine preferences of citizens. They may be captured and manipulated by narrow but powerful domestic groups which seek the protection of their interests.[230] International rules, including the science-based requirements of the SPS Agreement, may help national governments to shield themselves from those pressures, guaranteeing that regulations do not favour a particular interest group. Equally, public preferences may result from ignorance and the lack of necessary information to make a reasoned judgement. Thus, science-based provisions of the SPS Agreement can actually enhance the quality of the democratic deliberation process on risks and possible management strategies. As explained by Howse:

> there is more to democracy than a visceral response to popular prejudice and alarm; democracy's promise is more likely to be fulfilled when citizens, or at least their representatives and agents, have comprehensive and accurate information about risks, and about the costs and benefits associated with alternative strategies for their control.[231]

Thirdly, science may be seen as actually re-enforcing the democratic values (in the sense of considering the interests of those who are affected by particular

[228] Ibid, 763–4. [229] Ibid, 763.

[230] Ernst-Ulrich Petersmann, 'Trade Policy as a Constitutional Problem. On the "Domestic Policy Functions' of International Trade Rules"', (1986) 41 *Aussenwirtschaft* 405, 426.

[231] Howse ('Democracy') 2330.

governmental action) on the international level by establishing the external accountability of states.[232] The requirement of scientific justification, as well as other obligations of the SPS Agreement, limit the discretion of national governments and remove some capricious regulations. The pragmatic argument should be probed against the history of SPS disputes. Out of six decided SPS cases, three were fully complied with by the losing parties, in one case the compliance was partial (*EC – Biotech Products*) and one still needs to be decided as to the compatibility of the implementing measures (*US/Canada – Continued Suspension*). Therefore, *EC – Hormones* may constitute an exception in the SPS field rather than the general rule. In this context it is also worth noting that a future decision in the new *Suspension* dispute may actually resolve the problem of compliance in the *Hormones* dispute. At the same time, there is no hard statistical data which would show any decrease in public support for an international trading system resulting from the outcomes of the SPS disputes. Last but not least, it may also be argued that the possibility of the compensation and the suspension of concessions, instead of compliance, serve as a safety valve. A Member may always opt for non-compliance and keep a measure which was found to be incompatible with WTO obligations, in force.

As far as the objectivity of the science is concerned, it seems that the WTO case law has progressively acknowledged the limits which are inherent in the scientific methods. The Appellate Body did not subscribe to any orthodox and simplified vision of science. It recognized that science is not a static or determined set of knowledge but is rather a constant process of inquiry on the nature and behaviour of natural things, characterized by continuous verification of subsequent hypotheses. Such an understanding of science has far-reaching practical implications. As science is not perceived as constituting a monolithic view, divergent scientific opinions may provide a valid basis for the SPS measure. Consequently, 'the call for a scientific basis may be satisfied by one or multiple mutually exclusive sciences'.[233] Likewise, as explained recently by the Appellate Body, a panel, when evaluating whether the SPS measure is based on sufficient scientific evidence (or risk assessment), limits its examination to the plausibility of the evidence that constitutes a foundation for a measure. The panel is not expected to assess the correctness of scientific and technical determinations made on a domestic level but only to examine whether they are defensible (ie whether they hold the minimum epistemic threshold and whether a measure constitutes a reasonably justified choice). In the same vein, WTO Members are not required to rely on quantitative risk assessment and qualitative evaluation is a fully legitimate option. The case law also recognized that the government might act from the perspective of prudence and precaution and differ in its assessment of insufficiencies.[234]

[232] Scott (*Commentary*) 43–4, citing R Keohane, who defined external accountability as 'accountability to people outside of the acting entity, whose lives are affected by it'.

[233] Atik ('Science') 749.

[234] Cf Chapter 5, sections 5.4.2 and 5.4.5.

Although one has to agree that some science-based standards developed by the SPS case law may compromise the choice of WTO Members as to their ALOP,[235] this seems to be a problem of the correct interpretation of the SPS Agreement rather than science itself. As discussed in this chapter, the proportionality requirement, the concept of negligible risk, an overly high specificity threshold for risk assessment, and its impact on the possibility of regulating low-level risk or relying on minority scientific opinions, are all standards that were developed by the case law. They are not explicitly provided by the SPS Agreement and may easily be corrected by the future decisions of the WTO dispute settlement bodies. Indeed one may see some considerable improvements in the recent report of the Appellate Body in *US/Canada – Continued Suspension*.

As far as the concept of negligible risk is concerned, the Appellate Body, when discussing the applicable standard of review, noted that the panel was not authorised to determine whether the risk from consumption of meat from cattle created with hormones was appreciable. This holding seems to suggest that once risk is more than just theoretical, the panel is not permitted to conduct any further analysis on the extent of such risk. If my interpretation of the Appellate Body report is correct, this will preclude in the future the application of the concept of negligible risk as well as any strict proportionality analysis in the context of science-based provisions. Similar change is visible with respect to specificity requirements in the context of complex and low-level risk. The Appellate Body expressly observed that the specificity requirement is qualified by the expression 'as appropriate to circumstances'. This means that methodological difficulties posed by the nature and characteristics of a particular risk may affect the expected level of specificity. As a consequence, the EC was not required to establish in its risk assessment a direct causal link between the possibility of adverse health effects and specific hormone residues and it was sufficient to show that such residue is one of the contributing factors to overall risk. In the same vein, it was not necessary to isolate a contribution that might be attributed to oestradiol-17b but rather to assess the overall risk resulting from exposure to different sources of hormones, including naturally occurring hormones. This seems to come very close to the acceptance of general studies and indirect evidence as a legitimate ground for subsequent regulatory action.

Although the Appellate Body did not address the problem of specificity in the context of minority scientific opinion, the new standard of review which was proposed for examination of factual determinations in SPS disputes bears a direct relevance on the treatment of minority views. The Appellate Body several times stressed that WTO Members may rely on such opinions, while the task of the panel is only limited to the assessment of minimum epistemic thresholds and does not extend to the examination of their correctness. Nor is the panel permitted to weigh different scientific evidence in order to establish which view is

[235] See also, Chapter 6, section 6.4.1(b) and Chapter 7, section 7.1.

better. Once it is determined that a particular claim is at least scientifically defensible, even if improbable, the task of the panel ends. Even though the new standard of review has not yet been tested in practice, the language of the report indeed allows us to be optimistic about the future course of the SPS case law. If one combines this with the new standards that were developed in the context of Article 5.7 (eg accepting the existence of the normative dimension in the determination of insufficiency of scientific evidence),[236] there are good grounds for concluding that the SPS Agreement is actually able to provide a workable mechanism that seriously takes into account the complex nature of science and scientific risk assessment and does not compromise the legitimate regulatory choices of WTO Members.

In this context it is also worth noting that the criticism which is raised against the use of science-based disciplines usually does not offer any convincing alternatives.[237] Widely opening up the scientific test provided by the SPS Agreement for other considerations such as cultural factors or risk perception may produce even more arbitrariness in the dispute settlement process and probably reduce the predictability of the outcome of potential disputes. Last but not least, even if one subscribes to the view that the WTO rules should be limited to the detection of protectionism rather than promoting domestic rationality, science may still be a useful instrument. The problem of over-inclusiveness may be mitigated to a great extent (although it cannot be removed entirely) by proper application of science-based standards of the Agreement. In fact, those decisions which are conventionally identified by the critics as examples of excessive intrusiveness of the agreement (*EC – Hormones* and *EC – Biotech Products*) seem to find their proper solution under the new *Suspension* standards. Note that there are good chances that the EC will be able in the subsequent compliance proceeding to show that its SPS measures on hormones-treated beef are compatible with the SPS Agreement. As far as the *Biotech* report is concerned, the application of the *Suspension* reasoning would have probably resulted in reversing the decisions of the panel on many important issues (eg particularly with respect to the determination of insufficiency of scientific evidence in the context of safeguard measures maintained by EC Member States).[238] Consequently, while the problem of over-exclusion does not disappear completely, in practice it may prove to be less serious than was predicted in the literature.

236 Cf Chapter 5, section 5.4.2.

237 Tracey Epps, *International Trade and Health Protection. A Critical Assessment of the WTO's SPS Agreement* (2008) 186 (noting that 'most critics of a science-based requirement for justification of trade-restrictive health regulations do not suggest a credible alternative means of sifting out protectionism from genuine health regulations').

238 Cf Chapter 5, section 5.4.2(b).

5

Precaution and the SPS Agreement

5.1 Introduction

Obviously, there are situations where science is not able to provide risk managers with any conclusive answers. The results of scientific research may be contradictory or unreliable. Sometimes, uncertainty remains as to the existence of the hazard, the extent of potential damage, or its probability. As discussed in Chapter 1, risk assessors use different techniques to cope with the problem of uncertainty. However, in some cases the extent of uncertainty is such that it simply prohibits the performance of a meaningful risk assessment. Consequently, risk managers may be faced with the dilemma of whether to act on the basis of inconclusive evidence, or await more certain data. One answer is provided by the precautionary principle. The principle permits, or sometimes even mandates, protective actions even before there is a complete scientific proof of risk.

The SPS Agreement, although it does not refer to the principle, also contains special rules which are intended to address such situations. Contrary to the precautionary principle, these disciplines are not triggered by the presence of uncertainty but rather by the existence of insufficiency of scientific evidence which is usually characterized in the literature as a more demanding criterion.[1] In addition, the case law has identified a number of other SPS provisions which embody the idea of precaution. On the other hand, all attempts of the parties to SPS disputes to directly invoke the precautionary principle in the context of the Agreement were met with hostility by the WTO dispute settlement bodies.

The above description gives a picture of complex interrelations that exist between the precautionary principle and the disciplines of the SPS Agreement. This chapter will attempt to simplify that picture. In the first part I concentrate on the precautionary principle itself, trying to identify its meaning. Against that background, the second part analyses the relevant SPS case law in order to assess its treatment of the principle. The third part considers the disciplines of the Agreement that were characterized as precautionary. A special emphasis is put on Article 5.7, a central precautionary provision of the Agreement. This part of the

[1] For example, Jacqueline Peel, 'A GMO by any Other Name...Might be an SPS Risk!: Implications of Expanding the Scope of the WTO Sanitary and Phytosanitary Measures Agreement' (2007) 17(5) European Journal of International Law 1009, 1017.

discussion is divided into two separate sections. The first addresses the problem of applicability of Article 5.7 and its relationship with other provisions of the SPS Agreement. The second looks at the substance of the article and analyses its obligations in detail. The last part attempts to draw overall conclusions on the position of the precaution under the SPS Agreement.

5.2 The concept of precaution

The body of the literature on precaution and the precautionary principle is immense.[2] This includes contributions from legal scholarship but also from political science and economy. Evidently, the size and purpose of this chapter do not allow me to present a detailed discussion on all aspects of the principle. The goal here is more modest, as I only intend to provide the reader with a useful background against which the disciplines of the SPS Agreement may be analysed and assessed.

The precautionary principle appeared for the first time in German environmental law.[3] The objective of the principle was to provide a rationale for risk regulatory actions, even if the causal relationship between the source of the harm and potential damage was not clear.[4] In the years that followed, the principle has become one of the major tools used by regulators in the management of uncertainties in the area of environmental, and gradually health, risk regulation, both on the national and international levels. The principle aims to provide the regulator with guidance as to how to react to uncertain (in practice mainly technological) risks, when the occurrence of risk, and even the existence of hazard, remain uncertain.[5] The principle provides that regulators, rather than awaiting certainty with respect to the actual existence of risk, should act in anticipation of potential environmental and human health harm, in order to prevent it. Having said this, it must also

[2] For a more in-depth discussion on the precaution and precautionary principle, see David Freestone and Ellen Hey (eds), *The Precautionary Principle in International Law* (1996); Timothy O'Riordan and James Cameron (eds), *Interpreting the Precautionary Principle* (1994); Georges Kremlis, Yiorgos Balias, and Antonis Sifakis (eds), *The Precautionary Principle* (2004); Cass R Sunstein, *Laws of Fear, Beyond the Precautionary Principle* (2005); Elizabeth Fisher, Judith Jones, and Rene von Schomber (eds), *Implementing the Precautionary Principle: Perspectives and Prospects* (2006); Nicolas de Sadeleer (ed), *Implementing the Precautionary Principle: Approaches from the Nordic Countries, EU and USA* (2007).

[3] The principle (*Vorsogeprinizp*, which can be translated as the 'forecaring principle') was included in the 1974 Federal Emission Protection Act. Some scholars, however, find the elements of the precautionary principle as early as the 1906 US Federal Food and Drugs Act (see eg, Jonathan B Wiener, 'Whose Precaution After All? A Comment on the Comparison and Evolution of Risk Regulatory Systems' (2003) 13 Duke Journal of Comparative and International Law 207, 211).

[4] For detailed discussion of the meaning and function of *Versogeprinizp*, see Sonja Boehmer-Christiansen, 'The Precautionary Principle in Germany—Enabling Government' in Timothy O'Riordan and James Cameron (eds), *Interpreting the Precautionary Principle* (1994).

[5] See eg, Christopher D Stone, 'Is There a Precautionary Principle?' (2001) 31 Environmental Law Reporter 10790, 10794.

be recognized that in the context of precaution and the precautionary principle, one can only speak about potential risks, since the notion of risk retains the connotation of something that can be measured and hence managed. However, the essence of the precautionary principle lies in the fact that it intends to assist the regulators in managing those situations where (identified or unidentified) uncertainties prevent any meaningful assessment of risk. This element distinguishes precaution from prevention. The last one is broadly recognized as a principle of international environmental law and requires addressing risks before they materialize into harms. Despite obvious similarities (both concepts concentrate on *ex ante* actions), the precautionary principle intends to address not so much the risks as the potential risks, that is risks whose existence cannot be established because of scientific uncertainty.

There is no agreement as to when the precautionary principle appeared for the first time on the international level. Some elements of the precaution can be found as early as 1982 when the UN General Assembly adopted the World Charter for Nature (eg by requiring special precautions to be taken in order to prevent discharge of radioactive or toxic wastes). The other environmental instruments adopted in the 1980s also employed the precautionary approach or principle[6] (the 1985 Vienna Convention for the Protection of the Ozone Layer, and the 1989 Montreal Protocol on Substances that Deplete the Ozone Layer[7]). The first roughly precise articulation of the precautionary principle came with the 1992 Rio Declaration which provided that 'where there are threats of serious or irreversible damage, lack of full scientific certainty shall not be used as a reason for postponing cost-effective measures to prevent environmental degradation'.[8] This definition has quickly become the most widely accepted and referred formulation of the principle. Since then, a number of other multilateral instruments have used the precautionary clauses.[9]

[6] There is an intense discussion in the literature on the character of precaution. Some scholars claim that it has acquired a status of the principle, others prefer to use the notion of precautionary approach. The disagreement is not only about semantics, as characterizing precaution as an approach or principle reflects a deeper disagreement over its status in international law; cf the literature cited in n 1.

[7] Montreal Protocol to the Vienna Convention on Substance that Deplete the Ozon Layer, Montreal, 16 September 1987, 30 ILM 541.The Preamble to the Protocol states that it is '[d]etermined to protect the ozone layer by taking precautionary measures to control equitably total global emissions of substances that deplete it, with the ultimate objective of their elimination on the basis of developments in scientific knowledge, taking into account technical and economic considerations and bearing in mind the developmental needs of developing countries'.

[8] Rio Declaration on Environments and Development, 14 June 1992, UN A/Conf 151/5/Rev 1 (1992), reprinted in 31 ILM 876 (1992) at principle 15. Note, however, that there are authors who label the Rio Declaration as precautionary approach rather than precautionary principle (Olivier Godard, 'The Precautionary Principle, the Environment and International Trade: Sovereignty and Collective Preferences in Question', Ecole Polytechnique, Centre National de la Recherche Scientifique, Cahier No (2006–05) 1).

[9] These include: the United Nations Framework Convention on Climate Change (1992), the Convention on Biological Diversity (1993), the Cartagena Protocol on Biosafety (2000) and the Stockholm Convention on Persistent Organic Pollutants (2001). Besides, there are also a number of

The concept of precaution, which at the beginning was only used in the field of national and international environment law, with time has also appeared in other areas, notably in the field of human life and health protection. The Cartagena Protocol on Biosafety (CPB) is a good example. Although the major concern of the Protocol is the protection of biodiversity, it also touches upon the protection of human health.[10] The same is true for the Stockholm Convention on Persistent Organic Pollutants which asked for a precautionary approach with regard to the protection of human life and health. This new scope of the principle was also confirmed in the more recent definition that was proposed by a group of scholars and ecological NGOs (the so-called 'Wingspread Statement'). The statement particularly provides that 'when an activity raises threats of harm to human health or the environment, precautionary measures should be taken even if some cause and effect relationships are not fully established scientifically'.[11]

If one compares the formulation of the principle in different international instruments, it becomes immediately apparent that there is no one uniform definition of the precautionary principle. With a certain degree of generalization one may say that 'the essence of the precautionary principle is that positive action ... may be required before the existence of a risk has been scientifically established'.[12] However, when it comes to details, the uniformity disappears. First, different definitions of the principle provide for different thresholds for its application. The Rio Declaration refers to threats of serious or irreversible damage. This means that only those threats which can be characterized as serious or irreversible can trigger the application of the principle.[13] The other instruments provide for a much lower threshold. The Biosafety Protocol, for example, only speaks about 'potential adverse effect of a living modified organism' without qualifying its seriousness and probably presuming the irreversible character of threats connected with GMOs. The Convention on Biological Diversity is somewhere in the middle as it refers to 'threats of significant reduction or loss

regional legal instruments that provide for the precautionary principle or precautionary approach (eg the Bamako Convention (1991), the OSPAR Convention (1992), and the Treaty Establishing the European Community).

[10] See eg, Article 2.2 of the CPB, which provides that '[t]he Parties shall ensure that the development, handling, transport, use, transfer and release of any living modified organisms are undertaken in a manner that prevents or reduces the risks to biological diversity, taking also into account risks to *human health*' (emphasis added).

[11] Wingspread Statement on the Precautionary Principle, 25 January 1998, reprinted in Carolyn Raffensperger and Joel Ticker (eds), *Protecting Public Health and the Environment: Implementing the Precautionary Principle* (1999); cf also, Regulation (EC) No 178/2002 of the European Parliament and of the Council of 28 January 2002 laying down the general principles and requirements of food law, establishing the European Food Safety Authority and laying down procedures in matters of food safety (OJ L031 of 1 February 2002).

[12] Jan Bohanes, 'Risk Regulation in WTO Law: A Procedure-Based Approach to the Precautionary Principle' (2002) 40 Columbia Journal of Transnational Law 323, 331.

[13] Similar language is contained in the Climate Change Convention, which provides that threat of damage needs to be serious or irreversible (Article 3.3) and the United Nations Fish Stocks Agreement (1995) where the threshold is seriousness (Article 6.1–7).

of biological diversity'. The least demanding is the Communication of the EC Commission which proposes recourse to the precautionary principle if a potential risk is simply above the ALOP.[14]

Secondly, different versions of the principle seem to require different thresholds of scientific uncertainty. The Rio Declaration speaks about the lack of full scientific certainty (note that such an expression is in itself ambiguous as science never provides full certainty[15]); the Wingspread Statement speaks about 'some cause and effect relationships [that] are not fully established scientifically' (which and to what degree?); while the World Charter for Nature speaks about adverse effects that are not fully understood. The EC Communication goes even further and requires the performance, prior to adoption of any measures based on the precautionary principle, of a scientific risk assessment (as complete as possible).[16] Only after such an assessment, if uncertainty is identified, may a risk manager rely on the precautionary principle. Although all the above definitions presume the existence of some positive scientific evidence which would authenticate potential risks, they are silent on the required threshold (ie to what extent a risk manager should prove the existence of potential risks). In practice, it may turn out that the judgment on uncertainty is to a large extent discretionary and depends on the body making the decision. Another group of definitions gives even more discretion to regulators as they only require the existence of doubt as to the safety of a product or activity.[17] If interpreted liberally, such formulations allow for the introduction of measures with respect to any kind of situation as some residual doubts on safety almost always remain. On the other hand, the precautionary principle does not address the situation of informed ignorance which is defined by Godard as a situation 'when after having used all resources of existing scientific knowledge it is not possible to give a face to a theoretic possibility that an unknown hazard exists'.[18]

Thirdly, some versions of the principle require cost-benefit analysis when applying the principle. Again, the Rio Declaration may serve as a good illustration, as it asks for 'cost-effective measures'. The scope and the relevance of such cost-benefit analysis for the final decision is not clear. Does it include evaluation of opportunity costs? How is one supposed to balance costs against effectiveness? Does a cost-effective requirement operate as a weak or strong form of proportionality? The Rio Declaration does not answer these questions. The Biosafety

[14] European Commission, Communication from the Commission on the Precautionary Principle, 2 February 2000, COM (2000) 17.

[15] Cf John N Hathcock, 'The Precautionary Principle—An Impossible Burden of Proof for New Products' (2000) 3(4) AgBioForum 255, 256 (noting that 'absolute proof of safety is not achievable because it would require the proof of a negative, a proof that something (risk) does not exist').

[16] EC Communication ('Precautionary Principle') 13–14; note that the CPB takes the same route.

[17] Final Declaration of the First European 'Seas at Risk' Conference, Annex 1, Copenhagen 1994.

[18] Godard ('Precautionary Principle') 2; but see Chapter 1, section 1.4.4.

Protocol speaks only about taking into account technical and any economic considerations when contemplating precautionary measures. This is a somewhat weaker requirement which only calls for consideration of these factors without requiring the regulator to reflect them in a final measure. At the other end of the spectrum, there is the Montreal Protocol which does not contain any comparable language and the precautionary measures arguably do not need to be mitigated by economic considerations.

Fourthly, it is also submitted in the literature that one of the essential elements of the precautionary principle is a particular allocation of the burden of proof that requires the proponent of a particular activity (or substance) to prove its safety.[19] Again, depending on the definition, different standards may be identified. Some definitions are silent on that issue while others lack consistency in the required threshold. Thus, the precautionary principle may ask for a no-harm threshold, meaning that the proponent of the activity needs to show that the particular activity or substance would cause no harm (which again at its extreme is impossible).[20] More lenient versions would require the level of 'certainty' of no harm. That level may be relational and depend on the seriousness and irreversibility of the harm and the extent of scientific uncertainty (ie the higher the uncertainty and irreversibility the higher the burden of proof required).

Fifthly, more recent definitions of precaution also recognize that the principle implicitly provides for a proportionality test. For example, Godard identifies four elements that influence the application of the precautionary principle: possible damage, safety goals, direct and opportunity costs of a measure, and the scientific consistency of hypotheses of hazards.[21] Depending on those variables, a particular measure may be more or less stringent. The same line is taken by the EC Commission which conceptualized the precautionary principle as requiring proportionality, albeit weak, between desired ALOP and a measure.[22]

Sixthly, the precautionary principle may be formulated as an imperative or as a mere signpost for risk managers. The principle as an imperative requires action in the face of uncertainty (and/or threats of serious or irreversible damage),[23] while in the second case it simply gives a justification for regulatory action, without compelling risk managers to act in the first place.[24]

Seventhly, some formulations presuppose the provisional character of the measures taken on the basis of precautionary principle, while the others do not

[19] For example, Lucas Bergkamp, *European Community Law for the New Economy* (2003) 179.

[20] Wiener ('Whose Precaution?') 210.

[21] Godard ('Precautionary Principle') 10.

[22] EC Communication ('Precautionary Principle') 18.

[23] See eg, Declaration of the Second International Conference on the Protection of the North Sea, 25 November 1987.

[24] EC Communication ('Precautionary Principle') 15–16 (noting that 'the Commission or any other Community institution has broad discretionary powers, notably as regards the nature and scope of the measures it adopts' and 'responding does not necessarily mean that measures always have to be adopted. The decision to do nothing may be a response in its own right').

have such a limitation. In the first case, a measure may be maintained as long as scientific uncertainty persists. The second group of definitions assumes that complexities of the world may be not reducible over time and therefore a precautionary measure may be of permanent character.[25]

This heterogeneity of definitions has led some scholars to argue that 'the precautionary principle is an idea rather than a clearly defined concept, much less a guide to consistent policy making'.[26] This group prefers to speak about precautionary approach rather than principle.[27] As noted in the literature the distinction between the approach and the principle has important legal consequences. The approach 'describes... a legal technique to deal with uncertainties, scientific or otherwise, while a "principle" can bind the actor (legislator, administration or judiciary) to apply those techniques'.[28] There are others who, although they do not question the ontological potential of the precautionary principle, note that 'the principle' is still evolving and has not yet crystallized into a rule of customary international law (or general principle of law).[29] This group also warns that 'much of the evidence for the precautionary principle comes from regional agreements that may not represent a global consensus'.[30] Some see this crystallization process as a trend towards less strict formulation of the precautionary principle.[31] The older versions of the principle appear to provide almost unlimited discretion to risk managers and set very low requirements for its applicability whereas newer versions seem to be more restricted. As observed by Applegate, 'those aspects of the principle that remain uncertain... have been systematically pushed in the direction of a weaker principle'.[32] On the other hand, the proponents of the principle claim that the heterogeneity of definitions results from the fact that the use of the precautionary principle is context-dependent and there is no formulation that suits all situations (eg a stronger version may appear in the regional arrangements among the homogenous states).[33] Moreover, they point out that a lack of

[25] See eg, the CPB. However, it was also argued that all versions of the precautionary principle are implicitly conditioned by development of scientific research (John S Applegate, 'The Taming of the Precautionary Principle' (2002–03) 27(13) William & Mary Environmental Law & Policy Review 13, 32); cf also Chapter 1, section 1.4.2.

[26] Giandomenico Majone, 'What Price Safety? The Precautionary Principle and Its Political Implications' (2002) 40(1) Journal of Common Market Studies 89, 93.

[27] Cf Appellate Body Report, *EC – Hormones*, para 60.

[28] Marie-Claire Cordonier Segger and Markus W Gehring, 'The WTO and Precaution: Sustainable Development Implications of the WTO Asbestos Dispute' (2003) 15(3) Journal of Environmental Law 289, 291.

[29] Patricia W Birnie and Alan E Boyle, *International Law and the Environment* (2002) 98.

[30] John O McGinnis, 'The Appropriate Hierarchy of Global Multilateralism and Customary International Law: The Example of the WTO' (2003–04) 44 Virginia Journal of International Law 229, 235.

[31] Applegate ('Taming') 23.

[32] Ibid, 16 (enumerating the following elements that have changed over time: triggering factors, threshold amount of information, levels of uncertainty, possible response, and requirement to revise a measure).

[33] Cf OECD Joint Working Party on Trade and Environment, Uncertainty and Precaution: Implications for Trade and Environment, COM/ENV/TD(2000)114/FINAL, 10; see also

clarity does not necessary preclude a norm from becoming a rule of international customary law.[34] Consequently, they argue that the precautionary principle may be qualified as a rule of customary international law (or general principle of law).[35]

Under international law, in order to be recognized as part of international customary law a particular rule has to meet two criteria. First, a rule needs to be followed by states in a constant and uniform fashion. Thus, divergences in the states' conduct would speak against qualifying the rule as customary law. At the same time, the required uniformity is not understood rigorously and general observance appears to be sufficient.[36] The practice also needs to be universal, in the sense that it has to be followed by a majority of states and in particular by those states whose interests are affected by the practice. In consequence, the identity of the states is more important than their overall number.[37] This also means that the express and constant opposition of a particular state to the rule results in the inapplicability of the rule to that state (the so-called persistent objector principle),[38] although the inaction of a state does not amount to effective opposition. The state practice also has a temporal dimension as it should be present over some time. In principle, a long period is required but there are equally instances when a rule of international customary law emerges after a relatively short period of time.[39] What seems to be more important in this context is the strength of the adherence to the custom. State practice may be deduced from several sources, including records of treaties' ratification, participation in treaty negotiations, national legislation and the domestic court practice, statements by national officials, pleadings of states before international tribunals, and so on.[40] The second condition which needs to be met requires states to follow the rule from a sense of legal obligation (*opinio iuris sive necessitatis*).[41] Thus, what the second condition calls for is a 'subjective belief' on the part of states that certain conduct is

Harald Hohmann, *Precautionary Legal Duties and Principles of Modern International Environmental Law* (1994) 339–40.

[34] David Freestone, 'The Road to Rio: International Environmental Law After the Earth Summit' (1994) 6 Journal of Environmental Law 193, 213 (giving the example of the principle of self-determination, which is recognized as a rule of customary international law despite the fact that no precise formulation exists).

[35] See eg, James Cameron and Juli Abouchar, 'The Status of the Precautionary Principle in International Law' in Freestone and Hey (eds) (*Precautionary Principle*) 52; Owen McIntyre and Thomas Mosedale, 'The Precautionary Principle As a Norm of Customary International Law' (1997) 9 Journal Environmental Law 221, 235.

[36] Hugh Thirlway, 'The Sources of International Law in International Law' in Malcolm D Evans (ed), *International Law* (2003) 127–8.

[37] Rebecca M Wallace, *International Law* (2005) 11; also cf the North Sea Continental Shelf Case, 3 ICJ Reports (1969) para 73.

[38] Wallace (*International Law*) 12.

[39] The rule which establishes a sovereign right of states over natural resources located in their continental shelf is an example of rule of customary law which developed relatively fast.

[40] Philippe Sands, *Principles of International Environmental Law* (2003) 144.

[41] Cf Restatement (Third) of Foreign Relations Law of the United States (1986) para 102.

required from them as a matter of law. *Opinio iuris* can be established on the basis of statements of officials, expressions of beliefs regarding the acts of international organizations, and so on.[42] International agreements which embody a particular rule may either serve as a proof of the existence of a customary rule or alternatively they may contribute to its development.

Qualifying a rule as a recognized general principle of law is even more difficult. The function of this category is to fill the gaps which inevitably exist in international law, by identifying those rules which are broadly shared by all states. Regrettably, there is no common understanding of the general principle of law. The mainstream approach is to define it as those rules in various systems of municipal law which are shared by all states or at least a majority of them.[43] As in the case of rules of customary law, there is no agreement on the number of required states and their relative importance. Some authors take a pragmatic view and argue that some municipal legal systems are more important than the others. However, there are also those who submit, on the basis of the principle of equality, that no one domestic system has a superior character. Similarly, there is no agreement as to the required compatibility in the formulation of the rule between different municipal legal orders. As a consequence, 'this particular source of law is of less practical importance in determining the rights and obligations of States'.[44]

The rules of international customary law as well as recognized general principles of law are clearly relevant in the context of WTO law. Some argue that they may be directly applied by WTO dispute settlement bodies in the dispute between WTO Members. Probably the most important proponent of this approach is Pauwelyn who submits that the defending party may invoke a rule of international law (including international customary law but also a bilateral treaty to which both parties are bound) as a part of its defence.[45] According to Pauwelyn 'if... the relevant conflict rule demonstrates that the other rule of international law overrides or even invalidates the WTO rule, then the WTO rule cannot be applied... irrespective of whether the WTO treaty itself includes an exception or justification for the measure at hand'.[46] Despite the very strong theoretical foundation, this proposition remains rather academic and it does not find a clear reflection in the WTO jurisprudence, including the SPS case law. This is also a reason why I will not discuss it in the context of the SPS Agreement.[47] Secondly, international customary rules may become relevant in the process of

[42] Sands (*Principles*) 146.

[43] Thirlway ('Sources') 131. The examples of general principles of international law include the principle of estoppel, states responsibility for the acts of its agents, and principles of reparation (see Wallace (*International Law*) 24).

[44] Ibid, 132.

[45] Joost Pauwelyn, 'The Role of Public International Law in the WTO' (2001) 95 American Journal of International Law 535, 566.

[46] Ibid.

[47] For a more detailed discussion, see Joost Pauwelyn, *Conflict of Norms in Public International Law: How WTO Law Relates to Other Rules of International Law* (2003) 440–86.

interpretation of WTO provisions. As discussed in Chapter 2, the WTO case law recognizes Article 31 of the VCLT as providing customary rules of interpretation of public international law.[48] At the same time, Article 3.2 of the DSU mandates dispute settlement bodies to interpret the provisions of WTO law in accordance with such rules. Article 31.3(c) of the VCLT is particularly important in this context as it requires an interpreter to take into account any relevant rules of international law applicable in the relations between the parties. Both international customary law and the recognized general principles of law are rules of international law,[49] which are by their nature applicable between WTO Members. Consequently, there are three distinctive situations where the precautionary principle may potentially affect the interpretation of WTO law: the principle can be considered as part of international customary law or it may be regarded as a general principle of law. The third option emerges when a precautionary principle is included in another international agreement that meets the conditions of Article 31.3(c) of the VCLT.

Note also that there are two distinctive problems with the application of the precautionary principle (either directly or as interpretative guideline) in the context of WTO law, and more specifically the SPS Agreement. First, this requires prior determination of the status of the principle (ie whether it can be qualified as a rule of customary international law or general principle of international law). Secondly, if the answer is positive, one has to determine to what extent and how this rule may have an impact on the disciplines of the Agreement. As far as the second element is concerned, the additional difficulty arises from the heterogeneity of the formulations of the principle. Which version should be relevant in the context of the SPS Agreement? This problem is well exemplified in the parties' submissions as none of them has provided any clear and coherent definition of the principle.

5.3 Direct relevance of the precautionary principle in the context of the SPS Agreement

It has become a common practice in SPS disputes to claim that the precautionary principle is relevant for the interpretation of different provisions of the SPS Agreement. On the other hand, the WTO settlement bodies have consistently held that the status of the principle in international law remains unsettled. This means that it could not be taken into account when interpreting WTO rules. At the same time, both the panels and the Appellate Body, without identifying the precautionary principle as a rule of international law, also held that the

[48] See eg, Appellate Body Report, *US – Gasoline*, 17; Appellate Body Report, *India – Patents (EC)*, para 46; Appellate Body Report, *Japan – Alcoholic Beverages II*, 10.

[49] Article 38 of the Statute of the International Court of Justice.

idea of precaution was actually reflected in the number of provisions of the SPS Agreement.

EC–Hormones was the first case where the relevance of the precautionary principle within the SPS Agreement was addressed. The position of the EC remains somewhat obscure. On the one hand, it argued that the precautionary principle had become a rule of customary international law or at least a general principle of law.[50] According to the EC, the application of the principle in the context of the SPS Agreement would produce the following consequences: (i) there is no need for consensus between scientists on the possibility and magnitude of risk; (ii) WTO Members may perceive and evaluate risk in a different manner; (iii) Article 5.1 to 5.2 of the SPS Agreement does not require any particular risk assessment and does not 'prevent Members from being cautious in their risk assessment exercise'.[51] On the other hand, the EC failed to explain how under point (ii) the precautionary principle would affect the specific requirements provided in the SPS Agreement. It is also unclear whether the EC saw the principle as directly applicable under the SPS Agreement or rather as an interpretive guideline in accordance with Article 31.3(c) of the VCLT.

The *Hormones* panel, when discussing the precaution and its relevance under the SPS Agreement, did not really address the specific arguments advanced by the EC and broadly stated that the principle could not override the explicit language of the SPS Agreement. Consequently, it seems that the panel understood the EC as arguing for direct application of the principle. At the same time, the panel noted that the principle found its explicit reflection in the SPS Agreement and identified Article 5.7 as embodying the idea of precaution.[52] In principle, the Appellate Body agreed with the panel. First, it observed that the status of the precautionary principle as a general principle or rule of customary international law was unclear. Although the Appellate Body refrained from taking the position on that issue, it found that the question on the status of the principle in international law was an abstract problem without any consequences for the dispute. It also added that 'the precautionary principle has not been written into the *SPS Agreement* as a ground for justifying SPS measures that are otherwise inconsistent with the obligations of Members set out in particular provisions of that Agreement'.[53] This closely corresponded with the approach of the panel, as the Appellate Body also saw the issue of the precautionary principle primarily in terms of the conflict between different rules on international law rather than as a problem of interpretation.[54] Finally, the Appellate Body also held that without clear textual formulation in the SPS Agreement, the precautionary principle did

[50] Appellate Body Report, *EC–Hormones*, para 16.

[51] Ibid, paras 16 and 24.

[52] Panel Report, *EC–Hormones (US)*, para 8.157.

[53] Appellate Body Report, *EC–Hormones*, para 124.

[54] Cf Robert Howse, 'The Use and Abuse of Other "Relevant Rules of International Law" in Treaty Interpretation: Insight from WTO Trade/Environment Litigations', IILJ Working Paper 2007/1, 7.

not 'relieve a panel from the duty of applying normal (ie customary international law) principles of treaty interpretation'.[55] As already mentioned, those rules are reflected in Articles 31 and 32 of the VCLT.

The approach of the Appellate Body is confusing. Leaving aside the problem of direct applicability of the precautionary principle (as the position of the EC on that issue was also unclear), the conclusion reached by the Appellate Body appears to be contradictory. Note that, on the one hand, it refused to assess the status of the precautionary principle under international law. Yet on the other hand, it instructed the panel to apply normal rules of treaty interpretation. As mentioned above, Article 31 of the VCLT mandates to take into account in the process of interpretation relevant rules of international law applicable in the relations between the parties. This also includes both rules of customary international law and general principles of law. Consequently, the status of the precautionary principle in international law could hardly be seen as an abstract question without relevance for the outcome of the interpretation.

Despite the fact that the Appellate Body refused to recognize the relevance of the precautionary principle in the context of the SPS Agreement, it did address some of the specific concerns raised by the EC. Thus, the Appellate Body agreed with the EC that a measure could be based on both mainstream and minority scientific opinion as the SPS Agreement does not require a consensus between scientists on the possibility and magnitude of risk before an action can be taken by a WTO Member.[56] The Appellate Body also confirmed that to some extent WTO Members may perceive and evaluate risk in a different manner. Specifically the governments may 'act from perspective of prudence and precaution where risks of irreversible, eg life-terminating, damage to human health are concerned'.[57] This also means that Article 5.1 to 5.2 does not prevent WTO Members from being cautious in their risk assessment. On the other hand, both the panel and the Appellate Body emphasized that the SPS Agreement requires from Members a particular risk assessment, namely the assessment which complies with the definition of Annex A(4).

The same problem resurfaced 10 years later in *EC – Biotech Products*. The EC again argued that the precautionary principle was a general principle of international law.[58] The main part of the EC argument was, however, of a very general nature with only limited references to any specific obligations of the SPS Agreement. This may suggest that either the EC saw the precautionary principle as an element that was relevant in the interpretation of all provisions of the SPS

[55] Appellate Body Report, *EC – Hormones*, para 124.

[56] For a detailed discussion on the position of scientific minority opinions, see Chapter 4, section 4.4.2(b).

[57] Appellate Body Report, *EC – Hormones*, para 124; for a more detailed discussion, see section 5.4.5 below.

[58] Panel Report, *EC – Biotech Products*, para 4.523 (as noted by the panel, the EC used somewhat imprecise language since the theory and practice of international law refers to rules of international customary law and general principles of law recognized by nations).

Agreement, or that it was unsure how the principle may affect the SPS rules.[59] Nevertheless, closer analysis allows the identification of at least some implications which, according to the EC, resulted from recourse to the principle. First, the EC argued that the precautionary principle was to be taken into account when assessing undue delay with respect to specific GMO applications as provided in paragraph 1(a) of Annex C.[60] Secondly, the EC claimed that the precautionary principle required recognizing Article 5.7 as an autonomous right and not an exception.[61] Thirdly, the EC expected a panel to adopt a deferential standard of review under Articles 2.2/5.1 and Article 5.7,[62] which would ensure greater discretion for national governments in their assessment of scientific evidence.

The panel essentially followed the reasoning of the Appellate Body in *EC – Hormones*. First, it noted that there was still disagreement on the status and the meaning of the precautionary principle in international law. The panel recognized that the principle was reflected in numerous international environmental agreements as well as frequently applied in the domestic context of some WTO Members. However, the panel also observed that the content of the principle remained unclear, while a number of authors questioned its status.[63] On that basis, it concluded that 'prudence suggests . . . not [to] attempt to resolve this complex issue' and added that there was no need to decide on this issue for the purpose of the case.[64] At the same time, the panel also accepted, without making a reference to the principle, some of the specific claims made by the EC. In particular, it found that Article 5.7 was indeed an autonomous right and that the burden of making a prima facie case of inconsistency rested on the complainants.[65] It also held that WTO Members enjoyed a certain level of discretion under Article 5.1, meaning that they could respond to uncertainties identified in a risk assessment in different ways, for example by adopting more precautionary (strict) measures than other Members. Although the panel did not make any findings with respect to Article 2.2, it seems that its approach would have been the same. Finally, the panel agreed with the EC that the language of the first clause of Annex C(1)(a) did not exclude the application of a prudent or precautionary approach. This flexibility is implied in the notion of reasonable time which permits WTO Members to take time that is reasonably needed to determine whether a relevant SPS requirement is met. Those Members that prefer the precautionary approach may request some additional information or clarifications.[66] This, however, does not mean that there are no limits under the SPS Agreement. On the contrary, a WTO Member cannot endlessly delay its decision only because it relies on the

[59] Cf ibid, para 4.339.

[60] Ibid, para 4.1242 (Annex C(1)(a) requires that 'procedures [control, approval and inspection] are undertaken and completed without undue delay and in no less favourable manner for imported products than for like domestic products').

[61] Ibid, para 4.523.

[62] Eg ibid, paras 7.2620 and 7.3065.

[63] Ibid, para 7.88.

[64] Ibid, para 7.89.

[65] See section 5.4.4 below.

[66] Panel Report, *EC – Biotech Products*, para 7.1522.

precautionary approach. Instead, a Member is obliged to take a decision at a certain moment in time which may be based either on a risk assessment as provided by Article 5.1 or be provisional in accordance with the disciplines of Article 5.7 (if it considers that there is insufficient evidence to complete the approval process).[67] Consequently, the precautionary approach could not provide a justification for delay in completing the approval procedure.[68]

Both reports leave the reader with ambiguous feelings. From the legal perspective, finding the status of the precautionary principle as an abstract question, without any relevance to the dispute, is deeply unsatisfactory. As explained above, if the principle has become a part of customary international law, the WTO dispute settlement bodies are obliged to take it into account when interpreting the provisions of the SPS Agreement. The approach of the SPS jurisprudence can probably be explained by political considerations. Express denial of the precautionary principle as a rule of international customary law could undermine the legitimacy of the organization as such as it would constitute yet more proof of favouring international trade over any other values. On the other hand, recognizing the principle as relevant could be unacceptable for some Members of the WTO (eg the US). These concerns, however, do not change the assessment of the Appellate Body approach from the DSU point of view as it fails to meet the standard provided by Article 31 of the VCLT. Arguably, it was possible to handle that issue in a more skilful manner. For instance, the Appellate Body could recognize the precautionary principle as an evolving principle of international law which has not yet attained the status of a customary rule of international law. Such an approach would allow the avoidance of analysing the impact of the precautionary principle on the disciplines of the SPS Agreement (as an evolving principle, it will be irrelevant from the perspective of the VCLT) and might also be politically acceptable for the parties of the dispute. (Although the principle would not be relevant in *EC – Hormones* this will not preclude its application in a future dispute.) At the same time, one has to recognize that in practice, both the Appellate Body and the *Biotech* panel endorsed the interpretation of specific SPS provisions that appears to take the principle into account. Accepting, in the context of Article 5.1, minority scientific opinions as a legitimate basis for SPS measures or recognizing flexibility that is inherent in the assessment of risk (eg permitting the application of the precautionary approach in response to identified uncertainties) seem to reflect the idea that is incorporated into the precautionary principle. Consequently, it seems that both the Appellate Body and the panel de facto treated the principle as a norm of international customary law, without, however, openly admitting that fact. It is at least disputable whether any different result (ie less stringent) would be achieved with the explicit recognition of the principle in the context of the SPS Agreement.

[67] Ibid, para 7.1525.

[68] Ibid, para 7.1529.

The other legal problem addressed by the panel in *EC – Biotech Products* was the recourse to the precautionary principle through the provisions of the CBD and the CPB. The panel, after adopting a very narrow interpretation of Article 31.3(c) of the VCLT, found that the expression 'applicable in the relations between the parties' required the treaty (ie the CPB) to be applicable to all WTO Members in order to become relevant for the purpose of interpretation.[69] The panel based its finding on three arguments. First, it made textual argument and found that the VCLT used the expression 'parties' in reference to the parties of a treaty and not to a dispute.[70] Secondly, the panel emphasized that its approach was motivated by the desire to 'ensure . . . or enhance . . . the consistency of the rules of international law applicable to these States'.[71] However, this consistency was understood as concerning only WTO law. Thirdly, it referred to the consent argument and stressed that the parties could not be bound by the rules which they did not expressly accept.[72]

This position was, however, mitigated as the panel noted a few lines later that its interpretation did not exclude the possibility of considering rules of other treaties which are applicable only to the parties of a particular dispute, if all of them argue that a particular WTO agreement should be interpreted in the light of these other rules of international law.[73] The structure of the panel reasoning seems be logically flawed. Its earlier requirement that all WTO Members have to be parties to a treaty logically excludes the possibility of accepting such a treaty as relevant even if it binds all parties to the dispute and those parties express their desire to use such rules in the interpretation of WTO provisions. Note also that this second finding of the *Biotech* panel is incompatible with at least two arguments that were advanced in support of the narrow interpretation of Article 31.3(c) of the VCLT (ie textual basis in the VCLT and consistency in the application of WTO law).

In this context, it is also worth noting that the arguments put forward by the panel were not entirely persuasive. As noted in the literature, the text of the VCLT is ambiguous and may equally support the conclusion that the expression 'parties' as used in Article 31.3(c) may refer to the parties to a dispute.[74] McGrady particularly notes that this expression is used in the VCLT in at least three different ways: as parties to a dispute, all parties to a treaty, and some subset of the parties. There are also 12 instances where the reference to 'all parties' is made. If the notion of 'parties' as used in Article 31.3(c) indicate all the parties, the second expression would become superfluous.[75] This leads to the conclusion, contrary to the

[69] Ibid, para 7.70; these findings also correspond closely to the approach taken by one of the GATT panels (see, GATT Panel Report, *US – Tuna/Dolphin II*, para. 5.19).

[70] Panel Report, *EC – Biotech Products*, para 7.68.

[71] Ibid, para 7.70.

[72] Ibid, para 7.71.

[73] Ibid, para 7.72.

[74] Howse ('Use and Abuse') 24.

[75] Benn McGrady, 'Fragmentation of International Law or Systemic Integration of Treaty Regimes: EC – Biotech Products and the Proper Interpretation of Article 31(3)(c) of the Vienna Convention on the Law of Treaties' (2008) 42(4) Journal of World Trade 589, 594–5.

panel's finding, that the text of Article 31.3(c) does not unequivocally determine the meaning of the term. Secondly, as was convincingly argued by Pauwelyn,[76] WTO agreements mainly consist of bilateral (or reciprocal), rather than collective, obligations.[77] The object of WTO law (trade which remains bilateral business), its origin (concessions negotiated on the bilateral basis), objective (trade liberalization which is not genuine collective interest), and the enforcement procedure (bilateral enforcement, including bilateral suspensions) indicate that WTO obligations are predominantly of a bilateral character. The SPS Agreement is not the exception. Although the SPS Agreement 'sets out minimum standards that WTO Members must meet, irrespectively of their actual or potential trade impact in (bilateral) relation with other WTO Members'[78] this does not change the bilateral character of its obligations. The breach of the SPS Agreement affects one or more WTO Members and not the collective obligations owed to the entire WTO membership. As summarized by Pauwelyn 'the mere fact that they [SPS obligations] are regulatory in nature—and thus substantially the same bundles for all WTO Members—does not suddenly transform them into constitutional-type norms in the interest of some "global common" transcending the sum total of individual interests of states concerned'.[79] Consequently, if one subscribes to the view that WTO rules (including SPS rules) are bilateral obligations, divergence in interpretation does not create the problem that the panel was concerned with. Countries, as actors in international relations, may take up different obligations arising from different treaties and conventions. To separate one set of bilateral rights and obligations and disregard others (vis-à-vis another country) means to interfere with the sovereign decisions of the states that agreed to be bound by a number of different international rules.[80] This observation is also connected with the third argument used by the panel (ie requirement of consent). It seems that the panel did not sufficiently analyse its argument. Note that *EC – Biotech Products* was only technically a single dispute between four parties. Legally speaking there were three disputes (*EC v Canada*, *EC v US*, and *EC v Argentina*) merged into one proceeding. The DSU makes clear that a single panel examining different complaints has to organize its examination and present its findings to the dispute settlement body (DSB) in such a manner that the rights which the parties to the

[76] Joost Pauwelyn, 'A Typology of Multilateral Treaty Obligations: Are WTO Obligations Bilateral or Collective in Nature?' (2003) 14(5) European Journal of International Law 907.

[77] Cf Chios Carmody, 'WTO Obligations as Collective' (2006) 17(2) European Journal of International Law 419.

[78] Pauwelyn ('Typology') 938.

[79] Ibid, 939–40; McGrady also notes that Article 31.3(c) lays down general interpretative rules as well as interpretative rules for resolution of disputes. This means that the phrase 'the parties to a dispute' would not be appropriate in the context of the provision.

[80] Cf also International Law Commission, *Fragmentation of International Law: Difficulties Arising from the Diversification and Expansion of International Law, Report of the Study Group of the International Law Commission*, A/CN.4/L.682 (2006), para 472 ('although this creates a possibility of eventually divergent interpretation that would simply reflect the need to respect (inherently divergent) party will').

dispute would have enjoyed had separate panels examined the complaints, are in no way impaired.[81] Thus, the panel should rather have addressed the question of consent in relation to the EC and each of the complainants. Although none of the complainants was a party to the CPB, both Canada and Argentina were parties to the CBD. Moreover, if one conceptualizes WTO obligations as bilateral, the problem of consent only becomes relevant with regard to parties to the specific dispute.

The panel also forgot to discuss an additional argument that is sometimes advanced in the literature and seems to support the panel's position. Some scholars are concerned that a broad reading of Article 31.3(c) of the VCLT will result in great uncertainty as to the nature of WTO rights and obligations and will reduce the precedential character of Appellate Body rulings. Although this is an important objection, it should not be overstated. The case law shows only a small number of instances where the relevance of extraneous treaty rules was argued by the parties.[82] This, in turn, could mean that the risk of reducing the precedential character of Appellate Body decisions may be limited.[83] The risk is further minimized through the application of traditional mechanisms elaborated by the WTO case law, such as recognizing particular provisions of the extraneous treaty as a reflection of customary international law, or considering a treaty norm as a form of legal dictionary.[84] Under such interpretation, the particular rule is applicable to all WTO Members.

The conclusion reached by the panel is also problematic as it transforms WTO law into a quasi self-contained regime. Note that the narrow interpretation of Article 31.3(c) practically excludes the possibility of taking into account any international treaty when interpreting the SPS Agreement (or WTO law if the general case law will subscribe to this approach). There are no legal instruments with the same composition as the WTO (eg some WTO Members, such as Hong Kong, are not even states but separate custom units). These are quite serious consequences. As recently noted by the International Law Commission, such an approach is 'contrary to the legislative ethos behind most of multilateral treaty-making and, presumably, with the intent of most treaty-makers'.[85] One would intuitively expect the panel to consider alternative approaches and engage in a discussion on their advantages and disadvantages. Note that there are at least two other options. First, one may interpret the term 'parties' as parties to the dispute (the broad approach). In such cases, the WTO dispute settlement bodies will be obliged, on the basis of Article 31.3(c) of the VCLT, to consider

[81] Article 9.2 of the DSU.

[82] Of course, it may be argued that the litigation strategies of parties result from the approach of the dispute settlement bodies.

[83] Michael Lennard, 'Navigating by the Stars: Interpreting the WTO Agreement' (2002) 5(1) Journal of International Economic Law 17, 38.

[84] Cf Chapter 2, section 2.3.

[85] Cf International Law Commission (*Fragmentation*) para 472.

rules of other international treaties in the process of interpretation of WTO rules. As discussed above, the textual basis does not exclude such an interpretation. Moreover, other arguments which were advanced by the panel are at least debatable whereas a number of considerations such as the bilateral character of WTO obligations appear to support the broad approach. Secondly, one may also take a middle ground and accept, as relevant in the process of interpretation, those rules which reflect the common understanding between the parties of a particular international agreement, irrespective of whether such rules are legally binding on all parties to the treaty.[86] Note that the middle ground position guarantees the consistency of WTO rules and maintains the precedential character of the Appellate Body decision. At the same time, it improves the unity of international law as it allows the adoption of an integrated interpretative approach. Although both alternatives pose their own problems (eg the first approach reduces the precedential character of Appellate Body rulings, while the second undermines the role of consent), the *Biotech* panel was not interested in any deeper analysis that would consider the broader consequences of its approach and simply promoted a narrow reading of Article 31.3(c) as the only defensible option.

The relevance of the precautionary principle was also examined in *Japan – Agricultural Products* and *Japan – Apples*. In the first case, Japan argued on its appeal that the principle mandated the panel to apply a deferential standard of review when assessing the rational relationship between scientific evidence and a measure as provided by Article 2.2. (Japan labelled the standard used by the panel as 'actual causal link'[87] as opposed to a more precautionary approach that would ask for a merely rational relationship.) At the same time, Japan did not specify what threshold would be appropriate for the assessment of such a relationship. Due to this lack of specificity in the Japanese argument, it is not surprising that the Appellate Body rejected it and simply repeated that Article 2.2 required a standard of rational relationship that was actually applied by the panel. At the same time, the Appellate Body recalled its observations from *EC – Hormones* that the precautionary principle had not been written into the SPS Agreement as a ground for justifying measures which are otherwise inconsistent with its provisions.[88]

In the second case, Japan argued that the application of the precautionary principle in the context of the SPS Agreement obliges the panel to 'take into account "the need of caution emphasized by the experts" with respect to the phytosanitary measure aimed at preventing the entry of fire blight into Japan'.[89]

[86] McGrady ('Fragmentation') 592; Howse ('Use and Abuse') 35. Note, however, that it may be argued that such a reading of Article 31.3(c) is incompatible with the text of the VCLT. It also raises the problem of a party's consent. This drawback may be somewhat overcome by examining the extent to which that other treaty was implicitly accepted or at least tolerated by other parties.

[87] Appellate Body Report, *Japan – Agricultural Products*, para 9.

[88] Ibid, paras 81–3.

[89] Appellate Body Report, *Japan – Apples*, para 50.

That caution, according to Japan, required the panel to recognize the possibility of entrance of fire blight together with the importation of apple fruits, despite the absence of evidence confirming the validity of the pathway. In other words, Japan argued that the application of the precautionary principle would require a panel to give more weight to those expressions of caution. The Appellate Body disagreed and stressed that the experts, rather than indicating uncertainties relating to the entrance of fire blight through the importation of apples from the United States, warned against the complete removal of the control system.[90] The Appellate Body also added that the panel actually took those cautious opinions into account and was aware of the effect of total removal of phytosanitary controls.[91] It seems that Japan's argument was somehow misplaced. Japan wanted to translate the caution expressed by the experts advising the panel into evidence of the existence of uncertainty, which in turn would legitimize, on the basis of the precautionary principle, the Japanese regimes for apples. However, as noted by the Appellate Body, the experts cautioned against the removal of all phytosanitary controls rather than identified areas of uncertainty. That seems to be a separate issue relating more to the problem of how Japan might comply with the WTO ruling (ie remove only some measures or all of them) than an actual justification for the existing system. The report also suggests that despite the inability of Japan to prove the exact pathway for fire blight, some control measures would have been acceptable under WTO law as a reflection of a prudent and precautionary approach. Consequently, what was condemned was the strictness of the measure and not a precautionary approach as such.

In order to clarify the content of the precautionary principle, the EC Commission also issued a special communication explaining in detail its approach to the principle and attempting to promote a certain 'common understanding of how to assess, appraise, manage and communicate risks that science is not yet able to evaluate fully'.[92] The precautionary principle was conceptualized as a risk management tool which allows decision makers to take decisions in the face of scientific uncertainty.[93] The Communication was clearly influenced by the language of the SPS Agreement. According to the Communication, measures adopted on the basis of the precautionary principle are required to be proportional to the ALOP (weak proportionality, meaning that a measure cannot be disproportionate to the desired level of protection), non-discriminatory, consistent with other measures,

[90] Ibid, para 236.

[91] Ibid, para 237 (citing the part of the panel report which provides that 'even if the scientific evidence before us demonstrates that apple fruit is highly unlikely to be a pathway for entry, establishment and spread of fire blight within Japan, it does suggest that some slight risk of contamination cannot be totally excluded.... none of the experts were comfortable with the notion of eliminating "in one step" all phytosanitary controls, taking into account Japan's island environment and climate').

[92] EC Communication (Precautionary Principle) 3.

[93] The Communication distinguishes, for example, between the precautionary principle and the elements of cautions which are applied by scientists in the process of risk assessment (ibid).

including measures taken on the basis of a proper risk assessment, and subject to a subsequent review. Although the Communication indicates that such measures should be based on examination of the potential costs and benefits, it does not introduce any form of strong proportionality test, in the sense of balancing between a measure and potential economic costs. The Communication requires prior identification of potentially negative effects resulting from a product (phenomenon or process) and subsequent assessment of available scientific data. On that basis, one may decide that scientific evidence does not allow the assessment of risk with sufficient certainty. The Communication acknowledges that certain types of uncertainty can be addressed in the process of risk assessment by adopting the so-called prudential approach (eg use of safety factors). At the same time, the Communication stressed that in some situations the deficiency of scientific data is such that the application of the prudential approach is insufficient. This is, according to the Communication, a situation which justifies recourse to precautionary principle.[94] The Communication does not explain who ultimately takes a decision as to whether a body of evidence is sufficient or insufficient (albeit in some parts it suggests that this is a risk manager prerogative).[95] Interestingly, the Communication stipulates that the precautionary principle is only triggered if, after evaluation of scientific evidence, it appears that the desired level of protection for the environment or a specific population group could be jeopardized. Thus, contrary to the other formulations of principle, the Communication does not refer to serious and irreversible damage, arguably making it easier to rely on the principle.[96]

Although the Communication is definitely helpful in clarifying the content of the principle and its possible application in the context of the SPS Agreement, it fails to answer a number of important questions. First, it is difficult to understand why the regime provided by the SPS Agreement (including Article 5.7) is not sufficient to address the problems articulated in the Communication. The description of the principle is generally compatible with the language of the Agreement. The Communication requires prior scientific assessment of potential risks (as full as possible), condemns arbitrary decisions, and calls for proportionality, consistency, and non-discrimination. The recourse to the precautionary principle is available when scientific information is insufficient, inconclusive, or uncertain, to such extent that it cannot be addressed within risk assessment, and when there are some indications of the possible effects on human health or the environment. Note that all these requirements closely correspond to the disciplines of the SPS Agreement, particularly those of Article 5.7. Arguably, the possible difference may relate to uncertainty. As will be discussed in more detail in section 5.4.2, some types of

[94] Ibid, 15.

[95] Ibid, 17 ('an assessment of... the uncertainties of the scientific evaluation should be considered by decision-makers when determining whether to trigger action based in the precautionary principle').

[96] Cf Majone ('What Price Safety?') 94.

uncertainty are covered by the SPS Agreement, while others fall outside its ambit. Therefore, one may argue that the Communication suggests some broader inclusion of uncertainties. If so (and since the Communication does not specify any required threshold with regard to uncertainty), the recourse to the principle could be practically unlimited and arbitrary. If this is not the case, the Communication fails to explain how the proposed precautionary principle differs from any risk decision which is almost always taken without complete scientific certainty. This brings us to the second issue. Surprisingly, the Communication actually nowhere defines the precautionary principle. It enumerates the triggering factors (albeit also without clearly specifying their content), it describes the rules of its application but, at the same time, it does not elucidate the exact legal meaning of the principle. This seems to be somewhat surprising in a document which aims to clarify the content of the principle. Thirdly, the Communication frequently employs the notion of risk with reference to potential risk, that is 'risk' which has not yet been verified. This creates a false impression that the precautionary principle addresses something actually existing (which might be the case in some instances). However, as already discussed, the principle is concerned with potential risks, a possibility of which can be neither excluded nor confirmed as a probable outcome. At the end of the day one gets a similar impression as when reviewing the arguments of the parties in SPS disputes. The normative content of the principle remains only vaguely defined while its potential impact on the SPS disciplines is far from clear.

5.4 Precaution in the SPS Agreement

The Appellate Body in *EC – Hormones* noted that the precautionary principle found its explicit reflection in Article 5.7 of the SPS Agreement. It also agreed with the EC that this fact did not exhaust the relevance of the precaution, and other provisions of the SPS Agreement could also be seen as embodying this idea. This section of the chapter will analyse the relevant provisions in more detail. The main emphasis will be on Article 5.7 which appears to be a central precautionary rule in the context of the SPS Agreement.

An SPS measure will conform to Article 5.7 only if four cumulative requirements are met: (i) there is insufficiency of scientific evidence for performance of risk assessment; (ii) an SPS measure is adopted on the basis of available pertinent information; (iii) a WTO Member seeks to obtain additional scientific data necessary for a more objective assessment of risk; and (iv) a provisional SPS measure is subject to review within a reasonable period of time.[97] The Appellate Body, when analysing Article 5.7, also distinguished between obligations that are concerned with the adoption of an SPS measure (the first sentence) and those which relate

[97] See Appellate Body Report, *Japan – Agricultural Products*, para 89; Appellate Body Report, *Japan – Apples*, para 176; Panel Report, *US – Continued Suspension*, para 7.593.

to the maintenance of a measure (the second sentence).[98] In another case, the Appellate Body also observed that all four elements 'must be interpreted keeping in mind that the precautionary principle finds reflection in this provisions'.[99] This corresponded with its earlier statement made in the context of Article 2.2 that 'a panel... should, bear in mind that responsible, representative governments commonly act from perspectives of prudence and precaution where risks of irreversible, eg life-terminating, damage to human health are concerned'.[100] The extent of the impact that the principle may have on the interpretation of the requirements of Article 5.7 remains, however, obscure.

5.4.1 Article 5.7 in the context of the SPS Agreement

Before going into the substance of Article 5.7, it is worth analysing its position in the broader framework established by the SPS Agreement. A proper qualification of Article 5.7 will ultimately determine the rights and obligations that are enjoyed by WTO Members in situations of insufficiency of scientific evidence. The analysis which is presented below is divided into two parts. The first addresses the general question of Article 5.7 applicability, focusing on the relationship between Articles 5.7 and 2.2/5.1 to 5.3. The second part discusses the position of Article 5.7 vis-à-vis other SPS provisions. The aim of this part is to identify those disciplines of the SPS Agreement which are relevant for SPS measures falling under Article 5.7.

(a) Applicability—Relationship between Articles 2.2/5.1 and 5.7

As discussed in Chapter 4, section 4.2, Article 2.2 of the SPS Agreement mandates WTO Members to base SPS measures on scientific principles and prohibits their maintenance without sufficient scientific evidence. Recall also that the SPS case law conceptualized Article 5.1 as a 'specific application of the basic obligations contained in Article 2.2',[101] which 'may be seen to be marking out and elaborating a particular route leading to the same destination set out in Article 2.2'.[102] In other words Article 5.1 provides a more detailed way of complying with the requirement of a scientific basis as incorporated in Article 2.2. Article 2.2 itself allows WTO Members to deviate from science-based disciplines in all situations that fall into the ambit of Article 5.7 ('except as provided for in paragraph 7 of Article 5'). Consequently, Article 5.7 may be seen as the exemption from the requirements established by Articles 2.2[103] and 5.1.[104]

98 Appellate Body Report, *Japan – Agricultural Products*, para 89.
99 Appellate Body Report, *US – Continued Suspension*, para 680.
100 Ibid, para 181.
101 Panel Report, *Japan – Agricultural Products*, para 82.
102 Panel Report, *Australia – Salmon*, para 8.52.
103 Appellate Body Report, *Japan – Agricultural Products*, para 89; but see Panel Report, *Australia – Salmon*, para 8.57 (referring to Article 5.7 as to the exception); see below section 5.4.4 below.
104 Panel Report, *EC – Biotech Products*, para 7.2998.

The text of the Agreement shows that Articles 2.2 and 5.1 and Article 5.7 establish two distinctive sets of obligations which are applicable to different situations. Under Articles 2.2/5.1, an SPS measure needs to be based on scientific principles and cannot be maintained without sufficient scientific evidence (ie in the form of scientific risk assessment). On the other hand, Article 5.7 permits, in certain situations which are characterized by insufficiency of scientific evidence, adoption and maintenance of measures that are not based on scientific evidence/formal risk assessment. Regrettably, this simple distinction was unnecessarily complicated by the SPS case law.

The relationship between these two sets of obligations was discussed comprehensively in the *Biotech* dispute. The panel followed the approach of the previous case law[105] and held that Article 5.7 is applicable in every case where relevant scientific evidence is insufficient.[106] In other sections of the report, the *Biotech* panel noted that 'first it would need to be established that... measures were adopted in respect of situations where relevant scientific evidence was insufficient'.[107] According to the panel, such an approach was required by the textual structure of Article 5.7 which provides the 'if-then' rule of conditionality (if p, then q).[108] On this basis, the panel concluded that in all cases where the scientific evidence is insufficient a WTO Member may adopt a provisional measure. This finding allowed the panel to rightly reject the EC argument that the demarcation line between Articles 2.2/5.1 and 5.7 was the provisionality of the SPS measure.[109] In this context, the panel also introduced a distinction between applicability and consistency. According to the panel, the first is concerned with the question of what triggers the application of Article 5.7, whereas the second relates to the conditions which must be met before finding that the SPS measure complies with the requirements of a particular article.[110] Provisionality was seen by the panel as one of the consistency requirements and not as a determinant of applicability.

The panel also discussed whether Article 5.7 is to be regarded as an exception from the general obligation of Articles 2.2/5.1 or rather an independent right of WTO Members. Although this issue was primarily relevant for the allocation of the burden of proof between the parties to the dispute, it also had an implication

[105] Appellate Body Report, *Japan – Apples*, para 184.

[106] Panel Report, *EC – Biotech Products*, para 7.2939 (noting that 'Article 5.7 is applicable whenever the relevant condition is met, that is to say, in every case where relevant scientific evidence is insufficient').

[107] Ibid, para 7.2944.

[108] Ibid, para 7.2939 (panel understood the if–then rule of conditionality as follows: 'if a certain condition is met (*in casu*, insufficiency of relevant scientific evidence), a particular right is conferred (*in casu*, the right provisionally to adopt an SPS measure based on available pertinent information').

[109] Ibid, para 7.2939; in this context the panel also noted that 'if the provisional adoption of an SPS measure had been intended as a condition for the applicability of Article 5.7, the first sentence of Article 5.7 would, in our view, have opened with a different phrase, such as "In cases where a Member provisionally adopts an SPS measure"' (ibid, para 7.2940).

[110] Ibid, para 7.2942.

for the question of applicability. The panel decided to base its analysis on the test that was developed by the Appellate Body in *EC – Tariff Preferences*. Pursuant to the test, a particular provision may be characterized as an independent right rather than exception if it meets three cumulative conditions: (i) one provision permits what would otherwise be inconsistent with an obligation of another provision; (ii) there is a reference between these two provisions; and (iii) one of the provisions suggests that its obligation is not applicable to a measure.[111] The *Biotech* panel found that Article 5.7 'permits, in certain circumstances, behaviour... that would otherwise be inconsistent with... the obligation of Article 2.2'.[112] Two other conditions were also met, as the panel was able to identify the existence of reference ('except as provided for in paragraph 7 of Article 5') and the exclusion of applicability ('except as'). In support of this conclusion, the panel also referred to other cases where the same relationship was identified. After finding Article 3.1/3.3 of the SPS Agreement as one of the examples, the panel observed that the language used in Articles 3.1 and 2.2 was substantially the same ('except as otherwise provided' v 'except as provided for'). Since Article 3.3 was characterized in the case law as the right of a WTO Member, the consistent interpretation of the SPS Agreement required the same conclusion with respect to Article 5.7.[113] Consequently, it was for the complainants to demonstrate that the EC SPS measures were inconsistent with at least one of the four requirements set forth in Article 5.7. The same approach was followed with regard to the relationship between Article 5.1 and 5.7. The panel again referred to the test of *EC – Tariff Preferences* and found that Article 5.7 permitted in certain circumstances behaviour otherwise inconsistent with Article 5.1,[114] there was a reference between two provisions,[115] and an implied suggestion that only one provision might be applied to a single measure.[116] This led the panel to conclude, similarly to the case of Articles 2.2/5.7, that the initial burden of proof was borne by complainants.[117]

[111] The Appellate Body particularly held that 'in cases where one provision permits, in certain circumstances, behaviour that would otherwise be inconsistent with an obligation in another provision, and one of the two provisions refers to the other provision,... the complaining party bears the burden of establishing that a challenged measure is inconsistent with the provision permitting particular behaviour *only* where one of the provisions suggests that the ***obligation is not applicable*** to the said measure. Otherwise, the permissive provision has been characterized as an exception, or defence, and the onus of invoking it and proving the consistency of the measure with its requirements has been placed on the responding party' (Appellate Body Report, *EC – Tariff Preferences*, para 88) (emphasis on *only* in the original; other emphasis added).

[112] Panel Report, *EC – Biotech Products*, para 7.2968.

[113] Ibid, para 7.2967.

[114] Ibid, para 7.2993 (basing its finding on the textual analysis of both provisions).

[115] Ibid, para 7.2994 (finding, on the basis of textual analysis of the phrases 'more objective risk assessment' and 'insufficiency of scientific evidence', that Article 5.7 contains implicit reference to Article 5.1).

[116] Ibid, paras 7.2995–6.

[117] Ibid, para 7.3000 (finding that 'in cases where a complaining party alleges that an SPS measure is inconsistent with Article 5.1, it is incumbent on the complaining party, and not the responding party, to demonstrate that the challenged measure is inconsistent with at least one of the four requirements set forth in Article 5.7').

The panel report up to this moment is more or less clear.[118] Following the logic of applicability, one would expect that after establishing the applicable provision, non-compliance with one of its requirements would not remove a measure from the scope of such a provision. This would also be in line with the distinction made by the *Biotech* panel between applicability and consistency. As rightly observed by the EC, the applicability 'does not and cannot depend on whether or not… [a measure] is consistent with one or other substantive provisions of that Agreement'.[119] Thus, finding that evidence is sufficient should lead to the applicability of the regime provided by Articles 2.2/5.1. In such a case, Article 5.7 will be inapplicable rather than violated. The opposite is also true: finding that Article 5.7 is applicable makes the disciplines of Articles 2.2/5.1 irrelevant for the assessment of a measure. The WTO Member is still obliged to comply with other requirements of Article 5.7 but failure to meet one of them leads only to the violation of Article 5.7 and not to its inapplicability. Under such understanding of applicability, deciding what serves as a triggering factor is of prime importance. If it is provisionality, as argued by the EC, the mere fact that a measure is provisional will allow a Member to escape the disciplines of Articles 2.2/5.1. On the other hand, if the demarcation line is determined by insufficiency of scientific evidence, much more is required. This is probably the scenario the panel had in mind when it rejected the EC argument.

The panel approach was, however, different. After discussing the applicability of Article 5.7 and making a distinction between applicability and consistency, it added in the footnote that 'when we refer to applicability of Article 5.7 we address the issue of whether or not the right conferred by the first sentence of Article 5.7 is, in principle, available to a Member. In a specific case a Member must of course, satisfy the various requirements set forth in Article 5.7 if it wishes to benefit from the right conferred by Article 5.7'.[120] The expressions 'in principle' and 'benefit from the right conferred by Article 5.7' appeared to be of crucial importance. The panel, elaborating on these terms, noted that 'if a challenged SPS measure was adopted and is maintained consistently with the four cumulative requirements of Article 5.7, the situation is "as provided for in paragraph 7 of Article 5" (Article 2.2), and the obligation in Article 2.2 not to maintain SPS measures without sufficient scientific evidence is not applicable to the challenged measure'.[121] This meant to the panel that the violation of any requirement provided in Article 5.7 removes a measure from its scope and results in the applicability of Article 2.2. In

[118] But cf section 5.4.4.

[119] Second Submission of the EC in *EC – Biotech Products*, 19 July 2004, DS291, DS292, DS293, para 83.

[120] Panel Report, *EC – Biotech Products*, fn 1807.

[121] Ibid, para 7.2974; cf also the other paragraph of the panel report where it found that the ordinary meaning of the clause 'except as provided for in paragraph 7 of Article 5' in Article 2.2 indicates that Article 2.2 would be applicable in a situation where a measure meets some, but not all, of the requirements of Article 5.7 (ibid, para 7.2975).

other words, Article 2.2 is not applicable only if all four requirements set out in Article 5.7 are met by a measure.

The above reasoning is disappointing. First, it makes redundant the whole distinction between applicability and consistency. A measure needs to be consistent with all the requirements of Article 5.7 in order to be covered by Article 5.7. This also makes the very extensive discussion as to what constitutes a demarcation line between Articles 2.2 and 5.7 (provisionality as argued by the EC, or insufficiency of scientific evidence as ultimately found by the panel) meaningless. Does it really matter if a violation of any conditions of Article 5.7 results in the applicability of Article 2.2? Even if provisionality was a triggering factor, an SPS measure would need to comply with all other requirements provided by Article 5.7. Failure to meet one of them (including insufficiency of scientific evidence) would send a measure back to Article 2.2 and make Article 5.7 inapplicable.

Secondly, the logic of the panel's reasoning is even more surprising if one looks at the case law referred to by the panel. The Appellate Body, when examining the relationship between Article 3.1 and 3.3 in *EC – Hormones*,[122] reversed the panel's finding that 'the EC by maintaining without justification under Article 3.3 SPS measures which are not based on existing international standards, acted inconsistently with 3.1 of the SPS Agreement'.[123] This conclusion is important as it indicates that these two provisions simply apply to different situations.[124] Using similar language, the panel in *EC – Tariff Preferences* confirmed that 'Article 3.3 excludes the application of Article 3.1 of the SPS Agreement. Where a Member has projected for itself a higher level of sanitary protection than would be achieved by a measure based on international standards, Article 3.3 applies and Article 3.1 does not apply at all'.[125] This clearly shows that violation of Article 3.3 does not lead to the infringement of Article 3.1 as they simply govern different situations.

It is not clear why the panel did not follow the logic of Article 3.1/3.3 in *EC – Biotech Products*. Perhaps the panel was uncomfortable with the fact that existing case law permits the examination of conformity with Article 5.7 without prior assessment of its applicability. Note that the panel in *Japan – Agricultural Products*, when analysing a measure under Article 5.7, assumed insufficiency of scientific evidence and focused its analysis on the third condition (obligation to seek to obtain the additional information necessary for a more objective assessment of risk). When it found that the condition was not met it concluded that a Japanese measure violated Article 2.2 of the SPS Agreement.[126] On appeal, Japan argued

[122] The relationship which served as a model for the examination of Article 2.2 and 5.7 in *EC – Biotech Products*.

[123] Appellate Body Report, *EC – Hormones*, para 253.

[124] Ibid, para 104; as noted by the Appellate Body, Article 3.1 'simply excludes from its scope of application the kind of situations covered by Article 3.3'.

[125] Panel Report, *EC – Tariff Preferences*, para 7.50.

[126] Panel Report, *Japan – Agricultural Products*, paras 8.59 and 8.61.

that the phrase 'except as provided for in paragraph 7 of Article 5' referred only to the first sentence of Article 5.7 (which includes the requirement of insufficient scientific evidence). The Appellate Body rejected this argument[127] and upheld the finding that the SPS measure violated Article 2.2.[128] A similar approach was taken in *Japan – Apples*. The panel noted that it might have 'began ... examination with either the requirements of the first sentence or of the second sentence of Article 5.7'.[129] Although the Appellate Body expressed some reservations with respect to the allocation of the burden of proof in the panel analysis, it did not reverse the findings on applicability.

Thirdly, the *Biotech* panel's reasoning appears to be much more similar to another section of the Appellate Body report in *EC – Tariff Preferences*. It is worth citing the relevant passage in full. As noted by the Appellate Body, 'the Enabling Clause "does not exclude the applicability" of Article I:1 in the sense that, as a matter of procedure ... the challenged measure is submitted successively to the test of compatibility with the two provisions'.[130] The Appellate Body continued by saying that:

> [A]s a matter of final determination—or *application* rather than *applicability*—it is clear that only one provision applies at a time. This is what the Panel itself found when, after stating that 'as an exception provision, the Enabling Clause applies concurrently with Article I:1', it added 'and *takes precedence* to the extent of the conflict between the two provisions'.[131]

Note that this is exactly the type of analysis that was performed by the *Biotech* panel. The EC SPS measures were examined successively under two provisions: first under Article 5.1[132] and subsequently under Article 5.7. As a matter of final determination only one set of provisions applied at a time (ie Articles 2.2 and 5.1). If all the conditions of Article 5.7 had been met, Article 5.7 would have taken precedence over the requirements of Articles 2.2/5.1 and only this provision would apply. The surprising thing about this somewhat unconscious application of the reasoning by the *Biotech* panel is the fact that the analysis in *EC – Tariff Preferences* was concerned with the provisions that were characterized in terms of the general rule and the exception. This, however, goes against another finding of the *Biotech* panel which saw Article 5.7 as a positive obligation and not

[127] Appellate Body Report, *Japan – Agricultural Products*, para 90 (noting that 'Japan's proposition that the wording "except as provided for in paragraph 7 of Article 5" in Article 2.2 refers only to the first sentence of Article 5.7, and that a Member should, therefore, be allowed to claim exemption from the obligation under Article 2.2 when it fulfils the requirements of the first sentence, is without basis in the text of either Article 2.2 or Article 5.7').

[128] Ibid, paras 94 and 143.

[129] Panel Report, *Japan – Apples*, para 8.214.

[130] Appellate Body Report, *EC – Tariff Preferences*, para 102.

[131] Ibid (emphasis in original).

[132] See Panel Report, *EC – Biotech Products*, para 7.3396 (noting that 'by maintaining the challenged safeguard measures inconsistently with Article 5.1, the European Communities has, by implication, also acted inconsistently with the second and third requirements in Article 2.2').

as an exception to other provisions of the SPS Agreement (specifically Articles 2.2 and 5.1).[133]

Last but not least, the approach of the panel appears to be logically flawed. A hypothetical case may serve here as useful example. Imagine a situation where there is insufficient scientific evidence to assess a particular SPS risk. A WTO Member nevertheless adopts an SPS measure but fails to observe one of the remaining requirements of Article 5.7 (eg to review a measure in reasonable time or to seek to obtain additional scientific information). Under the panel analysis, such a measure will not be saved under Article 5.7 and will be sent back to Articles 2.2 and 5.1 (even if insufficiency still persists). Thus, a WTO Member will violate the obligation not to maintain an SPS measure without sufficient scientific evidence (Article 2.2) in a situation where there is no such evidence.[134] The same is true with respect to Article 5.1. As a consequence of the panel approach, WTO Members will violate the requirement of risk assessment even if the existing scientific evidence is insufficient to perform such an assessment.[135]

The more appropriate approach would be to see disciplines established by Articles 2.2/5.1, on the one hand, and these of Article 5.7, on the other hand, as two distinctive paths of compliance that are available to WTO Members. The first regime is applicable to those situations where there is enough scientific evidence to perform risk assessment. Consequently, a WTO Member is required to base its SPS measure on such evidence and corresponding risk assessment. The second path covers all those cases that are characterized by insufficiency of scientific evidence. In such situations a WTO Member may adopt provisional measures that comply with Article 5.7. The relationship between these two paths is one of exclusion, meaning that inconsistency of an SPS measure with a provision relevant for a particular path does not send a measure to the second regime. This means that where scientific evidence is insufficient, failure to meet some of the requirements of Article 5.7 does not result in the violation of Articles 2.2 (to base a measure on sufficient scientific evidence) and 5.1 (to have an appropriate risk assessment). Such an approach seems to be more logical compared to the current standard of the case law.[136] It preserves conceptualization of Article 5.7 as a right and is consistent with the test of *EC – Tariff Preferences*.

[133] Ibid, para 7.2969.

[134] The panel seems to recognize such a problem, without, however, drawing any conclusions ('in cases where the relevant scientific evidence is insufficient, eg, because none is available, a Member who wishes nonetheless to take a precautionary SPS measure could not meet the requirement in Article 2.2 to ensure that this measure "is not maintained without sufficient scientific evidence"': ibid, para 7.2983).

[135] Similarly, Andrew Lang, 'Provisional Measures under Article 5.7 of the WTO's Agreement on Sanitary and Phytosanitary Measures: Some Criticism of the Jurisprudence So Far' (2008) 42(6) Journal of World Trade 1085, 1093. Lang also notes that the current approach of the SPS case law may fail to provide clear guidance to WTO Members as to how to comply with the rulings of the dispute settlement bodies (ibid).

[136] Cf also, ibid, 1091 (noting that such approach finds support in the text of Article 5.7).

The recent decision in *US – Continued Suspension* also appears to support the above interpretation. The Appellate Body, when discussing the applicability of the respective provisions, observed that 'Article 2.2 excludes from its scope of application *situations in which the relevant scientific evidence is insufficient.* In such situation, the applicable provision is Article 5.7 of the SPS Agreement'.[137] Note that the Appellate Body did not mention other conditions of Article 5.7 as factors which would be relevant for determination of its applicability. This may suggest that the only criterion is insufficiency of scientific evidence. Yet, since the Appellate Body did not elaborate on this issue, any final conclusion in this respect will require further confirmation of future SPS case law.

(b) Relationship to other provisions of the SPS Agreement

A separate issue which requires examination is the relationship between Article 5.7 and other provisions of the SPS Agreement. Clarification of this matter should allow the identification of those disciplines of the SPS Agreement that are applicable to Article 5.7 measures. As discussed in the previous section, Articles 2.2 and 5.7 are described as mutually exclusive obligations (under the current approach of the case law only if all four conditions are met). What remains unclear is the scope of the exclusion, as Article 2.2 is actually composed of three distinct obligations. Pursuant to this provision, an SPS measure needs to be: (i) necessary; (ii) based on scientific principles; and (iii) not maintained without sufficient scientific evidence.

One option is to assume that measures which are covered by Article 5.7 are exempted from all obligations of Article 2.2. Another reading would call for a more limited exclusion, where some obligations of Article 2.2 are still applicable to 'provisional' safeguard measures. Regrettably, the text of the relevant provisions does not provide any decisive answer. As submitted by the EC in the *Biotech* case, the comma after the word 'evidence' may imply that the words that follow the comma exclude all the words up to the word 'evidence'.[138] On the other hand, one may also argue that a comma has a neutral effect. In particular, the positioning of the exemption clause at the end of the sentence and not at the beginning may suggest that it only refers to the last part of Article 2.2. The case law is equally ambiguous. Although the Appellate Body said that Article 5.7 of the SPS Agreement operates as a qualified exemption from the obligation of Article 2.2 not to maintain SPS measures without sufficient scientific evidence, without mentioning other obligations of Article 2.2,[139] this finding was made in the context of the analysis on sufficiency of scientific evidence. Its value may therefore be limited.

[137] Appellate Body Report, *US – Continued Suspension*, para 674 (emphasis added).
[138] Panel Report, *EC – Biotech Products*, para 4.755.
[139] Appellate Body Report, *Japan – Agricultural Products*, para 80.

I believe that there are good grounds to accept a narrow reading of the exclusion. Article 5.7 deals with measures that are adopted in the circumstances of insufficient scientific evidence. The language of Article 5.7 corresponds closely with the third obligation of Article 2.2 (not to maintain SPS measures without sufficient scientific evidence) and one may rationally assume that these two are mutually exclusive. This logic would also require reaching the same conclusion with respect to the Article 5.1 provision, which may be seen as a specific application of the basic obligations set out in Article 2.2. On the other hand, Article 5.7 does not address the question of necessity of SPS measures. Indeed, why should a measure of Article 5.7 type be applied beyond what is necessary to protect human, animal, or plant life or health?

The second obligation (ie to base SPS measures on scientific principles) is more problematic. Although Article 5.7 speaks only about the insufficiency of scientific evidence without making any reference to scientific principles, an overlap that exists between the second and the third obligations of Article 2.2 may indicate that the exclusion also covers the requirement of scientific principles.[140] As discussed in Chapter 4, section 4.2, the notion of 'scientific evidence' implies a certain methodological rigour that is also required under the obligation to base measures on scientific principles. Nevertheless, the ultimate answer will depend on how one conceptualizes disciplines of Article 5.7, in particular the requirement to base a measure on pertinent information. If, as I believe, pertinent information is a category that is primarily of scientific character, there are no reasons to reject the applicability of the second sentence of Article 2.2 to measures that fall into the ambit of Article 5.7. This would guarantee that the selection and the evaluation of pertinent information reflect certain methodological standards. Although these tasks may require a certain level of flexibility, this cannot be understood as carte blanche for WTO Members to disregard scientifically sound methods.

There is no case law which would explain the relationship between Article 5.7 and other provisions of the SPS Agreement. It was submitted in the *Biotech* case that the position of Article 5.7 (after subparagraphs 1 to 6) indicates a special nature of this provision and implies that the obligations of all preceding subparagraphs are inapplicable to provisional measures.[141] Although the panel did not address this issue and any conclusion may only be speculative, it seems that the EC overstated the importance of the structure of Article 5. Note that neither Article 5.5 nor 5.6 contains language similar to Article 2.2, which could suggest a mutually exclusive relationship. Moreover, none of these provisions is related to

[140] Similarly, Joanne Scott, *The WTO Agreement on Sanitary and Phytosanitary Measures. A Commentary* (2007) 112 (noting that 'in circumstances in which scientific evidence is . . . insufficient to conduct risk assessment, it will be hard to make an informed evaluation of the soundness of the underlying scientific principles').

[141] Panel Report, *EC – Biotech Products*, para 4.758 (the EC argued that 'the fact that Article 5.7 is positioned after Articles 5.1 to 5.6 (and not after Article 5.1) also confirms its nature as a special regime applying specifically to provisional measures').

the third sentence of Article 2.2 in the same way as Article 5.1. If one adds that national risk regulators almost always, including in cases of insufficiency of scientific evidence, have a choice between different measures, there are no compelling reasons why their choices should be exempted from the respective obligations of the SPS Agreement. At the same time, the specificity of the regime established by Article 5.7 will require a certain degree of flexibility as compared to measures that are based on risk assessment and scientific evidence. The lack of sufficient information on the nature and extent of risk can complicate a comparison with other situations (Article 5.5), whereas in the context of Article 5.6, it may be 'difficult to calibrate a measure if the extent of the risks or the availability or effectiveness of protection measures is unclear'.[142] This, however, should not lead to the conclusion that those disciplines are inapplicable.

5.4.2 Insufficiency of scientific evidence

As discussed above, the first condition serves a two-fold purpose. It triggers the applicability of Article 5.7 (provisionally or ultimately) but it also constitutes one of its consistency requirements. So what is actually meant by insufficiency of scientific evidence? According to the case law, insufficiency exists if 'a body of available scientific evidence does not allow, in quantitative or qualitative terms, the performance of an adequate assessment of risks as required under Article 5.1 and as defined in Annex A to the SPS Agreement'.[143] The case law also stressed that insufficiency of scientific evidence under Article 5.7 cannot be equated with a situation where relevant scientific evidence is found to be insufficient for the purpose of Article 2.2.[144] This is logical as Article 2.2 refers to evidence that supports a particular SPS measure, while Article 5.7 is concerned with the entire body of science relevant for the adoption and maintenance of a provisional measure.

This subsequent analysis concentrates on four problematic issues that emerge under the first condition of Article 5.7: (i) the relationship between insufficiency of scientific evidence and uncertainty; (ii) assessment of insufficiency; (iii) the temporal dimension which is implied in such an assessment; and (iv) the role of international standards for the evaluation of insufficiency.

(a) Insufficiency of scientific evidence v uncertainty

The first issue which requires clarification is the nature of the relationship between insufficiency of scientific evidence and scientific uncertainty. Should these two be regarded as distinctive concepts or are they just different names for the same phenomenon? The SPS case law has consistently held that insufficiency

[142] Panel Report, *EC – Biotech Products*, para 4.760.

[143] Appellate Body Report, *Japan – Apples*, para 179, Panel Report, *US – Continued Suspension*, para 7.608.

[144] Panel Report, *Japan – Apples*, paras 215–16.

of scientific evidence cannot be equated with uncertainty. According to the Appellate Body, those two concepts are not interchangeable and the existence of uncertainty as such does not trigger the application of Article 5.7.[145] In the words of the Appellate Body 'existence of unknown and uncertain elements does not justify a departure from the requirements of Articles 5.1, 5.2 and 5.3'.[146] On the other hand, the *Suspension* panel used more careful language when it said that 'the existence of scientific uncertainty does not automatically amount to the situation of insufficiency of relevant scientific evidence'.[147] This may suggest that, according to the panel, at least some types of uncertainty could be regarded as a form of insufficiency of scientific evidence.

Recall from Chapter 1, section 1.4.4 that uncertainty is not a uniform concept and may be divided into different subcategories. This, in turn, means that any meaningful comparison between insufficiency of scientific evidence and uncertainty needs to take into account this heterogeneity. Following the taxonomy developed by Klinke and Renn, one may distinguish in the first place uncertainty that results from the ontological character of science and the reality of probabilistic risk assessment. Other types of uncertainty are caused by variability problems, systemic and random measurement errors, indeterminacy, and lack of knowledge. The last category can be further divided into uncertainty that results from lack of observations/measurements, practical immeasurability, conflicting evidence/competing theories, and reducible or irreducible ignorance.[148]

As might be expected, the WTO case law has refused to recognize the first type of uncertainty as a concern that may be legitimately addressed under the SPS Agreement. The Appellate Body stressed in the context of Article 5.1 that risk needs to be ascertainable and 'theoretical uncertainty is not the kind of risk which . . . is to be assessed'.[149] As discussed in Chapter 4, section 4.4.1, theoretical uncertainty was defined as the kind of uncertainty that is 'inherent in the scientific method and which stems from the intrinsic limits of experiments, methodologies, or instruments deployed by scientists to explain a given phenomenon'.[150] The existence of such uncertainty does not trigger the application of disciplines of Article 5.7. This seems to be a proper approach. Extending the scope of Article 5.7 so as to include theoretical uncertainty will simply make the whole system established by the SPS Agreement inoperable. Since such uncertainty always remains, a WTO Member would be able to claim it in every case. Consequently, the disciplines of Articles 2.2/5.1 would never be activated.

As far as variability and systemic and random measurement errors are concerned, under the current jurisprudence WTO Members are expected to address

145 Appellate Body Report, *Japan – Apples*, para 184.
146 Appellate Body Report, *Australia – Salmon*, para 130.
147 Panel Report, *US – Continued Suspension*, para 7.631.
148 See Chapter 1, section 1.4.4.
149 Appellate Body Report, *EC – Hormones*, para 186.
150 Appellate Body Report, *Japan – Apples*, para 241.

them within the framework of the risk assessment process. To this end Members can use all standard techniques such as conservative assumptions, safety factors, worst-case scenarios, and so on.[151] The SPS Agreement also allows for performing a fully qualitative risk assessment.[152] Moreover, as noted by the *Biotech* panel 'the mere fact that relevant scientific evidence is sufficient to perform a risk assessment does not mean that the result and conclusion of the risk assessment are free from uncertainties (eg uncertainties linked to certain assumptions made in the course of the performance of a risk assessment)'.[153] A WTO Member may, therefore, legitimately account for such uncertainties when deciding on a subsequent course of action. As a consequence, the same risk assessment, reflecting a different appreciation of identified uncertainties, may equally support various SPS measures (eg a risk assessment which identifies important uncertainties may constitute a sufficient basis for a very strict measure).[154]

The uncertainty caused by conflicting scientific evidence or competing theories is also clearly recognized in the SPS case law. As observed by the Appellate Body in the *Hormones* dispute 'in some cases, the very existence of divergent views presented by qualified scientists who have investigated the particular issue at hand may indicate a state of scientific uncertainty'.[155] The *Suspension* panel also subscribed to this view by noting that the 'presence of divergent views on an issue could be a form of scientific uncertainty'.[156] This type of uncertainty in principle does not fall into the ambit of Article 5.7 and may be addressed under the standard disciplines provided by Articles 2.2/5.1. Consequently, WTO Members may adopt a measure on the basis of one of the competing scientific views, even if such an opinion is shared only by a minority group,[157] or otherwise factor such uncertainty into the conclusions of the risk assessment (eg by assuming the worst-case scenario).[158] Of course, if minority opinion shows the existence of insufficiency of scientific evidence, the route provided by Article 5.7 will be available for a WTO Member.

Lack of observations and measurement data, practical immeasurability, and indeterminacy seem to come close to the concept of insufficiency of scientific evidence. Uncertainty that is caused by a recognized lack of knowledge would be a typical example of insufficiency of scientific evidence. Identified indeterminacy resulting 'from a genuine stochastic relationship between cause and effect(s), apparently non-casual or noncyclical random events or badly understood nonlinear chaotic relationship'[159] may be regarded as qualitative deficiency of scientific

[151] Panel Report, *US – Continued Suspension*, para 7.635 (citing opinions of its experts on how scientific uncertainty is addressed in risk assessment).

[152] Cf Chapter 4, section 4.4.1(b).

[153] Panel Report, *EC – Biotech Products*, para 7.1525.

[154] Ibid.

[155] Appellate Body Report, *EC – Hormones*, para 194.

[156] Panel Report, *US – Continued Suspension*, para 7.632.

[157] Cf Chapter 4, section 4.4.2(b).

[158] Panel Report, *US – Continued Suspension*, para 7.632.

[159] Andreas Klinke and Ortwin Renn, 'A New Approach to Risk Evaluation and Management, Risk-Based, Precaution-Based, and Discourse-Based Strategies' (2002) 22 Risk Analysis 1071, 1079.

data. If important enough for assessment of a particular risk, it could also qualify as a form of insufficiency of scientific evidence. As will be discussed in the subsequent section, WTO jurisprudence has recognized that data which is inconclusive or which fails to explain the nature of a particular risk may equally trigger the application of disciplines of Article 5.7.

The problems posed by irreducible and reducible ignorance have not been addressed by the case law but one may assume that none of these categories could constitute a basis for a provisional SPS measure. Since the SPS case law holds that theoretical uncertainty does not amount to insufficiency of scientific evidence, *a contrario* one cannot accept uncertainty that results from irreducible or reducible ignorance. Again such an approach would be in line in with the general practice in the area of risk regulation.[160]

The approach of the WTO case law to different forms of uncertainty is summarized below.[161]

Table 1 Treatment of uncertainty in SPS case law

Type of uncertainty*	The approach of the SPS case law
1. theoretical uncertainty	does not constitute a legitimate ground for an SPS measure
2. variability	addressed in risk assessment through the application of safety factors, conservative assumptions, worst-case scenarios and other standardized techniques
3. systemic and random measurement error	may amount to insufficiency of scientific evidence
4. indeterminacy	may amount to insufficiency of scientific evidence
5. lack of knowledge:	
• lack of observations/ measurement data	may amount to insufficiency of scientific evidence
• practical immeasurability	may amount to insufficiency of scientific evidence
• conflicting evidence/ competing theories	WTO Members may base their measures on one of the competing sets of evidence or theories; it does not preclude performance of risk assessment
• reducible or irreducible ignorance	does not constitute a legitimate ground for an SPS measure

* typology of uncertainty adopted on the basis of Klinke and Renn/Van Asselt and Rotmans

160 Note, however, that some national systems may aim to address reducible ignorance by adopting some measures; that option could be particularly used if a specific field of scientific research is new and the regulators do not even know what can be expected from specific technology.

161 Note that in one of the disputes, Japan made a distinction between 'new uncertainty' and 'unresolved uncertainty'. According to Japan, new uncertainty arises when a new risk is identified. On the other hand, unresolved uncertainty is uncertainty that the scientific evidence is not able to resolve, despite accumulated scientific evidence. Note, however, that this type of uncertainty can be reduced to one of the categories described here—for instance indeterminacy (Appellate Body Report, *Japan – Apples*, para 183).

The above shows that, although uncertainty and insufficiency of scientific evidence are two distinctive concepts, there is also an overlap between them. Some types of uncertainty are of such a nature that they may be qualified as an instance of insufficiency of scientific evidence. This also means that some of its forms (ie lack of observations/measurement data, practical immeasurability, and indeterminacy) may actually trigger the application of Article 5.7. Of course it is not difficult to imagine that there will be disagreement between the parties to the dispute as to the classification of particular uncertainty. For instance in *EC – Biotech Products*, the issues which were considered by the EC as falling into the category of indeterminacy were regarded by all complainants as an example of merely theoretical uncertainty.[162] A similar situation may be found in *US – Continued Suspension* where the EC identified a number of uncertainties that, in its opinion, precluded the performance of risk assessment. On the other hand, the JECFA believed that all of them could be properly addressed through the application of safety factors.[163]

(b) *Assessment of insufficiency*

The case law does not establish any general rules on how to assess the sufficiency/insufficiency of scientific evidence, preferring a case-by-case approach.[164] The Appellate Body emphasized that insufficiency cannot be considered in the abstract but requires evaluation in the light of all the circumstances of the specific situation.[165] What is clear is that a WTO Member, before concluding that evidence is insufficient to perform a risk assessment, needs to consider all available data.[166] At the same time, a mere possibility of conducting further research or analysing additional information does not mean that scientific evidence is insufficient.[167] The case law also recognizes that insufficiency under Article 5.7 has both a quantitative and a qualitative dimension. The first category relates to the amount of scientific data on a particular SPS problem, whereas the second is more concerned with their conclusiveness. This dual character of insufficiency was clearly articulated by the Appellate Body when it said that insufficiency includes 'cases where the available evidence is more than minimal in quantity but has not led to reliable or conclusive results'.[168] The panel, in the same case, was even more explicit in holding that 'it is possible that, in a given situation, a lot of scientific research may have been carried out on a particular issue without yielding

[162] Cf eg, Panel Report, *EC – Biotech Products*, para 7.3297.

[163] Panel Report, *US – Continued Suspension, Annex F, 'Comments by the Parties on the Replies of the Scientific Experts, Codex, JECFA and IARC to Questions Posed by the Panel and Comments by the Parties on the other Parties' Comments'*, F-8.

[164] Panel Report, *EC – Biotech Products*, para 7.3238.

[165] Appellate Body Report, *Japan – Apples*, para 179.

[166] Appellate Body Report, *US – Continued Suspension*, para 677.

[167] Ibid, para. 702.

[168] Appellate Body Report, *Japan – Apples*, para 185.

sufficiently "relevant"—within the meaning of Article 5.7—or reliable evidence'.[169] The quality of scientific data plays, therefore, an independent and important (if not a leading) role in the assessment of insufficiency. Indeed, does the extent of available scientific data really matter if it does not provide reliable or conclusive results?

It is also important to realize that 'there is no scientific definition of terms such as acceptable, reasonable, and sufficient'.[170] The same set of information can be sufficient for one scientist to assess risk and insufficient for another. This depends not only on the available body of scientific information but also on other factors such as subjective judgments of the experts that reflect their attitudes toward particular risks (less or more cautious), values of the particular community in which the expert is acting, as well as the nature of the assessment itself. Note that risk assessment requires answering a number of questions which are not purely scientific. What weight should be assigned to positive and negative studies? What is a statistical significance?[171] Can uncertainties be addressed through conservative assumptions and safety factors or perhaps is additional research needed? There are no absolute answers to those questions, and responses may legitimately differ between experts. Assessing the sufficiency of scientific evidence is therefore not a straightforward scientific task but it also has a normative dimension.[172] As noted by a group of eminent scholars in the field of risk science, 'evidence deemed reliable enough to generate a sufficient risk assessment in one regulatory context may fail in other contexts because of the different concerns, risk frames, and particular circumstance'.[173] As a particular field of scientific research develops, an increasing number of scientists will agree that the available body of scientific information allows the performance of risk assessment.[174] However, if a particular research area is novel and complex, no uniformity will exist.[175] This normative nature of insufficiency was accurately captured by one of the experts advising the *Biotech* panel, '[w]hen *additional* scientific knowledge is needed to evaluate new GM crops, each nation's regulators and scientific advisory committees are placed in the difficult position of choosing between expediency and greater certainty. It is not always clear where

169 Panel Report, *Japan – Apples*, para 7.9.

170 Douglas Crawford-Brown, Joost Pauwelyn, and Kelly Smith, 'Environmental Risk, Precaution and Scientific Rationality in the context of WTO/NAFTA Trade Rules' (2004) 24(2) Risk Analysis 461, 467.

171 Carl F Cranor, 'Science Courts, Evidentiary Procedures, and Mixed Science Policy Decisions' (1993) 4(2) Risk: Issues in Health and Safety 113, 119.

172 Ibid, 118; note also that it is at least debatable whether scientists alone can decide on such normative choices.

173 David Winickoff, Sheila Jasanoff, Lawrence Busch, Robin Grove-White, and Brian Wynne, 'Adjudicating the GM Food Wars: Science, Risk, and Democracy in World Trade Law' (2005) 30 Yale Journal of International Law 81, 114.

174 Ibid.

175 This is particularly visible in the case of biotechnology (sometimes referred to as science at the frontiers), which is both novel and complex.

the distinction lies between what regulators "need to know" *vs.* what is merely "nice to know"'.[176]

The normative element in the assessment of insufficiency remained, however, unappreciated in the panels' reports in *EC – Biotech Products* and *US/Canada – Continued Suspension*. Both panels saw the concept of insufficiency of scientific evidence as *quasi* absolute term which is determined solely on scientific grounds. This approach is well reflected in their analysis on the relationship between ALOP and insufficiency. The *Biotech* panel rejected the EC's argument that sufficiency is a relational concept that is influenced, amongst other things, by the protection goals pursued by the regulator. After over-textual interpretation of Article 5.7, the panel found that insufficiency might have existed only if available scientific evidence did not allow the performance of quantitative or qualitative risk assessment as defined by Article 5.1 and Annex A.[177] This was understood by the panel as a purely technical exercise which is both objective and absolute. According to the panel, the ALOP was relevant only when deciding on the regulatory measure and not in the process of determining the insufficiency of scientific evidence.[178] The same was true for any identified imprecision in the risk assessment as well as scientists' level of confidence in the results.[179] This could be taken into account by a Member only when choosing and adopting appropriate SPS measures but not when deciding whether evidence is sufficient.[180] The panel in *US/Canada – Continued Suspension* took essentially the same approach. It saw the assessment of insufficiency as an objectively verifiable determination that takes place in a purely scientific process. In the words of the panel, 'the determination of whether scientific evidence is sufficient to assess the existence and magnitude of a risk must be disconnected from intended level of protection'.[181] This narrow conception of insufficiency also required the panel to exclude the consideration of problems relating to observance of good veterinary practice (ie misuse or abuse in the administration of hormones). According to the panel, these issues, as unscientific, were not relevant for the assessment of insufficiency.[182]

The tension between the panel's conceptualization of insufficiency and the reality of risk assessment (including a decision as to whether evidence is sufficient to perform risk assessment) is well reflected in the treatment of scientific information provided by the experts in the context of the *EC – Biotech Products*

[176] Panel Report, *EC – Biotech Products, Annex H 'Replies by the Scientific Experts Advising the Panel to Questions Posed by the Panel'*, para 14 (emphasis in original).

[177] Ibid, para 7.3243. [178] Ibid, para 7.3238.

[179] Ibid, paras 7.3240 and 7.3244.

[180] The panel also cited the Working Principles for Risk Analysis for Application in the Framework of the Codex Alimentarius, which provides that '[t]he report of the risk assessment should indicate any constraints, uncertainties, assumptions and their impact on the risk assessment. Minority opinions should also be recorded. The responsibility for resolving the impact of uncertainty on the risk management decision lies with the risk manager, not the risk assessors' (ibid, para 7.3240).

[181] Panel Report, *US – Continued Suspension*, para 7.612.

[182] Ibid, para 7.603.

dispute. As part of its examination, the panel was obliged to analyse the insufficiency of scientific evidence underlying safeguard measures adopted by some of the EU Members. The French measure on oilseed rape MS1xRF1 and the German measure on Maize Bt-176 may serve here as useful examples. The panel asked the experts to assess whether in the light of evidence submitted by the EC there was any reason to believe that the scientific evidence available to Germany and France had not been sufficient to undertake an appropriate assessment of potential risks to human, plant, and animal health, and the environment from the importation and use of GMOs.[183] The question also distinguished between two temporal moments which were relevant for the assessment of sufficiency: the date of adoption of national safeguard measures (1998 and 2000 respectively) and the date of establishing panels' terms of reference (2003).

With respect to the French measure, Dr Nutti, one of the experts, was confident that the scientific evidence was sufficient to perform a risk assessment. She particularly pointed out the positive assessment of the body that performed the risk assessment on the European level (EC Scientific Committee on Plants (ECSCP)) and the British competent authority (the authority responsible for the evaluation of the initial application for approval of oilseed rape MS1xRF1).[184] Another of the experts, Dr Andow, on the other hand, when asked the same question, identified a number of instances where the evidence was insufficient. His view was that the information on the dispersal of oilseed rape pollen was insufficient to complete an accurate assessment of dispersal probability,[185] the molecular characterization was insufficient to identify all transgene products,[186] the risk of evolution of resistance in weeds not related to the Brassicaceae was incompletely assessed, while the data on non-target effects were inadequate.[187] He also expressed his concern with the assumption of the ECSCP that proper agricultural practices were being observed.[188] A third expert, Dr Snow, agreed with Dr Andow and after evaluation of available evidence confirmed that 'France had valid reasons... to carry out more research to "supplement existing scientific knowledge and validate methods for managing the cultivation of genetically modified oilseed rape"'.[189] He also identified one particular instance of insufficiency of scientific evidence, that is the rate at which the transgene for glufosinate tolerance would spread to volunteer plants (including the seed bank) and related weeds.[190] The fourth expert, Dr Squire, added that 'there was insufficient knowledge... to predict accurately... what the rates of spread and cross pollination would be (GM to non-GM) if a large part of the rapeseed areas were GM'.[191] Thus, three out of four experts advising the panel confirmed that scientific evidence was indeed insufficient. The same opinion was expressed by

[183] Panel Report, *EC – Biotech Products, Annex H, 'Replies by the Scientific Experts Advising the Panel to Questions Posed by the Panel'*, H-171.

[184] Ibid, para 780.

[185] Ibid, para 785.

[186] Ibid, para 789.

[187] Ibid.

[188] Ibid, para 786.

[189] Ibid, para 791.

[190] Ibid, para 793.

[191] Ibid, para 795.

the Biomolecular Engineering Committee, a committee which assessed data on MS1xRF1 for the French competent authority. Note also that the above answers were provided in the context of a very specific question posed by the panel which clearly connected the concept of insufficiency of scientific evidence with the possibility of conducting risk assessment. This additionally indicates that the experts saw deficiencies in existing knowledge to be so serious as to potentially prevent France from performing such an assessment.

When addressing the same question with respect to the German safeguard measure on Maize Bt-176, Dr Nutti once again confirmed that the available data allowed for the performance of risk assessment.[192] The other expert (Dr Andow) was less optimistic. He noted that although the ECSCP concluded that the scientific information provided by the German authorities did not alter the original risk assessments, that opinion was only 'one possible scientific opinion that can be reached from the information available at the time' and 'new information also allows several other scientifically valid opinions'.[193] Note that this seems to resonate closely with the approach of the SPS case law that recognizes the existence of insufficiency of scientific evidence where the available data have not led to reliable or conclusive results.[194] Dr Andow also added that the ECSCP had not considered all the scientific perspectives, and in some cases had even ignored prevailing scientific opinions. According to him, the ECSCP opinion did not invalidate the scientific opinion of Germany.[195] Dr Andow concluded by saying that 'the scientific evidence available to Germany ... was not sufficient to permit it to undertake an appropriate assessment of potential risks to plants and the environment'.[196] Note also that Dr Andow was conscious that some gaps in scientific knowledge could have been addressed through traditional risk assessment techniques, such as using the worst-case scenario or conservative assumptions.[197] For example, he explicitly admitted that data available in 2003 on non-target risks, although insufficient, did not prevent Germany from performing a risk assessment.[198] In other cases, however, he did not have this reservation.[199] In

[192] Ibid, para 921.

[193] Ibid, para 933; cf the statement of Dr Snow who observed that 'several ... papers challenged widely held assumptions about pollen, gene flow, or herbicide-tolerant crops' (ibid, para 15).

[194] Cf Appellate Body Report, *Japan – Apples*, para 185.

[195] Panel Report, *EC – Biotech Products, Annex H, 'Replies by the Scientific Experts Advising the Panel to Questions Posed by the Panel'*, para 937.

[196] Ibid, para 938.

[197] Cf also the opinion of Dr Andow on sufficiency of scientific evidence relating to Maize MON 810, where he noted that 'it could be argued that a risk assessment could be conducted using scientifically justified worst-case assumptions' (ibid, para 998).

[198] Ibid, para 941. However, Dr Andow also made clear that even in 2003 'Germany could reasonably maintain that there is still insufficient information to know which non-target species may be at risk' and that 'molecular characterization of Event 176 is insufficient to conduct a risk assessment' (ibid).

[199] Ibid, paras 933–7 (this particularly included data on development of resistance and effects of Bt toxin in the soil).

consequence, according to Dr Andow, some gaps in the knowledge were of such character so as to preclude adequate risk assessment.

The above examples show that even among the experts advising the panel there was no uniformity of views. The panel, however, disregarded dissenting opinions and found that since the ECSCP was able to perform the risk assessment of Maize Bt-176 and oilseed rape MS1xRF1 and subsequently reviewed the concerns which had been raised by France and Germany, it was no longer possible for these countries to claim that there was insufficiency of scientific evidence.[200] Thus, the existence of risk assessment on the European level de facto precluded EU Member States from the opportunity to claim that scientific evidence was insufficient. If one takes this approach to its extreme, the logical conclusion is that if one WTO Member is able to perform the risk assessment, no other Member may rely on Article 5.7.[201] Regrettably, such an approach fails to recognize that sufficiency is a relative and not an absolute term. Consequently, what is sufficient for some scientists, may equally be insufficient for others. As discussed above, the assessment of sufficiency of scientific evidence is not simply a scientific task but also has a normative dimension where judgments of the experts reflect their attitude toward particular risks (less or more cautious) and values of a particular community play an important role.

The approach of the *Biotech* panel may also be questioned with regard to the applied standard of review. The panel apparently conducted a *de novo* review which aimed at establishing what, in the opinion of the panel, was the best science. The panel, in a rather arbitral manner, preferred some opinions of its own experts over the others. As showed above, in some cases, the views expressed by those experts were simply disregarded without any explanation as to why the panel considered them irrelevant. Although there is no case law which would clearly articulate the applicable standard of review for assessing the insufficiency of scientific evidence, it seems rational to assume that the standard developed in the context of Article 5.1 is also applicable here. If sufficiency of scientific evidence is to be determined by scientists, there are no reasons to apply a different standard under Article 5.7. Accordingly, a panel should only assess whether a domestic determination of insufficiency of scientific evidence is supported by coherent reasoning and respectable scientific views.[202] This task will require identification of insufficiencies and corresponding scientific views which 'puts into question the relationship between the relevant scientific evidence and the conclusion in relation to risk'.[203] Such views may be examined in order to assess whether they hold a minimum epistemic quality. As under Articles 2.2/5.1, a scientific view

[200] See eg, ibid, para 7.3300.

[201] Note that the EC and its Member States are separate sovereign entities (in fact all of them are WTO Members).

[202] Cf Chapter 4, section 4.5.

[203] Appellate Body Report, *US – Continued Suspension*, para 677.

may come from mainstream and minority circles.[204] This will require accepting that some WTO Members may conclude, on the basis of such minority opinions, that scientific evidence is insufficient for the performance of risk assessment. The fact that another WTO Member takes a different view (based on mainstream or other minority opinion) and considers evidence as sufficient would not have any consequences. Finally, a panel should evaluate whether the reasonings articulated on the basis of such views are objective and coherent. The correct approach under Article 5.7, therefore, should be limited to assessing the plausibility of scientific opinions on the insufficiency/sufficiency of scientific evidence rather than deciding which of such views is better.

The simplified view of insufficiency of scientific evidence which was applied by both panels was rightly rejected in the recent report of the Appellate Body. In particular, the Appellate Body recognized that ALOP may play a role in framing the scope and the methods used in a risk assessment, including scientific information to be examined.[205] This may 'require a WTO Member to perform certain research as part of its risk assessment that is different from the parameters considered and the research carried out in the risk assessment'[206] of other WTO Members or one of the standard-setting organizations. Of course this does not mean that assessment of insufficiency can be reduced to a subjective process that depends entirely on the entity making the evaluation of scientific evidence. Such examination remains a rigorous and objective process with ALOP being one of the relevant factors.[207] However, since the observation of the Appellate Body was very general, future SPS case law will need to clarify the exact impact of all those different elements on the final determination of insufficiency.

(c) *Time for assessing the insufficiency of scientific evidence*

Another problematic issue under the first condition of Article 5.7 is the identification of the temporal moment that is relevant for the assessment of insufficiency of scientific evidence. Is it the time when an SPS measure was adopted? Or rather the moment when a panel's terms of reference are fixed? Regrettably, the SPS case law has failed to establish any consistent and convincing interpretation.

The *Biotech* panel, on the basis of the distinction introduced by the Appellate Body between the first and the second sentence of Article 5.7, differentiated between the obligations of Article 5.7 that are concerned with the adoption of an SPS measure (the first sentence) and those that relate to the maintenance of a measure (the second sentence). On the basis of this distinction, the panel found that the insufficiency in the first sentence should be determined by reference to the time when an SPS measure was adopted.[208] In support of its interpretation, the panel also referred to contextual argument and found that a broad reading

204 Ibid. 205 Ibid, para 685.
206 Ibid. 207 Ibid, para 686.
208 Ibid, para 7.3253.

of the first sentence of Article 5.7 would make the second sentence redundant. Thus, according to the panel, the question of whether the scientific evidence is still insufficient is to be addressed under the second sentence of Article 5.7, as it may 'shed light on whether the Member... has complied with the requirement to "seek to obtain the additional information necessary for a more objective risk assessment" ' and 'whether the Member... has conducted a review of its provisional measure "within a reasonable period of time"'.[209] Since the phrase 'available pertinent information' is found in the first sentence of Article 5.7, its existence is also ascertained as of the day of adoption of an SPS measure.

The panel's argument, although textually correct, does not provide a pragmatic solution. First of all, it may be difficult to determine a state of scientific knowledge in a particular remote moment in the past. Obviously, the problem will be amplified as the gap between the adoption of a measure and a panel proceeding increases. It is also unclear how to address under the *Biotech* panel approach all those cases where initially false insufficiency turns out, after the adoption of a measure, to be real. Under which condition of Article 5.7 should one take those new scientific developments into account? The panel report does not provide any answer to this problem. Although it states in a different context that its interpretation 'leaves room for the possibility that even if at a given point in time relevant scientific evidence is sufficient to perform a risk assessment, a situation might subsequently arise where the relevant scientific evidence could be considered insufficient to perform a risk assessment as required under Article 5.1 and as defined in Annex A(4) of the SPS Agreement',[210] it is not clear how, under the panel's analysis, the change of circumstances with respect to the insufficiency of scientific evidence could be taken into account. The panel's interpretation also creates systemic problems with respect to the interrelationship between Articles 5.7 and 2.2. A particular measure may be found compatible with the sufficiency requirement of Article 2.2 (as the relevant moment for the assessment of sufficiency is a time when the panel's terms of reference are fixed) and with the insufficiency requirement of Article 5.7 (as the relevant moment is the adoption of a measure). If, as it is argued here, each provision applies to the exclusion of the other, which of them will be applicable in such a case?[211] The approach of the *Biotech* panel also appears to be inconsistent with the previous case law. The panel in *Japan – Apples*, when assessing sufficiency/insufficiency of scientific evidence, did not concentrate on any specific temporal moment. Particularly, it did not require its experts to assess the insufficiency of scientific evidence as of the time of the adoption of the Japanese SPS measure. More importantly, it explicitly rejected the argument of Japan that evidence which became available after adoption of the measure,

[209] Ibid, paras 7.3255–6.

[210] Panel Report, *EC – Biotech Products, Annex K*, 3.

[211] Of course this problem will not arise under the existing interpretation of the relationship between Articles 2.2 and 5.7.

should be disregarded for the purpose of Article 5.7.[212] The panel particularly said that:

> [W]e do not see in the text of Article 5.7, or of Article 2.2 for that matter, any reason to limit our assessment of the 'relevant scientific evidence' to evidence available before 1995. On the contrary, since Article 5.7 provides for an exception to Article 2.2, and an assessment of the compatibility of a measure with Article 2.2 is made at the time the matter is reviewed by the Panel, there is no justification for assessing any alleged provisional measure at a different date.[213]

Note that the panel made the above observation in the context of the first sentence of Article 5.7. Actually, the panel did not address any other conditions, except for insufficiency of scientific evidence. One may, therefore, assume that its findings were relevant only for the first condition of Article 5.7.

The more recent case law has followed the direction taken in the *Biotech* dispute.[214] The panel in *US – Continued Suspension* when assessing insufficiency of scientific evidence sought to determine the state of scientific knowledge for the date of adoption of Directive (EC) 2003/74 (a measure that was examined under Article 5.7). To this end, the panel specifically asked the experts to distinguish in their answers between scientific evidence which was available in 2003 as opposed to the current state of knowledge.[215] Although the experts in principle followed the panel's request,[216] a closer analysis of their answers also reveals that in some instances they cited, in support of their scientific opinions, materials that were published only after the adoption of the Directive.[217] This may suggest that even for the experts, it was difficult to precisely determine the state of scientific research at a particular moment in the past.

(d) International standards and insufficiency of scientific evidence

The last issue which requires clarification is the nature of the relationship between international standards and insufficiency of scientific evidence. To what extent does the existence of a relevant standard preclude a WTO Member from relying on Article 5.7? On the textual level, there is nothing in the SPS Agreement which would suggest that an international standard per se affects the applicability of Article 5.7. Panels are not relieved from the obligation to assess the insufficiency of scientific evidence simply because a particular risk is addressed by an international standard. On the other hand, it would be going too far to conclude that

212 To be precise, the specific date in that case was the date of entry into force of the SPS Agreement, as the SPS measure itself had been adopted earlier.

213 Panel Report, *Japan – Apples*, para 7.10.

214 The Appellate Body did not address this aspect of the panel report. Nevertheless, it also emphasized that the obligation of the first sentence of Article 5.7 had to be met before the adoption of a measure (Appellate Body Report, *US – Continued Suspension*, para 676).

215 See eg, Panel Report, *US – Continued Suspension, Annex D, Replies of the Scientific Experts to Questions Posed by the Panel*, D-37.

216 See eg, ibid, D-39.

217 Ibid.

the existence of such standards is without any relevance. Note that international standard-setting bodies, before adopting a standard, conduct a comprehensive scientific examination in order to assess a particular risk agent and related risks. As explained by one of the experts advising the panel in *US/Canada – Continued Suspension*, the Codex adopts standards only if it considers that the scientific database is complete and that there are no outstanding scientific issues.[218]

The impact of the standards on the availability of measures taken under Article 5.7 was recently extensively examined in *US/Canada – Continued Suspension*. The EC argued that there was insufficient scientific evidence for the assessment of certain hormones (ie progesterone, testosterone, zeranol, trenbolone acetate, and MGA) used for bovine growth promotion purpose. At the same time, the EC admitted that four provisionally banned hormones were also covered by international standards adopted by the Codex. In principle, the panel agreed with the EC that the existence of international standards could not be dispositive for the purpose of Article 5.7. However, this finding of the panel was qualified by its second observation. The panel believed that in order to find insufficiency of scientific evidence there needs to be a critical mass of new evidence and information which undermines previous findings. Notwithstanding a rhetorical observation of the panel that the test of critical mass should not result in the imposition of a special burden of proof on a defendant,[219] its application actually led to the establishment of a threshold that was very difficult to meet. The rather uncontroversial statement that 'suggesting hypothetical correlations or merely arguing that there could be more evidence on one concern or another should not be deemed sufficient to successfully claim that relevant scientific evidence has become insufficient',[220] was followed by a much more demanding requirement. According to the panel, the threshold could only be met by evidence which would bring into question fundamental principles of previous knowledge and evidence so as to make relevant previously sufficient evidence insufficient. Elaborating on this new standard, the panel explained that 'the new scientific information and evidence must be such that they are at the origin of a change in the understanding of a scientific issue'.[221] In support of its reasoning, the panel advanced two arguments. According to the panel, the existence of an international standard indicated that a risk assessment was performed within a particular international standard-setting body which implied existence of sufficiency as a matter of evidence. The relevance of this risk assessment was subsequently strengthened by the normative consideration which compelled the panel to recognize the importance attached to international standards by the SPS Agreement. Consequently, under the panel analysis, an international standard together with underlying risk assessment served not only as evidence of the fact but also as a normative benchmark determining the state of scientific knowledge. Interestingly, in the other section

[218] Ibid, D-18. [219] Ibid, para 7.646.
[220] Ibid, para 7.647. [221] Ibid, para 6.141.

of the report, the panel used more cautious language which is however difficult to reconcile with its previous findings. The panel, when discussing the relationship between the concepts of 'critical mass' and 'weight of evidence', observed that the test of critical mass was less demanding and could include situations:

> where the weight of the evidence has not shifted away from the existing prevailing knowledge, but where the new knowledge is qualitatively and quantitatively sufficient to create a situation where a Member can legitimately decide that the pre-existing scientific evidence is no longer sufficient to complete a risk assessment.[222]

The panel also saw a role for international standards in the allocation of burden of proof between the parties. After noting the existence of the consistency presumption for those measures which conform to international standards, it found that it was for the EC to bear the burden of proof with respect to identification of the issues for which scientific evidence was insufficient.[223] Surprisingly, the test of critical mass as well as reversed burden of proof was also applied to those cases (ie MGA) where intensive work has been performed at the international level, without leading to the adoption of a standard.[224] The panel particularly pointed out two risk assessments performed by the JECFA in 2002 and 2004, and the fact that a relevant draft of a standard was already at Step 7 of the Codex elaboration procedure. The panel also believed that the prominent role played by the JECFA in the field of international risk assessment had to find its reflection in the SPS Agreement. Nevertheless, this was rather a strange conclusion, since the introduction of the test of critical mass was justified by the panel through reference to the role which is played by international standards under the SPS Agreement. The SPS Agreement, however, in contrast to the TBT Agreement, does not recognize the relevance of the standards that are in the approval pipeline.[225]

It was not difficult for the panel, on the basis of the above conclusions, to find that evidence submitted by the EC did not call into question the fundamental precepts of previous knowledge or render previously sufficient scientific evidence insufficient.[226] In consequence, the panel held that the EC failed to establish the existence of insufficiency of scientific evidence for all five hormones.

The above analysis again reveals the general position of the panel on the nature of insufficiency of scientific evidence which is understood as an absolute category independent from domestic ALOP or subjective beliefs held by risk assessors. In the context of international standards, this conviction was reflected in the recognition of only a temporal and not a spatial dimension of insufficiency. According

[222] Ibid, para 6.143.

[223] Ibid, para 7.652; the panel advanced two other arguments in this context. First, it noted that its allocation of burden of proof was motivated by the general practice of the WTO dispute settlement bodies which required each party to prove its allegations. Secondly, the panel also mentioned the specific circumstance of the post-suspension case (ie the EC as a complainant).

[224] Ibid, para 7.813.

[225] Cf Chapter 3, section 3.4.

[226] See eg, Panel Report, *US – Continued Suspension*, paras 7.671 and 7.827.

to the panel, insufficiency of scientific evidence may only result from the evolution of science when new scientific evidence or information calls into question previous evidence.[227] The approach of the panel again fails to recognize that insufficiency of scientific evidence is not an absolute term and that the relevance which is attached to the lack of information may vary from assessor to assessor (and between countries). The panel's reasoning as a whole also resembles the approach taken by the panel in *EC – Biotech Products*. The *Biotech* panel, after identifying a relevant risk assessment on the European level, almost automatically found a sufficiency of scientific evidence that precluded the application of Article 5.7. Thus, both reports may be seen as forerunners of a new approach of the SPS case law under which it would be increasingly difficult to rely on disciplines of Article 5.7 if a relevant risk was assessed 'somewhere there'.[228] Taken to the extreme, this approach would require the rejection of availability of provisional measures taken under Article 5.7 whenever another WTO Member, including a complainant, was able to perform relevant risk assessment. This obviously interferes with the sovereign right of each WTO Member to establish its own ALOP. Preventing one WTO Member from relying on Article 5.7, when a risk assessment has been conducted somewhere else, deprives that Member of the opportunity to review scientific materials, to assess them, and, in consequence, to establish a level of protection which it deems appropriate.

The approach of the Appellate Body in *US/Canada – Continued Suspension* was entirely different. It is definitively more nuanced and more closely corresponds with the reality of risk assessment. The Appellate Body rejected at the outset the concept of critical mass as demanding a proof of a paradigm shift in the particular scientific field. According to the Appellate Body, such a threshold was far too high for the assessment of insufficiency of scientific evidence. Instead, the appropriate benchmark against which one should assess whether scientific evidence once considered to be sufficient became insufficient, was located somewhere between two extremes: incremental advances of science on one side and a paradigm shift on the other.[229] Although the Appellate Body did not determine its precise location,[230] it seems that such a benchmark is situated closer to the lower rather than the higher margin. In particular, new evidence does not need to invalidate earlier scientific data; it is sufficient that it 'cast[s] doubts as to whether the previously existing body of scientific evidence still permits . . . assessment of

[227] See eg, ibid, para 7.645.

[228] Cf also, Lang ('Provisional Measures') 1097.

[229] Appellate Body Report, *US – Continued Suspen*sion, para 703; incremental advances of science were defined are advances at the margins of knowledge.

[230] The Appellate Body cited some examples provided by Canada and the US that may qualify as evidence which makes a body of scientific data insufficient. This particularly includes discovery of a new pathway for a risk, new adverse effects or adverse effects at lower exposure levels, or identification of a new source of exposure. Arguably other less serious discoveries are also able to meet the required threshold (ibid, para 700).

risk'.[231] In this context, it is also worth noting that the Appellate Body did not subscribe to a one-dimensional perspective of insufficiency of scientific evidence that was preferred by the panel. The report clearly shows that the Appellate Body recognized that insufficiency may result not only from the scientific developments which took place after adoption of a standard[232] but also from a difference in the appreciation of available scientific information and data[233] as well as applicable ALOP.[234]

The Appellate Body also disagreed with the panel as to the value that should be attached to the JECFA's risk assessment (either the one that constitutes a basis for the standard or only its draft). According to the Appellate Body, the proper role of a standard (or, to be more precise, scientific data which supports underlying risk assessment) in the context of Article 5.7 is much more limited than what had been suggested by the panel. A standard should be regarded as evidence of specific scientific information which may help to establish a prima facie case of sufficiency of scientific evidence for the purpose of Article 5.7. Nonetheless, such evidence is not dispositive.[235] Consequently, scientific evidence that underlies an assessment performed within the relevant international standard-setting body constitutes only a factual issue and does not establish a normative benchmark.[236] In other words, international standards do not create any legal presumption of sufficiency for the purpose of Article 5.7.[237] There are two implications that arise from this conclusion. First, it is not important whether a standard was adopted or not. What matters in the assessment of insufficiency of scientific evidence are the scientific materials that were used in the process of adoption of a standard. Secondly, there is no reversal of burden of proof. It is always a claimant who needs to establish that scientific evidence is sufficient. Of course it may use the data that were relied on in the risk assessment produced in one of the sister organizations but this does not change the general rules on the allocation of the burden of proof.

5.4.3 Other obligations

As already mentioned, Article 5.7 provides a set of various requirements for measures that are adopted in the circumstances of insufficiency of scientific evidence. This includes the obligation to base a measure on pertinent information, to seek to obtain additional information, and to review a measure within a reasonable period of time. This section will discuss those requirements in more detail.

[231] Ibid, para 703. [232] Ibid, para 701.

[233] Ibid, para 695.

[234] Ibid, para 686 (noting that 'information from relevant international organizations may not necessarily be considered "sufficient" to perform a risk assessment').

[235] Ibid, para 697. [236] Ibid, para 707. [237] Ibid, para 697.

(a) Obligation to use pertinent information

The requirement to adopt a measure on the basis of available pertinent information has received only limited attention in the case law. Consequently, one may only speculate on the normative content of this obligation. The 'pertinent information' definitely requires a lower threshold of conclusiveness than scientific evidence. On the continuum from speculative information to scientific evidence, pertinent information is to be found somewhere between these two extremes. Nevertheless, it seems impossible to establish any abstract benchmark and the assessment of pertinent information has to be performed on a case-by-case basis. Arguably, a required threshold cannot be set at a high level as this may limit the possibility of invoking the disciplines of Article 5.7 and affect the right of WTO Members to establish a level of protection they deem appropriate.

In this context it is worth recalling that there is no specific moment when purely speculative claims transform into scientific information or evidence; it is rather the accumulation of different factors that makes particular information scientific. The scientific community uses diverse standards to assess the scientific value of evidence. As discussed in Chapter 4, section 4.2, these standards include confirmation by peer reviews, reproducibility of observations, consistency with relevant theories, the ability to explain the observations with a plausible hypothesis, the inability to disprove an underlying hypothesis with experiments, and reinforcement of observations through several independent kinds of evidence and ideas that are interconnected.[238] Pertinent information provides for a lower threshold than that required from scientific evidence and may include information that, for example, has not yet been subjected to peer review or has not been confirmed in other studies (reproducibility), or is not fully consistent with theory. General studies on a particular risk which according to the SPS case law do not satisfy the requirements of specificity of risk assessment, can probably be considered as pertinent information. The same is true with respect to mere suggestive and indirect evidence. It is also advocated in the literature that pertinent information should include substantive 'inputs from officially recognized public deliberation, . . . other information concerning public values such as consumer data and public attitudes'.[239] Without rejecting the value of such data, one may doubt that they will satisfy the requirements of Article 5.7. The SPS Agreement, including Article 5.7, strongly relies on the technical paradigm which perceives risk as a pure combination of probability and effect. In consequence, information that is relevant in the context of the Agreement is predominantly of a scientific character (in the sense of hard science such as toxicology or microbiology) or relating to the risk management practises in the exporting member (eg observance of GVP). Such reading fits the language of Article 5.7 which links in its second sentence

[238] Daniel M Byrd III and C Richard Cotheren, *Introduction to Risk Analysis: A Systemic Approach to Science-based Decision-making* (2005) 27.

[239] Winickoff et al ('Adjudicating the GM Food Wars') 115.

information with assessment of risk ('Members shall seek to obtain the additional information necessary for a more objective assessment of risk'). It may be rationally assumed that the nature of pertinent information (albeit not the quality) will be the same as in cases of regular risk assessment. A mere consideration of data relating to public values or consumer preferences does not meet this threshold. This conclusion also seems to find a support in the recent report of the Appellate Body where pertinent information was conceptualized through reference to possible existence of risk.[240]

The expression 'on the basis of' in Article 5.7 should probably be understood as 'based on', which is used in Article 5.1.[241] Besides the similar textual foundation, this conclusion is also justified by systemic similarities between these two provisions. Both articles require a rationale for an SPS measure, either in the form of a risk assessment or as a collection of pertinent information. The application of the Article 5.1 standard would require the existence of a rational or objective relationship between pertinent information and an SPS measure.[242] Using the language of Article 5.1, available pertinent information must reasonably support a provisional SPS measure. In other words, there needs to be an '*objective relationship* between two elements, that is to say, to an *objective situation* that persists and is observable between an SPS measure'[243] and pertinent information. The rational relationship has to be examined on a case-by-case basis, in which 'account is taken of all considerations rationally bearing upon the issue of the potential adverse health effects'.[244]

Article 5.7 also suggests where pertinent information is to be sought. This includes relevant international organizations as well as SPS measures applied by other WTO Members. It is not clear whether organizations other than those enumerated in Annex A may be considered as relevant for the purpose of Article 5.7. On the one hand, it is advisable that a WTO Member draws from the experience of as many international organizations as possible (eg WHO). On the other hand, the practice of the SPS Committee may indicate that only the sister organizations are to be considered as relevant.[245] As far as the second source of information is concerned, Article 5.7 uses somewhat imprecise language. It is not a measure itself that may provide pertinent information, but data that were used as a basis for its adoption. An SPS measure is rather an indicator of the existence of such data. Of course, information needs to be available in practice. The existence of

240 Appellate Body Report, *US – Continued Suspension*, para 678.

241 The case law decided under Article 2.2 (which also uses the phrase 'based on') may be only of limited help here. Panels, in their analysis under Article 2.2, concentrated more on the sufficiency of scientific evidence rather than on the requirement to be 'based on'.

242 Cf Appellate Body Report, *US – Continued Suspension*, para 678 (speaking about rational and objective relationship between the information concerning a certain risk and a Member's provisional SPS measure).

243 Appellate Body Report, *EC – Hormones*, para 189 (emphasis in original).

244 Ibid, para 194; see also Appellate Body Report, *Japan – Agricultural Products*, para. 79.

245 But see Scott (*Commentary*) 120.

information which is subject to, for example, confidentiality obligations, renders it unavailable.[246]

As already mentioned, the existence of pertinent information is assessed as of the day of the adoption of an SPS measure. According to the current practice, availability of new pertinent information which has became available after the adoption of an SPS measure will be addressed under the third and fourth conditions of Article 5.7.

(b) Obligation to seek to obtain additional information

Article 5.7 also requires WTO Members maintaining provisional measures to seek to obtain the additional information necessary for a more objective assessment of risk. It is clear that the obligation calls for specific conduct rather than a result. In the words of the Appellate Body, a WTO Member 'must make best efforts to remedy the insufficiencies in the relevant scientific evidence with additional scientific research or by gathering information from relevant international organizations or other sources'.[247] The case law also held that Article 5.7 does not establish any prerequisite with respect to additional information.[248] Of course, data need to be relevant for both the risk in question and the prospective risk assessment corresponding with identified insufficiencies.[249] Equally, the case law does not require following any particular collection procedure.[250] Nevertheless, since the obligation of Article 2.2 to base SPS measures on scientific principles seems to be fully applicable to measures taken pursuant to Article 5.7, the employed methodology has to be scientifically sound. Moreover, the use of the verb 'seek' indicates that WTO Members are required to take active steps. Therefore, a mere declaration in a measure that speaks about the obligation to pursue further research does not satisfy this condition. It is the actual conduct of a WTO Member which matters.

The phrase 'necessary for a more objective assessment of risk' also provides additional guidelines on the meaning of the third conditions of Article 5.7. First, it conceptualizes the examination of pertinent information as a form of risk assessment.[251] Of course, due to deficiencies in scientific knowledge, such risk assessment cannot be required to satisfy the requirements of Annex A(4) and Article 5.1. Secondly, the expression also signifies that the ultimate goal under the SPS Agreement is to perform risk assessment which fully meets its requirements, or at least which is closer to that standard than a mere consideration of

[246] Ibid, 121 (referring to the case law where the Appellate Body considered undisclosed confidential information as non-existent).

[247] Appellate Body Report, *US – Continued Suspension*, para 679.

[248] Appellate Body Report, *Japan – Agricultural Products*, para 92.

[249] Appellate Body Report, *US – Continued Suspension*, para 679.

[250] Appellate Body Report, *Japan – Agricultural Products*, para 92.

[251] Panel Report, *EC – Biotech Products*, para 7.2992.

available pertinent information.[252] This in turn suggests that Article 5.7 imposes the obligation that is in essence continuous and which requires performance of 'transitional' assessments as new evidence or pertinent information emerges.

In this context, it is interesting to ask to what extent the situation of a particular WTO Member is relevant for the assessment of its conduct under the third condition of Article 5.7. In other words, the question is whether available national recourses allocated to scientific research matter at all. Neither the SPS Agreement nor the case law provides any direct answer. It seems that it is neither possible nor feasible to establish any predetermined benchmark. I would rather suggest a reliance on a flexible standard of reasonablness, applicable on a case-by-case basis. Such a standard would require WTO Members to make genuine efforts in the search for additional scientific information, while taking into account limitations of domestic research capacities, national priorities in addressing SPS risks, and allocation of funds between different areas of scientific research relative to the costs that are borne by other Members in maintaining a measure.[253] On the other hand, the availability of information in the relevant international standard-setting bodies as well as the possible assistance of other Members and international organizations should also constitute relevant factors. Such an approach is even more desirable in the case of developing countries.[254] Note that the SPS Agreement recognizes and acknowledges in its Preamble that WTO Members from developing countries may encounter special difficulties in the formulation and application of SPS measures in their own territories. Although this consideration finds specific reflection in Article 10, there is nothing in the Agreement which would suggest that this provision exhausts the relevance of the special treatment of developing countries. The recent decision of the Appellate Body confirms this interpretation. The Appellate Body particularly noted that 'due account shall be taken of the special needs of developing Members on respect to their ability to procure the additional information for a more objective assessment of risk'.[255] Despite a certain inaccuracy by the Appellate Body in identifying the relevant legal basis for such conclusion,[256] one may consider this statement as a general recognition of the WTO dispute settlement bodies of the needs of developing countries in the context of the SPS Agreement that require special treatment.

[252] Ibid, paras 7.2988–9 (noting that such dynamic dimension of that obligation could be deduced from the use of the word 'more').

[253] Note, however, that under such an approach there is a risk of a panel second-guessing the regulatory choices made by a WTO Member.

[254] In practice, this will be mostly relevant for developing countries, as they often have only limited financial resources dedicated to scientific research in the SPS field.

[255] Appellate Body Report, *US – Continued Suspension*, fn 1398.

[256] The Appellate Body held that this observation finds support in Article 10.1 of the SPS Agreement. Note, however, that this provision provides only the obligation for importing Members to consider the special needs of developing countries. It does not say anything on the implementation of the disciplines established by the SPS Agreement by developing countries.

(c) Obligation to review a measure within a reasonable period of time

Article 5.7 implies that an SPS measure which falls within its scope needs to be provisional, as measures are subject to review that should take place within a reasonable period of time. The provisionality as such does not need to be explicitly stated in a measure; it is rather the conduct of a WTO Member that shows whether or not a measure is provisional. The case law holds that assessment of what constitutes a reasonable period of time requires a case-by-case approach, which leaves dispute settlement bodies with a considerable degree of discretion.[257] A similar description may be found in the observation made by the Appellate Body under Article 6.8 of the Anti-Dumping Agreement where it found that a reasonable period 'must be interpreted consistently with the notions of flexibility and balance that are inherent in the concept of "reasonableness", and in a manner that allows for account to be taken of the particular circumstances of each case'.[258] In *Japan – Agricultural Products* the Appellate Body also identified elements which may influence the assessment of a reasonable period of time in a particular case. This includes difficulties in gathering new information as well as characteristics of a provisional measure.[259] This should not be considered as a closed list, and other elements may also play a role in the determination of reasonable period of time.[260]

The first factor suggests that the notion of 'reasonable period of time' is related not so much to specific time limits but rather to the development of scientific knowledge. In other words, slow progress of scientific research in a particular domain resulting from objective difficulties may justify a longer period of time. It is not clear what may qualify as a difficulty which is relevant for the assessment of reasonable time. It is at least defendable that a situation of a particular Member, including not only its budgetary constraints and political priorities but also its development status, plays a role in such a determination. Indirect support for this interpretation may be found in the case law decided under Article 21.3 of the DSU,[261] which recognizes internal constraints, such as complexities of national regulatory processes. Arbitrators, in their determination of a reasonable period of time, tend to take into account the type of regulatory action

[257] Appellate Body Report, *Japan – Agricultural Products*, para 93; in *Japan – Apples*, the Appellate Body found that a period of eight years did not constitute a reasonable period of time.

[258] Appellate Body Report, *US – Hot-Rolled Steel*, paras 84–5; note also that this standard seems to be applicable to other WTO agreements, as the relevant passage of the Appellate Body report was also referred to under Article 21.3(c) of the DSU—see Award of the Arbitrator, *US – Hot-Rolled Steel*, paras 25–6.

[259] Appellate Body Report, *Japan – Agricultural Products*, para 93.

[260] This conclusion is supported by the language used by the Appellate Body, particularly the verb 'include', which denotes non-exclusive character of the enumeration (see Appellate Body Report, *Japan – Agricultural Products*, para 93).

[261] Article 21.3 of the DSU provides that 'if it is impracticable [for a WTO Member] to comply immediately with the recommendations and rulings [of the dispute settlement body], the Member concerned shall have a reasonable period of time in which to do so'.

which is required in order to bring a particular measure into conformity with WTO law. This means that compliance that can be accomplished through administrative measures allows for a shorter period of time[262] compared to those instances where legislative action is required.[263] Moreover, the *Korea – Alcoholic Beverages (Article 21.3)* arbitrator found that a reasonable period of time 'does not require a Member ... to utilize an extraordinary legislative procedure, rather than the normal legislative procedure'.[264] There are no compelling reasons for a different standard under Article 5.7. The language is exactly the same ('reasonable period of time'), while the nature of both obligations suggests that a more stringent standard under Article 5.7 would be inappropriate. Note that Article 21.3 of the DSU is concerned with measures which were already identified as being inconsistent with WTO law. On the other hand, measures which are taken under Article 5.7 of the SPS Agreement are presumed to be compatible with WTO obligations. Applying a more lenient standard under the DSU as opposed to Article 5.7 will be counter-intuitive.

The nature of the second factor identified by the Appellate Body which is relevant in the examination of a reasonable period of time is ambiguous. Does it require a panel to assess the necessity of an SPS measure? If so, this would be an unnecessary repetition of the obligations already included in the other provisions of the SPS Agreement (eg Article 2.2 first part, Article 5.6). The statement of the Appellate Body may also be understood as a form of balancing test which weighs a reasonable period of time against the stringency of a measure. If an SPS measure establishes an absolute ban, a relatively short period would be available, while a less strict measure could justify a longer time. Although logically correct, such a reading will be at odds with the function of Article 5.7 which provides for the possibility of maintaining provisional measures as long as the scientific evidence remains insufficient. This would also violate a WTO Member's right to establish the level of protection it deems appropriate as it would require the Member to amend its measures even if scientific evidence persists in being insufficient.

It is also submitted that in situations that are characterized by low certainty and low scientific consensus (ie novel risks relating to new technologies such as genetic modification) a longer, reasonable period of time may be required.[265] The EC raised a similar argument in the *Biotech* dispute arguing that 'GMO technology is still at the frontiers of science and its future consequences (compared to a case like fire blight) are highly uncertain'.[266] The EC also added that when assessment of long-terms effects was required, while the potential changes might have a permanent character, the reasonable period of time must be longer. The language

[262] Award of the Arbitrator, *EC – Hormones*, paras 25–6; Award of the Arbitrator, *Chile – Alcoholic Beverages*, para 38; Award of the Arbitrator, *Canada – Pharmaceutical Patents*, para 47.

[263] Award of the Arbitrator, *Chile – Alcoholic Beverages*, para 42.

[264] Award of the Arbitrator, *Korea – Alcoholic Beverages*, para 42.

[265] Winickoff et al ('Adjudicating the GM Food Wars'), 115–16.

[266] Panel Report, *EC – Biotech Products*, para 7.3230.

of Article 5.7 is definitely able to include such considerations. Nevertheless, the more probable scenario would be to address those elements when assessing difficulties in gathering new information rather than to introduce them as independent factors.

Article 5.7 does not specify what should count as a starting date for the calculation of a reasonable period of time. Intuitively, one would point to the moment when a provisional measure was adopted. However, in some cases what might seem to be an easy task becomes more complicated. A good example is provided by *US – Continued Suspension* where the parties disagreed with respect to the calculation method. The EC argued that its provisional measure had to be distinguished from the previous permanent measure which had been found illegal in *EC – Hormones*. According to the US position, the new provisional measure was only a re-labelled form of the previously existing permanent ban. In consequence, the reasonable period of time should have been assessed by reference to the adoption date of the original measure (ie early 1990s). In the end, the issue was not directly addressed in the final report. Nevertheless, the approach of the panel indicates that the interpretation proposed by the US was implicitly rejected as unnecessarily expansive. The panel, when determining insufficiency of scientific evidence, identified September 2003 as a relevant moment. Obviously, this has consequences for calculating the reasonable period of time. The SPS measure which is referred to under the fourth condition of Article 5.7 is exactly the same measure that is analysed within the first sentence of Article 5.7. This seems to be the correct approach.

5.4.4 Burden of proof

Article 5.7 was characterized in the SPS case law as an exception, a qualified exemption, an autonomous right, and a qualified right.[267] This heterogeneity in the nomenclature is probably a reflection of the confusion that persists in WTO jurisprudence as to the nature of Article 5.7. As far as the allocation of burden of proof is concerned, qualifying a particular provision as a right (or a positive obligation) requires a complainant to establish a prima facie case of incontinency between a measure and a relevant provision. On the other hand, in case of an exception, it is a defendant who needs to demonstrate that its measure can be justified under such an exception.

The panel in *Japan – Agricultural Products* held that the burden of making a prima facie case of violation of Article 5.7 was borne by the complainant. In accordance with the standard rules, after its successful establishment it was for a defendant to rebut the presumption of inconsistency.[268] Consequently, the

[267] Cf Appellate Body Report, *Japan – Agricultural Products*, para 80; Panel Report, *Japan – Apples*, para 7.10; Panel Report, *EC – Biotech Products*, para 7.2976.

[268] Panel Report, *Japan – Agricultural Products*, paras 4.187 and 8.48.

United States was expected to provide evidence showing that available scientific data were sufficient to perform a risk assessment. Once a prima facie case was established, it was for Japan to adduce evidence demonstrating the opposite.[269] This approach changed, however, with the panel's report in the *Apples* case. The panel reached the provisional conclusion that the Japanese measure was maintained without sufficient scientific evidence within the meaning of Article 2.2. Only then did the panel proceed with its analysis under Article 5.7.[270] It referred to the standard developed in *US – Shirts and Blouses* and held that it was for the party invoking a particular provision to bear the burden of making a prima facie case.[271] In other words, the panel saw Article 5.7 as a form of justification for otherwise WTO-inconsistent measures rather than a provision which establishes a positive obligation.[272] Since the allocation of the burden of proof was not challenged by Japan in the appeal, the Appellate Body was not required to address this issue. However, it is also worth noting the Appellate Body explicitly mentioned that this part of the panel interpretation was not appealed,[273] which might suggest a certain reservation with respect to the panel's allocation of the burden of proof.[274]

Again it was the *Biotech* panel which clarified this ambiguity. As already discussed in section 5.4.1, the panel recognized Article 5.7 as a positive obligation (in the panel's nomenclature a 'qualified right') rather than the exception to other obligations of the SPS Agreement.[275] This qualification of Article 5.7 required a panel to assign the complainants with the burden of showing a prima facie case of inconsistency of provisional measures.

The final conclusion of the *Biotech* panel on the allocation of the burden of proof under Article 5.7 is definitively correct. Assigning a complainant with the burden of making a prima facie case corresponds with the nature of Article 5.7. The SPS Agreement recognizes that a WTO Member may take any kind of regulatory action subject to certain requirements. If Articles 2.2/5.1 and 5.7 are conceptualized as providing two separate paths of compliance, it is logical that in both cases it should be a complainant that is responsible for making a prima facie case of violation. If scientific evidence is sufficient, this would require showing

[269] Ibid, para 8.58.

[270] This approach was understandable since Japan presented its arguments as an alternative, arguing that the panel should have examined a measure under Article 5.7 only if it found that Article 2.2 was violated (Panel Report, *Japan – Apples*, para 8.203).

[271] Ibid, paras 7.26 and 8.222.

[272] Note that the allocation of burden of proof adopted in *Japan – Apples* was strongly criticized by WTO Members: see eg, Minutes of the DSB meeting of 10 December 2003, WT/DSB/M/160, paras 9–10, where both the US and Japan expressed their disappointment with respect to the panel's approach.

[273] Appellate Body Report, *Japan – Apples*, fn 316.

[274] Cf Panel Report, *EC – Biotech Products*, para 7.2980.

[275] Ibid, para 7.2960; note that in *EC – Biotech Products* the panel introduced the notion of 'qualified rights' while the previous case law refers to 'autonomous right'. Despite different terms it seems that these two notions refer to the same concept.

that a WTO Member did not observe disciplines provided by Articles 2.2 and 5.1 to 5.3. A distinct situation exists if there is insufficient scientific evidence where a separate set of obligations applies to the regulatory actions of a WTO Member. In such a case, a complainant should establish that one of the conditions provided by Article 5.7 was not met by a measure of a defendant. Last but not least, the *Biotech* panel decision also needs to be welcomed as it improves the overall consistency of the SPS case law by clarifying who bears the burden of proof under Article 5.7.

What is, however, disappointing about the decision in *EC – Biotech Products* is the structure of the panel's reasoning.[276] Although the panel recognized Article 5.7 as a positive obligation (autonomous right), it came to this conclusion on very narrow textual grounds. The panel did not engage in a more normative analysis which would address the nature of both provisions (ie Article 2.2 as opposed to Article 5.7). The whole examination was based on the mechanical application of the test introduced by the Appellate Body in *EC – Tariff Preferences* and its outcome turned out to depend entirely on the interpretation of the expression 'except as provided for'. The panel held, without any elaborate discussion, that the phrase denoted exclusion rather than exception.[277] On that basis, the panel concluded that Article 5.7 was the autonomous right and not the exception.[278] Similar logic was employed with respect to the relationship between Articles 5.1 and 5.7. Here the panel was able to identify the existence of exclusion even without any textual basis in the relevant provisions. According to the panel, since Article 5.1 was a specific application of the obligation found in Article 2.2 it could not apply in situations where Article 2.2 is inapplicable.[279] Again, the panel did not examine the nature of both obligations which would support or oppose such a conclusion and was satisfied with its textual argument.

The *Suspension* panel confirmed that the burden of making a prima facie case under Article 5.7 is borne by a complainant.[280] In the specific circumstances of the post-suspension case, it was for the EC to show that the relevant scientific evidence regarding the five hormones was insufficient to perform a risk assessment.[281] In support of such allocation of burden of proof, the panel also adduced an additional argument. As discussed earlier, the panel believed that the existence of international standards for those hormones would in any case shift the

[276] For an illuminating critique of the panel's approach, see Tomer Broude, 'Genetically Modified Rules: the Awkward Rule—Exception—Right Distinction in *EC – Biotech*' (2007) 6(2) World Trade Review 215.

[277] Ibid, 219 (as noted by Broude 'the panel's textual logic is such that had the phrase "except for..." not been included in Art 2.2, it could not characterize Art 5.7 as a "right", because the text would not "suggest" that Art 2.2 is inapplicable').

[278] Cf ibid, 221–7 (noting the confusion in the *Biotech* panel report between different terms such as positive rule, exception, and autonomous right).

[279] Panel Report, *EC – Biotech Products*, paras 7.2995–96.

[280] Panel Report, *US – Continued Suspension*, para 7.652.

[281] Note that in a regular dispute it would be for a complainant to make a prima facie case with regard to sufficiency of scientific evidence for the purpose of risk assessment.

traditional burden of proof to a WTO Member maintaining a measure (*in casu* the EC).[282] The same approach was adopted with respect to those cases where intensive work had been performed within one of the sister organizations, even if a standard as such had not yet been adopted.[283] As a consequence of such an approach, the panel concluded that it could limit the 'review exclusively to the "insufficiencies" expressly identified by the European Communities in its submission'.[284]

Although the Appellate Body did not eventually change the outcome of the panel decision on the allocation of burden of proof, it based its findings on entirely different grounds. The duty of the EC to identify and substantiate its claims as to the insufficiency of scientific evidence under Article 5.7 came only as a result of specific circumstances of the case. There is no place and need to analyse here the particularities of such post-suspension proceedings.[285] It is worth noting that the Appellate Body held that it was for the EC to 'provide an adequate explanation of how the provisional ban taken under Article 5.7 rectifies the inconsistencies found in the *EC – Hormones*'.[286] This should particularly include the identification of those insufficiencies which prevented the EC from conducting risk assessments as required by Article 5.1.[287] What, however, seems to be more important in this context is the fact that the existence of international standards (or risk assessment performed by an international scientific body such as the JECFA) does not reverse the normal burden of proof under Article 5.7. In a traditional SPS dispute it will be for a complainant to establish a prima facie case of violation of Article 5.7, even if there is a corresponding international standard. Again, as already noted, the complainant may use the scientific data that constitutes the basis of a standard in order to establish a prima facie case of inconsistency, but this fact does not change anything in the allocation of the burden of proof.

5.4.5 The precautionary principle and other provisions of the SPS Agreement

The case law confirms that Article 5.7 does not exhaust the relevance of the precautionary principle under the SPS Agreement and there are other provisions which also reflect the idea of precaution. The Appellate Body in *EC – Hormones* identified the sixth paragraph of the Preamble together with Article 3.3, and noted that these provisions 'recognize the right of Members to establish their own appropriate level of sanitary protection which level may be higher (ie more cautious) than that implied in existing international standards, guidelines and

[282] Panel Report, *US – Continued Suspension*, para 7.652.
[283] Ibid, para 7.661 in connection with para 7.813.
[284] Ibid, para 7.653.
[285] For details on the allocation of burden of proof in post-suspension cases, see Appellate Body, *US – Continued Suspension*, paras 360–5.
[286] Ibid, para 716. [287] Ibid.

recommendations'.[288] This is a somewhat imprecise observation since the establishment of higher, or, using the words of the Appellate Body, more cautious, ALOP than reflected in international standards does not depend on the existence of uncertainty. The recourse to Article 3.3 is equally justified in the circumstances of high scientific certainty as to the existence of risk, and it may only reflect a different tolerance of a particular state for such risk. Arguably, the Appellate Body used the expression 'precaution' in the more colloquial sense of prudence understood as caution or circumspection to risk. The case law has also recognized the relevance of the precaution in the context of Article 2.2 when examining the sufficiency of scientific evidence. In the words of the Appellate Body, 'a panel...should, bear in mind that responsible, representative governments commonly act from perspectives of prudence and precaution where risks of irreversible, eg life-terminating, damage to human health are concerned'.[289] As already noted, this probably implies a willingness of the WTO dispute settlement bodies to grant national governments a certain additional margin of discretion when regulating irreversible risks. Consequently, the nature of risk has an impact on the degree of required certainty. Elaborating on this finding, the *Biotech* panel found that Article 5.1 equally has some built-in precautionary elements. In particular, the precaution may influence the assessment of an objective relationship that is required between risk assessment and an SPS measure. As explained by the panel, in some cases a Member which relies on the precautionary approach, and 'which confronts a risk assessment that identifies uncertainties or constraints, would be justified in applying (i) an SPS measure even though another Member might not decide to apply any SPS measure on the basis of the same risk assessment, or (ii) an SPS measure which is stricter than the SPS measure applied by another Member to address the same risk'.[290] This again confirms an additional margin of discretion enjoyed by WTO Members when confronted with uncertainties in their risk assessment. Of course, the margin is not unlimited; the uncertainties only affect the extent (and not the existence) of required objective relationship. The panel in *US – Continued Suspension* also subscribed to the same view that uncertainties identified in the process of risk assessment may have a direct influence on the regulatory discretion of WTO Members. Thus, the level of confidence attributed to the results of risk assessment may be affected by assumptions that were made in the course of risk assessment and Methodoligal constraints, and justify the adoption of more stringent measures than would otherwise be possible.[291] WTO Members may also follow the minority scientific views which may be more cautious than mainstream science. This gives additional room for manoeuvre, where precautionary considerations easily find their place.

[288] Appellate Body Report, *EC – Hormones*, para 124.
[289] Ibid, para 181.
[290] Panel Report, *EC – Biotech Products*, para 7.3065.
[291] Panel Report, *US – Continued Suspension*, para 7.633.

5.5 Conclusions

The case law has only partially succeeded in elucidating the normative content of the disciplines established by Article 5.7, and a number of important theoretical and practical issues still remain unclear. From a theoretical point of view, one may ask to what extent Article 5.7 is indeed a reflection of the precautionary principle. Since there is no uniform definition, the answer is not an easy one. It seems at least defendable to argue that Article 5.7 provides a specific formulation of the principle.[292] Note that it permits, similarly to other versions of the principle, regulatory action even without the existence of conclusive scientific evidence as to the nature of hazard and the extent of potential risk. The concept of insufficiency of scientific evidence, albeit different from the idea of uncertainty, in fact covers some of its types (eg uncertainty resulting from indeterminacy or lack of necessary measurement data). At the same time, the SPS version of the principle expresses it own particularities. With regard to some aspects it provides less demanding requirements than other formulations of the principle. For instance, it asks for prior positive scientific evidence (in the nomenclature of the SPS Agreement, pertinent information) which authenticates potential risks. This seems to go further than just the mere existence of doubts as to the safety of a product or technology, but it is not unique. Article 5.7 also presupposes the provisional character of precautionary measures by mandating their review once new scientific information becomes available. Other versions do not explicitly provide such temporal limitations. Surprisingly, there are also elements that are less demanding compared to other formulations of the precautionary principle. For instance, Article 5.7 does not require making any trade-offs between protective goals and potential costs to international trade. National regulation can be as expensive as the particular government wishes (of course subject to the four requirements provided by the provision). Other versions (eg the Rio Declaration) ask for 'cost-effective measures' which may imply the performance of some cost–benefit analysis before the precautionary measure is adopted. Article 5.7 does not contain any reference to threats of serious or irreversible damage and its application is possible if a potential risk is simply above ALOP. Again, such formulation gives a much wider discretion to domestic regulators as opposed to other versions of the principle.

On a practical level, the case law has not yet addressed some important issues that are relevant in the context of Article 5.7. Thus, the character of the relationship between Article 5.7 and other provisions of the SPS Agreement, including different obligations of Article 2.2, still remain unclear and one may only speculate on the requirements that are applicable to provisional measures. As far as the substance of Article 5.7 is concerned, the SPS jurisprudence had a chance to

[292] Cf Applegate ('Taming') 31.

clarify more systematically only its first condition, while the other requirements still await authoritative interpretation. For instance, further guidance is required as to the nature of pertinent information, the relevance of the development status of a Member, and domestic regulatory priorities in the assessment of the obligations to seek to obtain additional information for a more objective evaluation of risk, as well as the treatment of situations that are characterized by low certainty and low consensus when calculating a reasonable period of time.

Article 5.7 generally proved to pose particular difficulties for the SPS dispute settlement bodies. Thus, it took some time for SPS case law to elaborate what I consider to be a proper allocation of the burden of proof, which assigns to a complainant the duty of making a prima facie case of inconsistency with the disciplines of Article 5.7. The same is true for assessment of insufficiency of scientific evidence. The panels tended to see it as a purely scientific and absolute idea which could be separated from normative considerations that are country-specific (eg appropriate level of protection, and the cautious approach of domestic risk assessors to particular risks). This belief was translated into the excessive importance that was assigned to risk assessment existing on the supranational and international level. Similarly, panels recognized only the temporal and not the spatial dimension of insufficiency. If one takes this approach to its extreme, disciplines of Article 5.7 become unavailable whenever any WTO Member, including a complainant, is able to perform the relevant risk assessment. This not only affects the right of a defendant to establish its own level of protection but is also incorrect from the perspective of risk science. Assessing the sufficiency of scientific evidence is not a simply scientific task but also has a normative dimension where judgments of the experts reflecting their attitude towards specific risks, together with the values of a particular community, play an important role. The standard of review applied to examination of insufficiency was equally erroneous. The panel engaged in a *de novo* review, weighing different scientific positions and deciding what constituted the best science in a particular case. This also meant that minority scientific opinions, including those which were expressed by the panels' own experts, were frequently underestimated or even disregarded. The above developments of the case law were corrected by the Appellate Body in *US/Canada – Continued Suspension*. The normative dimension in the assessment of insufficiency of scientific evidence was finally accepted as a relevant factor. This finding was also combined with the recognition that other risk assessments (including those which come from sister organizations) may only serve as a source of scientific data that are helpful, but not dispositive, in the determination of insufficiency. One of the consequences of such an approach was denial of the test of critical mass that was promoted by the *Suspension* panel. Although the Appellate Body did not pronounce any standard of review that is applicable in the context of Article 5.7, its approach under Article 5.1 indicates that the role of the panel is limited to the assessment of whether the domestic determination of insufficiency of scientific evidence is supported by coherent reasoning and comes from respectable scientific sources. Among other things, this requires acceptance that the existence of insufficiency may be determined on the basis of minority scientific opinions.

At the same time, some of the interpretations proposed by the case law remain disappointing. Probably the most serious failure of the SPS jurisprudence is its approach to the relationship between the disciplines of Articles 2.2/5.1 and 5.7. The current practice confuses the issues of applicability and consistency by requiring measures to comply with all conditions provided by Article 5.7 as a prerequisite of its application. The approach which is postulated in this chapter calls for conceptualizing Articles 2.2/5.1 and 5.7 as providing two distinctive paths of compliance that are applicable to entirely different situations. When scientific evidence is sufficient, a WTO Member is required to follow the normative prescription of Articles 2.2 and 5.1. A different set of obligations applies in all those cases where scientific data are insufficient to perform a risk assessment. The relationship between these two paths is one of exclusion, and a failure to meet the requirements applicable to one path does not mean that an SPS measure must be considered in the other.

The interpretation of Article 31.3(c) of the VCLT which was proposed by the *Biotech* panel appears to be equally unsatisfactory. Although that issue is of a more general character since it was decided in the context of the discussion on the precautionary principle, it should also be mentioned here. The narrow reading proposed by the panel has very significant consequences as it transforms the WTO into a *quasi* self-contained regime. Regrettably the seriousness of potential implications was not translated in the panel report into a more elaborate discussion on the possible alternative approaches, their advantages and drawbacks. Instead the *Biotech* panel promoted a narrow reading of the VCLT as the only defensible option.

Despite the above failures (as well as some others that are mentioned here), the overall assessment of the case law relating to Article 5.7 is rather positive. The new standards introduced by the Appellate Body in *US/Canada – Continued Suspension* seem properly to address the major points of concern and leave a considerable degree of freedom to WTO Members when regulating uncertain risks.

6

Risk Management Disciplines and the SPS Agreement

6.1 Introduction

The core of the SPS Agreement consists in its science-based provisions such as Articles 2.2, 5.1, and 5.7. The Agreement is, however, not limited to these types of obligations and it also imposes a number of requirements on the domestic risk management process. Interestingly, this part of the SPS Agreement has received less attention in the legal scholarship. This chapter intends to restore the balance in the discussion by analysing the normative content of these disciplines in a more detailed way. Some of those provisions, such as the consistency requirement or equivalence, are equally innovative in the context of WTO law as science-based rules.

The chapter proceeds as follows. The first part examines the relevance of the distinction between risk assessment and risk management in the context of the SPS Agreement. In the second part, special attention is given to the Appellate Body report in the *Hormones* case which seems to establish a standard for subsequent case law. This part concludes that, despite the lack of textual basis in the Agreement, it is legitimate to speak about its risk management dimension. The third part concentrates on the relevant provisions of the Agreement. This particularly includes establishment of ALOP, options evaluation, implementation and application of SPS decisions, as well as their monitoring and review. The last section attempts to draw some overall conclusions on the risk management dimension of the SPS Agreement.

6.2 Risk assessment and risk management

As discussed in Chapter 1, risk assessment is conventionally defined as a process of a probabilistic estimation of the potential adverse health or environmental effects of a substance, process, action, or event, determined according to scientifically plausible methods. The goal of risk assessment is to provide risk managers with the information necessary for rational decision-making. On the other hand, risk management is defined as 'a process of identifying, evaluating, selecting and implementing actions to reduce risk'[1] and reflects the preferences of a particular society for an acceptable level of risk exposure. The same language is contained in the documents of the sister organizations. The Codex definition stipulates that risk management is a process 'distinct from risk assessment, of weighing policy alternatives, in consultation with all interested parties, considering risk assessment and other factors relevant for the health protection of consumers and for the promotion of fair trade practices, and, if needed, selecting appropriate prevention and control options'.[2] Risk management is also recognized as an integral part of pest risk analysis.[3] The IPPC defines it as 'an evaluation and selection of options to reduce the risk of introduction and spread of a pest'.[4] A similar definition is provided by the OIE Terrestrial Animal Health Code which also sees risk management as an indispensable part of risk analysis.[5] Despite some differences in the language used by the international standard-setting bodies, all of them recognize risk management as an element which is distinct from risk assessment.[6] Remember that the aim of this functional separation between risk assessment and management is to ensure the scientific integrity of risk assessment by differentiating scientific and other factual findings from value judgments. Of course, I do not claim here that risk assessment is an absolutely value-free process. Again, as already discussed in Chapter 1, the distinction should be seen as an attempt to minimize their values in the assessment rather than a magic formula that makes the whole process value resistant.

Remember also from Chapter 1 that risk management is conceptualized as a four-element process consisting of risk evaluation, options evaluation, implementation of decision, and monitoring and review.[7] The first phase comprises an evaluation of risk assessment results in order to determine whether any regulatory action is required.[8] If no risk has been identified by risk assessment or such risk is below ALOP, no further action is taken. Otherwise, in the second phase,

[1] Presidential/Congressional Commission on Risk Assessment and Risk Management *Final Report, Vol 1, Framework for Environmental, Health Risk Management* (1997) 1.

[2] Codex, *Procedural Manual* (18th edn, 2009) 73.

[3] ISPM No 2, Framework for Pest Risk Analysis (2007) 5.

[4] ISPM No 5, Glossary of Phytosanitary Terms (2005) 13.

[5] OIE, *Terrestrial Animal Health Code* (18th edn, 2009), Glossary (defining risk management as a 'process of identifying, selecting and implementing measures that can be applied to reduce the level of risk').

[6] Ibid, Article 2.1.1; Codex, *Procedural Manual* 73; ISPM No 2, Framework for Pest Risk Analysis (2007) 5.

[7] OIE, *Terrestrial Animal Health Code*, Article 2.1.6; Codex, *Procedural Manual*, 71; ISPM No 2, Framework for Pest Risk Analysis (2007) 10–12 (identifying three steps of risk management—option evaluation, selection, monitoring and evaluation, while risk evaluation is included in the phase of option evaluation).

[8] The Codex refers to this phase as 'preliminary risk management activities'. According to the *Procedural Manual*, this category, alongside risk evaluation, also covers activities such as identification of a food safety problem, establishment of a risk profile, ranking of the hazard for risk

risk managers determine available options and evaluate them with regard to their efficacy and feasibility (including social, economical, and environmental impact of proposed measures). Risk managers may modify ALOP during this phase if reaching such a level would be too expensive or otherwise too difficult. The third phase consists of implementation of a specific SPS measure. Such a measure is subject to an ongoing process of monitoring and review in order to ensure that it corresponds with the changing circumstances and new data.

The SPS Agreement does not refer to risk management. Nevertheless, the substantive analysis of its obligations reveals that its different provisions are clearly concerned with risk management actions.

6.3 What the Appellate Body said and did not say in the *Hormones* cases

The relevance of the distinction between risk assessment and management under the SPS Agreement was addressed in the first SPS case. The *Hormones* panel observed that Article 5 of the SPS Agreement was concerned with two aspects of the national decision-making process. The first was risk assessment, defined by the panel as 'a process of assessment of potential adverse effects (if any) related to a specific substance... together with the probability of occurrence of any such effects'.[9] The second aspect could be labelled as risk management and related to the determination of ALOP together with the subsequent implementation of SPS decisions. The panel characterized risk management as a non-scientific process which involves social value judgments.[10] According to the panel, 'once the risks have been assessed,... a Member will need to decide, on the basis of its own value judgments, whether it can accept these risks'.[11] The panel also added that the obligations that are particularly relevant during this phase were included in Articles 2.2 to 2.3 and 5.4 to 5.6 of the SPS Agreement,[12] as they determined how a WTO Member may react to risks identified in the risk assessment. This finding led the panel to conclude that the evaluation of risks connected with possible abuses of hormones in agricultural practice (non-observance of GVP) had to be taken into account only in the risk management phase.[13] On the same basis, the panel rejected some reports of the European Parliament,[14] as it found that they only evaluated other scientific documents and, consequently, constituted a

assessment and risk management priority, establishment of risk assessment policy for the conduct of the risk assessment, and commissioning of the risk assessment (Codex, *Procedural Manual* 71).

9 Panel Report, *EC – Hormones (US)*, para 8.92.

10 Ibid, para 8.160.

11 Ibid.

12 Ibid, para 8.96.

13 Ibid, para 8.146.

14 Ibid, para 8.109 (specifically rejecting consideration of the 1981 Nielsen Report, the 1985 Collins Report, the 1989 Collins Report and the 1989 Pimenta Report).

part of risk management. Again, all those reports analysed the problem of abusive administration of hormones in the American agriculture sector.[15]

The Appellate Body, when discussing the approach of the panel, used somewhat unfortunate language, and criticized the distinction that was introduced by the panel. It is worth citing the Appellate Body in full:

> [T]he Panel observed that an assessment of risk is, at least with respect to risks to human life and health, a 'scientific' examination of data and factual studies; it is not, in the view of the Panel, a 'policy' exercise involving social value judgments made by political bodies. The Panel describes the latter as 'non-scientific' and as pertaining to 'risk management' rather than to 'risk assessment'. We must stress, in this connection, that Article 5 and Annex A of the SPS Agreement speak of 'risk assessment' only and that the term 'risk management' is not to be found either in Article 5 or in any other provision of the SPS Agreement. Thus, the Panel's distinction which it apparently employs to achieve or support what appears to be a restrictive notion of risk assessment, has no textual basis. The fundamental rule of treaty interpretation requires a treaty interpreter to read and interpret the words actually used by the agreement under examination, and not words which the interpreter may feel should have been used.[16]

More than 10 years later, the problem of the distinction between those two phases of the risk analysis model resurfaced in *US/Canada – Continued Suspension*. The EC argued that risk assessment as provided by Article 5.1 also included a risk management stage which involved weighing policy alternatives in light of the results of risk assessment, and selecting and implementing appropriate measures.[17] The panel disagreed and explained that the term 'risk management' was not used in the Agreement. According to the panel, the standard that was established in the first *Hormones* case did not provide a sufficient grounding for the interpretation proposed by the EC.[18] On that basis, the panel concluded that it was only obliged to examine whether the SCVPH opinions (alleged risk assessment of the EC) identified the potential of adverse effects on human health resulting from residues of oestradiol-17β in beef meat when applied in accordance with GVP. Only if the answer was positive, would the panel consider the impact of misuses and abuses of hormones in agricultural practice. The Appellate Body again criticized the panel. In the view of the Appellate Body, the approach of the panel resulted in the same restrictive notion of risk assessment that had been rejected in *EC – Hormones*. Consequently, the panel erroneously sought to re-establish the rigid distinction between risk assessment and risk management, a conclusion that was not supported by the language of the SPS Agreement.[19]

[15] Interestingly, the Appellate Body subsequently found that those reports did not constitute either scientific evidence or risk assessment, and consequently could not serve as a basis for an SPS measure (Appellate Body Report, *EC – Hormones*, para 207).

[16] Ibid.

[17] Panel Report, *US – Continued Suspension*, para 7.519.

[18] Ibid, para 7.520.

[19] Appellate Body Report, *US – Continued Suspension*, para 542.

Did the Appellate Body really reject the concept of risk management in the SPS Agreement? Note that in both *Suspension* reports, without counting the annexes, the term appears 47 times. That seems to be quite a lot for a concept that is supposedly irrelevant in the context of the Agreement.

A closer analysis of the *Hormones* and *Suspension* reports indicate that in both cases the Appellate Body was more concerned with the specific consequences resulting from the panel's approach rather than the distinction itself. In particular, one of the problematic outcomes of the panel's reasoning was the exclusion from risk assessment of 'all matters not susceptible of quantitative analysis by the empirical or experimental laboratory methods commonly associated with the physical sciences'.[20] However, as further explained by the Appellate Body 'the risk that is to be evaluated... is not only risk ascertainable in a science laboratory operating under strictly controlled conditions, but also risk in human societies as they actually exist, in other words, the actual potential for adverse effect on human health in the real world where people live and work and die'.[21]

It is not difficult to agree with the findings of the Appellate Body in *EC – Hormones* that the definition of risk assessment under the SPS Agreement is sufficiently broad to cover assessment of risks resulting from the non-observance of GVP.[22] In fact, this can be done only in the risk assessment process, either as a separate evaluation or as an element of the original risk assessment. Risks resulting from the abusive use of hormones in agricultural practice are in principle the same as other types of risks and in order to be managed have to be assessed in the first place. The risk management phase would only address the questions of whether and how to deal with pre-identified risks. It may therefore be argued that what the Appellate Body did not like in the panel analysis was a narrow conceptualization of risk assessment and not the distinction as such. Consequently, one may see the approach of the Appellate Body not as a general rejection of the concept of risk management within the SPS Agreement, but rather as a dismissal of restrictive, and in fact incorrect, formulation of risk assessment. The same is true for the second *Hormones* case. Since the evaluation of potential abuses in administration of hormones by American farmers was a part of the EC risk assessment,[23] the panel methodology, which asked for a separate examination of such risks and risks related to hormones when applied in accordance with GVP, was clearly erroneous. These findings of the Appellate Body, however, do not per se undermine the relevance of the distinction between the two phases of the risk analysis model. Moreover, the Appellate Body, when reversing the panel, stressed that the panel erred in making a *rigid* distinction rather than by merely differentiating between risk assessment and management.

[20] Appellate Body Report, *EC – Hormones*, para 187. [21] Ibid.
[22] Cf Chapter 4, section 4.4.1(h).
[23] Appellate Body Report, *US – Continued Suspension*, para 548.

The approach of the Appellate Body also appears to correspond with the modern understanding of those concepts. As discussed in Chapter 1, section 1.4.3, there seems to be a consensus that strict separation between risk assessment and management is not possible. The relation between these two is more dynamic and iterative than provided by earlier models of risk analysis. Some risk management considerations are accepted as indispensable factors of the risk assessment process. Default assumption, framing, application of safety factors, or a choice of research methods are all elements that cannot be explained by simple reference to science. Despite this, the majority of domestic and international institutions responsible for assessment or management of different types of risks still recognize the importance of separating risk assessment from its management. The value of the distinction lies in continued attempts to control political aspects that are inevitably involved in a risk assessment in order to ensure, to the greatest possible extent, the integrity and quality of scientific advice. Again, this seems to be reflected in the approach of the Appellate Body which did not endorse the EC position that Article 5.1 actually included both risk assessment and management phases. In fact, the Appellate Body in another section of the report clearly held that risk assessment remains a rigorous and objective process in 'which possible adverse effects are evaluated using scientific methods'.[24]

There are some additional arguments which support the recognition of risk management within the SPS Agreement.[25] First, this would facilitate reconciling the text of the Agreement with the documents of the international standard-setting bodies operating in the SPS field. As noted above, all sister organizations rely on the risk analysis paradigm which distinguishes between risk assessment and management. The same is true for most of the modern national risk regulatory systems. In fact the precise identification of the SPS provisions that are concerned with risk management may actually facilitate domestic compliance. A Member that follows a risk analysis model will be able to determine more easily which SPS obligations are applicable to specific steps that are normally taken within its risk management process. Secondly, the distinction may help to avoid in the future some of the mistakes of the previous case law, such as characterizing risks as negligible or non-appreciable which are clearly subjective judgments belonging to risk management phase judgements. Thirdly, the analysis of the SPS Agreement, as provided below, shows that risk management disciplines can quite easily be identified in the Agreement. These provisions establish a number of different

[24] Ibid, para 534.

[25] Contra, David Winickoff, Sheila Jasanoff, Lawrence Busch, Robin Grove-White, and Brian Wynne, 'Adjudicating the GM Food Wars: Science, Risk, and Democracy in World Trade Law' (2005) 30 Yale Journal of International Law 81, 98. The authors criticize the distinction between risk assessment and management, claiming that it is 'neither feasible nor appropriate to separate science policymaking into a purely technical phase and a political phase'. Without deciding on this broad issue, it seems that the above argument is based on the very narrow understanding of risk assessment (ie purely quantitative laboratory assessment of risk), which does not find a support in SPS practice.

requirements on risk and options evaluation, implementation, monitoring, and review—activities that are conventionally described as elements of the risk management process. Last but not least, it also seem that the current academic discussion tends to concentrate on risk assessment disciplines, while the rest of the SPS Agreement does not receive proper attention. Thus, reintroducing a distinction between risk assessment and management may also be helpful in highlighting the importance of other provisions of the SPS Agreement. Note also that the introduction of the concept of risk management to the SPS Agreement remains neutral for WTO Members. It does not change the allocation of the rights and obligations, but rather it helps to conceptualize SPS disciplines in a clear and more consistent manner.

6.4 Risk management provisions in the SPS Agreement

The SPS Agreement sets different requirements with respect to national risk management activities. First of all, it confirms the right of each WTO Member to set its own ALOP.[26] That right is, however, not absolute as Members have to avoid arbitrary or unjustifiable distinctions in the levels they consider to be appropriate in different but related situations.[27] Moreover, Members are expected to take into account the objective of minimizing potential negative trade effects when establishing ALOPs.[28] The SPS Agreement also requires WTO Members to evaluate in their quarantine risk assessment the efficiency of different SPS alternatives. Irrespective of the type of assessment, countries are obliged to supply both the WTO Secretariat and other Members with relevant information relating to contemplated SPS measures and to take into account their comments and observations. A final measure cannot be applied beyond what is necessary to protect human, animal, or plant life or health and be more trade-restrictive than required to achieve ALOP.[29] The Agreement suggests two specific ways that may help to ensure the necessity of a measure and its least restrictiveness, through application of regionalization and equivalence provisions. WTO Members also need to review their measures as scientific knowledge develops in order to guarantee that they are based on up-to-date information.

When one compares the above obligations with the structure of risk management, it becomes clear that the SPS Agreement imposes a number of disciplines on how risks are handled. It sets certain requirements on the establishment of ALOP, evaluation of different options, implementation of the risk-reducing actions, and monitoring and review of SPS measures. The subsequent part of

26 Recital No 5, Preamble the SPS Agreement, and Articles 3.3, 4.1, 5.4–5.6 and Annex A(5).

27 Article 5.5 of the SPS Agreement.

28 Article 5.4 of the SPS Agreement.

29 Article 5.6 of the SPS Agreement.

the chapter will address each of the risk management phases, comparing it with relevant SPS provisions and corresponding case law.

6.4.1 Establishing the appropriate level of protection (ALOP)

ALOP is defined in Annex A of the SPS Agreement as the level of protection deemed appropriate by a WTO Member introducing an SPS measure to protect, within its territory, human, animal, or plant life or health. ALOP determines how much risk is acceptable for a particular Member and can be understood as the highest level of risk that a country is prepared to tolerate. The Preamble to the SPS Agreement confirms that the Agreement does not require WTO Members to change their ALOP.[30] Along the same lines, the definition of ALOP which is included in Annex A of the SPS Agreement characterizes it as the independent choice of a WTO Member (ie 'deemed appropriate by the Member'). Other provisions of the SPS Agreement also suggest that ALOP should be seen as a sovereign right of Members.[31] This explicit language is reflected in the case law, which confirms that WTO Members have considerable discretion in setting their level of SPS protection.[32] This implies that they may 'trade off health values from other social values (including economic welfare)'.[33]

The SPS Agreement implicitly requires positive determination of ALOP,[34] either quantitatively (eg 1:1,000) or qualitatively (eg low, moderate, high, conservative).[35] This is, however, a relatively undemanding obligation as a mere statement of a WTO Member on its ALOP in a panel proceeding seems to satisfy the requirement of positive determination.[36] Due to the indirect character of the obligation, lack of (or imprecision in) a determination does not constitute a violation of the SPS Agreement. In such a case ALOP 'may be established by panels on the basis of the level of protection reflected in the SPS measure actually applied'.[37] Consequently, failure of a WTO Member to clearly articulate its ALOP does not impair the ability of a panel to examine a measure under those provisions of the SPS Agreement which require pre-assessment of a level of SPS protection (eg Article 5.5).

(a) ALOP and SPS measures

One has to distinguish between an ALOP and an SPS measure. The former is an objective, while the latter is an instrument used to attain or implement that

[30] Recital No 6, Preamble to the SPS Agreement.

[31] See eg, Article 3.3, which allows WTO Members to introduce and maintain measures that result in a higher level of SPS protection than would be achieved if based on the international standards; other provisions include Articles 4.1, 5.4–5.6 and Annex B(3) of the SPS Agreement.

[32] Appellate Body Report, *EC – Hormones*, para 124; Appellate Body Report, *Australia – Salmon*, para 199.

[33] Jeffrey Atik, 'The Weakest Link: Demonstrating the Inconsistency of "Appropriate Levels of Protection" in Australia – Salmon' (2004) 24(2) Risk Analysis 483, 484.

[34] Appellate Body Report, *Australia – Salmon*, para 206.

[35] Ibid. [36] Ibid, para 197. [37] Ibid, para 207.

objective. As noted by the Appellate Body, 'determination by a Member of the "appropriate level of protection" logically precedes the establishment or decision on maintenance of an SPS measure'.[38] Thus, under the SPS Agreement, ALOP is a benchmark that is external to an SPS measure. A measure itself also reflects a certain level of protection (ie an import ban reflects a zero-risk level). In the majority of cases these two are the same, meaning that an SPS measure simply reflects an ALOP. Further, two levels will be regarded as equal when a Member fails to make its determination or does it with insufficient precision. In this case, ALOP will be established by a panel on the basis of the level of protection reflected in an SPS measure. On the other hand, in some cases both levels may differ. As observed by the Appellate Body 'to imply the appropriate level of protection from the existing SPS measure would be to assume that the measure always achieves the appropriate level of protection determined by the Member. That clearly cannot be the case.'[39] Consequently, the level which is reflected in an SPS measure can be lower than ALOP. Interestingly, the opposite situation, when a level of protection reflected in a measure is higher than ALOP, is also possible. For example, in the *Salmon* case, Australia described its ALOP as conservative while the relevant SPS measure was based on a zero-risk approach.[40] Such a situation is, however, exceptional as it makes the establishment of a prima facie case of violation of Articles 2.2 and 5.6 much easier. Note that under both provisions the relevant benchmark for assessment of SPS measures is ALOP and not the level of protection that is reflected in a contested measure. I will come back to this issue in the section of the chapter that is dedicated to Article 5.6.

(b) ALOP and scientific disciplines

Scholars initially seem to agree that the process of establishing ALOP was subject to certain requirements of the SPS Agreement, such as the consistency obligation of Article 5.5, but not to science-based criteria.[41] This common belief was brought into question by the panel decision in *Japan – Apples*, which introduced the concept of negligible risk together with a strict proportionality test. Moreover, as discussed in Chapter 4, there are also some other scientific standards that may affect the discretion of WTO Members when establishing their ALOP.

The first situation is rather simple and probably not so controversial. Remember that risks that may be legitimately regulated under the SPS Agreement have to be ascertainable. Purely hypothetical risks are covered neither by Articles 2.2/5.1 nor

[38] Ibid, paras 200–1. [39] Ibid, para 203.

[40] Ibid, para 197.

[41] Jacqueline Peel, 'Risk Regulation under the WTO SPS Agreement: Science as an International Normative Yardstick?', Jean Monnet Working Paper No 2002/04 14; Joost Pauwelyn, 'Does the WTO Stand for "Deference to" or "Interference with" National Health Authorities When Applying the Agreement on Sanitary and Phytosanitary Measures (SPS Agreement)' in Thomas Cottier and Petros C Mavroidis (eds), *The Role of the Judge in International Trade Regulation: Experience and Lessons for the WTO* (2003) 175.

5.7. If one recognizes the limitations of science, it is not difficult to imagine that some of these hypothetical risks may turn out in the future to be very real ones. Secondly, as discussed in Chapter 4, section 4.4.1(e), the WTO jurisprudence requires from risk assessment a relatively high level of specificity. This threshold seems to be particularly problematic for low-level risks which are difficult to measure, either quantitatively or qualitatively, and easily escape traditional assessment methods. If general studies showing the presence of some risk related to a particular substance do not satisfy the specificity threshold, it may appear that in low-risk situations the right to decide on its ALOP is rather illusory. The specificity requirement also seems to create a hurdle for minority scientific opinions as a basis for SPS measures. One of the features of such opinions is their lower level of conclusiveness and limited empirical basis. They are frequently based on suggestive but not definitive evidence showing little more than a theoretical possibility of risk rather than a concrete causal relationship between a specific risk agent and an adverse effect. Consequently, in many cases they simply cannot provide highly particularized studies that would clearly indicate the existence of risk. Thirdly, as mentioned above, the SPS case law has conceptualized the relationship between scientific evidence and an SPS measure as a kind of proportionality test. If a measure is strict (presumably aiming at zero risk level), while the risk is negligible (which does not mean that it does not exist) a panel may conclude that there is no rational relationship between these two. Lack of such a relationship indicates that a measure is maintained 'without sufficient scientific evidence' and violates Article 2.2 of the SPS Agreement. Again, if a WTO Member is entitled to establish its ALOP with respect to any kind of identifiable risk, it is difficult to accept such an approach. The right to establish ALOP includes the objective of zero risk policy for all kind of risks (including small or 'negligible' ones). Precluding a Member from taking strict measures to address low-level risks effectively deprives a Member of the right to regulate such risks.

There are also those provisions of the SPS Agreement which, depending on their interpretation, may or may not interfere with the right of WTO Members to establish ALOP. First, the notion of probability which appears in the definition of quarantine risk assessment can be read as requiring quantitative evaluation of risk. This would obviously prevent Members from regulating those quarantine risks that cannot be reduced to quantitative figures. Although the panel report in *Australia – Salmon (Article 21.5 – Canada)* indicates that this is not necessarily the case, some uncertainty as to the future direction of the WTO jurisprudence remains.[42] Secondly, interpreting rules provided by Article 5.3 and Annex A(4) as autonomous provisions may also affect the right of WTO Members to establish their ALOP. Since the SPS case law requires the existence of an objective relationship between risk assessment and an SPS measure, while the evaluation of economic consequences may constitute an element of the quarantine risk

[42] See Chapter 4, section 4.4.1(b).

assessment (or quarantine risk assessment that is concerned with risks to animal life and health), one may argue that a final measure has to be rationally related to economic considerations as well. This would imply the existence of some kind of proportionality test that calls for balancing the restrictiveness of a quarantine measure against its economic costs (in terms of restriction to international trade).[43]

As discussed in Chapter 4, the recent decision of the Appellate Body appears to introduce new standards, which exclude the application of the strict proportionality test in the context of Article 2.2.

(c) Minimizing negative trade effects (Article 5.4)

Article 5.4 stipulates that WTO Members, when determining an ALOP, should take into account the objective of minimizing negative trade effects. One would intuitively accept that Article 5.4 provides a procedural obligation that requires WTO Members to consider potential trade effects resulting from contemplated ALOP. This would imply that negative trade effects need to be considered but not necessarily reflected in the final level of protection which is deemed appropriate by a particular Member.

The approach of the case law went in a different direction. The *Hormones* panel, after following an over-textual approach, held that 'this provision of the SPS Agreement does not impose an obligation'.[44] The panel justified this conclusion by reference to the language of Article 5.4 which uses the word 'should' and not 'shall'. Although it also added that the objective reflected in Article 5.4 could be taken into account as part of the analysis under other SPS provisions, it did not specify any relevant article. However, since the determination of an ALOP is in principle not mitigated under the SPS Agreement by economic considerations, whether a WTO Member actually took into account the objective of minimizing negative trade effects is irrelevant, and the panel's statement appears to have no legal consequences. The interpretation adopted by the panel is also questionable in the light of general standards of interpretation of WTO law. A reading of Article 5.4 which deprives it of any legal relevance, can hardly be seen as compatible with the principle of effective treaty interpretation that requires giving meaning to every provision of the Agreement. Conceptualizing Article 5.4 as a procedural rule would help to avoid these deficiencies. At the same time, such an approach would respect the textual basis of the article and ensure that the right to establish ALOP is not compromised by economic factors. Note also that a procedural reading may actually improve the quality of national decision-making by promoting a regulatory dialogue which openly considers costs of regulation in terms of the impact on international trade.

[43] See Chapter 4, section 4.4.1(i).

[44] Panel Report, *EC – Hormones (Canada)*, para 8.169; Panel Report, *EC – Hormones (US)*, para 8.166.

Not surprisingly, Article 5.4 has received almost no attention in the subsequent case law. Complainants, probably discouraged by the findings of the panel in *EC – Hormones*, have never again argued the violation of Article 5.4. The recent decision of the Appellate Body in *US/Canada – Continued Suspension* may, however, give some promise for the revival of its disciplines. As already discussed, the report appears to take a more procedural approach to the assessment of SPS measures compared to the previous practice. This, in turn, may encourage greater recourse to those provisions of the SPS Agreement which provide procedural obligations. In addition, the Appellate Body in one of the passages in its report explicitly recognized Article 5.4 as a rule which may have an impact on the extent of the freedom that is enjoyed by WTO Members when establishing ALOP.[45] Unfortunately this observation was made on the margin of the main discussion and its importance may be limited in practice.

(d) Weak consistency (Article 5.5)

The right to establish ALOP is not absolute. As we have seen, Article 5.4 requires WTO Members to consider the negative trade effects when adopting ALOP. Some scientific standards that were developed by the case law also appear to limit the discretion of WTO Members in setting their levels of protection. Another relevant provision of the SPS Agreement is Article 5.5 which requires WTO Members to address similar or the same risks in a consistent fashion. At the same time, inconsistency is still possible if it is not arbitrary or if it may be justified. Consequently, Article 5.5 does not impose what is sometimes imprecisely labelled a requirement of full consistency. This is rather a goal to be achieved in the future as Article 5.5 only prohibits those inconsistencies that are arbitrary and unjustifiable. In the words of the Appellate Body, 'governments establish their appropriate levels of protection frequently on an *ad hoc* basis and over time, as different risks present themselves at different times, it is difficult to require a full consistency'.[46] Bearing this observation in mind, for sake of simplicity, this chapter will refer to Article 5.5 as the consistency requirement.

The substantive obligations of Article 5.5 are accompanied by a procedural rule that asks WTO Members to cooperate with the SPS Committee for the purpose of developing guidelines for the practical implementation of the concept of consistency (Article 5.5 second sentence). On that basis, the SPS Committee adopted the Guidelines to Further the Practical Implementation of Article 5.5 ('Guidelines').[47] The document is discussed briefly in the subsequent section.

The Appellate Body identified three elements that have to be demonstrated in order to establish the violation of Article 5.5. This includes: (i) existence

[45] Appellate Body Report, *US – Continued Suspension*, fn 1088.

[46] Appellate Body Report, *EC – Hormones*, para 213.

[47] SPS Committee, Guidelines to Further the Practical Implementation of Article 5.5, 18 July 2000, G/SPS/15.

of different ALOPs in different but comparable situations; (ii) arbitrary or unjustifiable character of such differences; and (iii) discrimination or disguised restriction on international trade resulting from that distinction.[48] The test is of a cumulative nature and Article 5.5 is violated only if all its conditions are met. At the same time, the Appellate Body described the third element of the test as the most significant one.[49] This may look somewhat contradictory. If the test is of a cumulative character, all its elements are equally important. But the statement of the Appellate Body can be read in a different way, indicating the overall function of Article 5.5. This would mean that Article 5.5 is concerned in the first place with the elimination of discriminatory and protectionist measures rather than the imposition of a certain technocratic rationality on domestic regulators. This narrow function of Article 5.5 is clearly exemplified in another report where the Appellate Body observed that 'discrimination between Members, including their own territory and that of other Members within the meaning of Article 2.3, first sentence, can be established by following the complex and indirect route worked out and elaborated by Article 5.5'.[50]

Before going into details of the test, it is worth noting the ambiguous effects of Article 5.5 on trade liberalization. The *Australia – Salmon* dispute shows that compliance with the consistency obligation may be achieved not only through withdrawal of a measure which was found to be inconsistent with Article 5.5, but also by tightening the requirements with respect to those situations (products) that were used as comparators. In an attempt to comply with the recommendation of the original report (successfully as it appeared later), Australia extended its ban from salmon products (original SPS measure) to non-salmonids, including in particular herring for use as bait and live ornamental finfish (comparators).[51] That clearly shows that successful challenges of the measure under Article 5.5 may actually result in additional restrictions on trade rather than a freer flow of goods. This outcome troubled some of the scholars. As noted by Atik 'if [the] ultimate finding of Article 5.5 inquiry... is that the salmon import ban was a disguised restriction it is difficult to imagine how imposition of restriction on other products (for sake of consistency) lessens this conclusion'.[52]

Although the above may seem to be at odds with idea of free trade, it probably needs to be accepted as an unavoidable consequence. The *quasi* absolute character of the right of WTO Members to establish ALOP requires the allowance of the existence of false negatives (wrongly exonerating measures as compatible with Article 5.5) rather than false positives (wrongly condemning measures as violating Article 5.5). Moreover, this situation does not seem to be qualitatively different from the possibility which exists under the traditional non-discrimination standards

48 Appellate Body Report, *EC – Hormones*, para 214.
49 Ibid.
50 Appellate Body Report, *Australia – Salmon*, para 252.
51 Panel Report, *Australia – Salmon (Article 21.5 – Canada)*, para 7.91.
52 Atik ('Weakest Link') 483.

of GATT 1994. Note that a WTO Member instead of simply removing a measure that was found incompatible with, for example, Article III:4 may introduce the same treatment that was found to be discriminatory to like domestic products. As far as Atik's argument is concerned, it is also worth recalling another passage of the panel's report which was probably overlooked. The panel particularly noted that there was some positive evidence which indicated 'that its [Australia's] new regime on imports of salmonids does *not* result in discrimination or a disguised restriction on international trade inspired to avoid import competition, but is rather a quarantine measure to protect Australia against diseases'.[53] In particular, Australia imposed less demanding requirements with respect to the import of salmon from New Zealand; a product that was one of the most competitive on the Australian market, but was also coming from a country that had a low disease status. If the main reason that stood behind the import restriction was protectionism how does one explain fairly open market access for very competitive New Zealand products?

Of course, there is a danger that the interpretation proposed by the panel may discourage future complainants from pursuing it claims under Article 5.5. A pragmatic litigation strategy would suggest identifying those situations which involve products that are important for the major trading partners of a defendant or alternatively for its domestic interest groups. In case of defeat, it could be more practical for a defendant to lower the requirements imposed by a contested measure rather than to move upwards with respect to those products that they were compared with.

The analysis under the first condition of the Article 5 test is twofold. Initially, it is necessary to select a 'different situation' which would be a case against which the comparison is made. Although both situations have to be different, they also need to express some common element(s). This seems to be a logical conclusion because 'if the situations proposed to be examined are *totally* different from one another, they would not be rationally comparable and the differences in levels of protection cannot be examined for arbitrariness'.[54] The case law refrained, however, from stating in general terms what makes two situations comparable and preferred a case-by-case approach. Nevertheless, a closer look at the SPS report allows the identification of some factors which make the two situations comparable. This particularly includes:

- presence of the same substance (eg hormone) or pathogen;[55]
- presence of the same adverse health effect (eg carcinogenicity);[56]
- risk of entry, establishment or spread of the same or a similar disease;[57]

[53] Panel Report, *Australia – Salmon (21.5 – Canada)*, para 7.106.
[54] Appellate Body Report, *EC – Hormones*, para 217.
[55] Panel Report, *EC – Hormones (US)*, para 8.176; see also SPS Committee, Guidelines (5.5), para A2.
[56] Panel Report, *EC – Hormones (US)*, para 8.176.
[57] Appellate Body Report, *Australia – Salmon*, para 146.

- same or similar potential biological and economic consequences that are associated with a risk.[58]

Note also that the test, which was developed in the case law, allows the comparison of different products as long as the common element(s) is involved.[59] Given the case-by-case approach, the above list is in no way exhaustive and other factors may also be helpful when establishing that two situations are comparable.[60] At the same time, it is not clear whether the scope of points (a) and (b) was deliberately limited by the *Hormones* panel. Note that the Appellate Body in *Australia – Salmon* saw both parity and similarity as sufficient to establish that two situations are comparable, while the *Hormones* panel presumably required a higher threshold by demanding the existence of the same substance or the same adverse health effect. This is probably an unintended omission.

The case law has not settled whether a source of a risk matters in the determination of comparability. That is particularly important with respect to the distinction that is frequently made between man-made and naturally occurring risks. Note that neither ordinary people nor regulatory practice treat these two types of risk as comparable. As discussed below, lay people tend to consider man-made risks as more dangerous, while regulators, responding to differences in risk perception, frequently adopt distinctive regimes for these two types of risk. The approach taken in *EC – Hormones* (and subsequently in *Australia – Salmon*) suggests that under the SPS Agreement the source of a risk is irrelevant in determining whether two situations are comparable. Both the panel and the Appellate Body accepted the comparison between synthetic hormones used for growth promotion purposes with hormones occurring endogenously in meat and other foods.[61] This confirms the strong reliance of the WTO adjudicating bodies on the technical paradigm which conceptualizes risk in terms of probability and harm, with the source of risk and the public perception being of secondary (if any) importance. The Guidelines (5.5) also take a similar stance and provide that 'categorizing risks as "similar" must include a comparison of both the relevant likelihoods and the corresponding consequence',[62] without indicating the source of the risk as a relevant factor. Note, however, that the popular perception

[58] Ibid.

[59] Different panels compared the levels of protection in the following situations: natural hormones used as growth promoters as opposed to the same hormones when used for therapeutic or zootechnical purposes, natural hormones for growth promotion as opposed to the same hormones occurring endogenously in meat and other foods, natural and synthetic hormones for growth promotion as opposed to carbadox and olaquindox (*EC – Hormones*), salmon products as opposed to uncooked Pacific herring, cod, haddock, Japanese eel, and plaice for human consumption, salmon products as opposed to uncooked Pacific herring, Atlantic and Pacific cod, haddock, and European eel for human consumption, salmon products as opposed to herring in whole, frozen form for use as bait, and salmon products as opposed to live ornamental finfish (*Australia – Salmon*).

[60] It was also argued that the comparison can be made only between affirmative measures, or in other words 'occasions where the member has actively regulated' (Atik ('Weakest Link') 485).

[61] See eg, Appellate Body Report, *EC – Hormones*, paras 216–18.

[62] SPS Committee, Guidelines (5.5), para A4.

of risk (including the distinction between man-made and naturally occurring risks) together with the corresponding regulatory practice may be also taken into account under the second condition of Article 5.5. I will come back to this issue in the pages that follow.

Once the different situation is identified, one has to determine ALOPs with respect to each situation that was identified in the first step. Following the general approach of the case law, it may be assumed that it is for each WTO Member to identify its ALOP in both instances (original situation and comparator). The panel's intervention will only be justified if ALOP was not established or if it was determined without sufficient precision.

The second element of the test requires a complainant to show that the difference in the levels of protection in two comparable situations is arbitrary or unjustifiable. Since the case law does not distinguish between two parts of the alternative, both elements have to be considered together. Regrettably, the dispute settlement bodies have failed to develop any clear criteria for the evaluation of arbitrariness and lack of justification. As rightly observed by Hurst, the approach of the case law shows that 'the inquiry into whether differences are arbitrary or unjustifiable is purely subjective; the result will turn entirely on the body making the evaluation'.[63] This corresponds, however, with the approach of the general GATT 1994 case law which was also unsuccessful in setting any plain standards and preferred a case-by-case examination.[64]

The examination of the case law reveals that making a prima facie case with respect to the second element is a rather undemanding task.[65] Consequently, a defendant needs to be prepared to present evidence which will show that identified differences are not arbitrary, or to provide some other convincing justification. What is then relevant under the second condition of Article 5.5? One element that seems to justify the difference between two risk situations is the extent of regulatory action. In other words, the more governmental intervention is required, the less are the chances that differences in levels of protection in the comparable situations will be arbitrary or unjustifiable.[66] This meant, in the specific circumstances of the *EC – Hormones* case, that unlimited-residue level of protection with regard to hormones occurring naturally in foods, as opposed to no-residue level of protection against natural hormones used for growth promotion, was not arbitrary. According to the Appellate Body, requiring the EC to apply the same level of protection for hormones naturally occurring

[63] David R Hurst, 'Hormones: European Communities—Measures Affecting Meat and Meat Products' (1998) 1(9) European Journal of International Law (<http://207.57.19.226/journal/Vol9/No1/sr1g.html> last visited 25 July 2009).

[64] See eg, Appellate Body Report, *US – Shrimp*, para 150.

[65] See eg, Panel Report, *EC – Hormones (US)*, para 8.193; Panel Report, *Australia – Salmon*, para 8.139.

[66] But see, Atik ('Weakest Link') 485 (arguing that those two situations are not comparable but rather arbitrary and without justification). Note, however, that the Appellate Body finding was made in the context of the second element of Article 5.5 and not the first.

in food 'entails such a comprehensive and massive governmental intervention in nature and in the ordinary lives of people as to reduce the comparison itself to an absurdity'.[67]

Another factor which may serve as a justification, or which may show that the distinction is not arbitrary, is higher risks related to substance (product) under examination as opposed to the substance (product) being compared.[68] This is a correct approach as the level of risk is usually connected with the applicable ALOP (higher risk requires higher ALOP). Note also that the category of 'higher risk' may either relate to the possibility/likelihood of adverse effects or to the magnitude of potential damage.[69]

The Appellate Body also identified the character of a measure as a relevant factor that may explain divergences in ALOPs.[70] In consequence, difference in the levels of protection with respect to the administration of hormones for growth promotion purposes and their administration for therapeutic and zootechnical purposes was considered not to be arbitrary or unjustifiable.[71] According to the Appellate Body, the frequency and scale of the treatment (ie therapeutic use takes place on a small scale while zootechnical use, although it may relate to entire herds, normally occurs only once a year) were specifically relevant in this context as well as the mode of administration of hormones (ie hormones for therapeutic and zootechnical purposes are administered by a veterinarian or under the responsibility of a veterinarian).

As already mentioned, the SPS case law presumably disregards differences in risk perception (depending on the sources of risk) as a relevant factor for determination of comparability. Having said that, it seems that the existing case law does not exclude a more permissible interpretation of the second condition of Article 5.5. At the same time, there are important reasons for accepting differences in risk perception as a possible justification. Remember from Chapter 1 that popular perception of risk may differ from the assessment of technical experts. Risks that are highly ranked by experts can be disregarded by laypersons. The opposite is also true; risks which, according to the experts are unlikely, can generate great public concern. The heuristic availability, controllability, perceptibility, and source of risk as well as potential benefits are all important elements which influence the assessment of risk by laypersons. As many of those factors are culturally related, it is not surprising to see differences in risk perception between different countries. For instance, Europeans seem to be more concerned with the potential risks connected with genetic modifications, while Americans tend to overestimate

[67] Appellate Body Report, *EC – Hormones*, para 221.

[68] Panel Report, *EC – Hormones (US)*, para 8.193, Panel Report, *Australia – Salmon*, para 8.134.

[69] Joanne Scott, *The WTO Agreement on Sanitary and Phytosanitary Measures. A Commentary* (2007) 148.

[70] But cf ibid, 147 (interpreting the approach of the Appellate Body in terms of easiness of control and not intended application).

[71] Appellate Body Report, *EC – Hormones*, para 225.

(in a technical sense) cancer risks.[72] Without deciding whether perception of risk by laypersons is a psychological deficiency or rather a rival rationality to the scientific discourse,[73] it seems advisable to accept it as a possible justification. As perceived risk needs to be genuine (otherwise a measure will fail the scientific tests provided by Articles 2.2 and 5.1), there are no compelling reasons which would speak against such an approach. This would also correspond with the function of Article 5.5 which intends to detect discrimination and protectionism through identification of inconsistencies in risk regulation rather than to impose any technocratic rationality under which legitimate responses to risk are determined solely on the basis of probability and adverse effects. Moreover, the SPS Agreement itself seems to recognize risk perceptions as legitimate grounds for differences in ALOPs. For example, Article 5.5 instructs the SPS Committee, when developing the guidelines relating to consistency, to take into account the exceptional character of human health risks to which people voluntarily expose themselves. The text of Guidelines (5.5), elaborating on this requirement, speaks about the identification of those situations that justify acceptance of the lower level of protection for similar or the same types of risk. The Guidelines also enumerate some risks that may require such special treatment (eg consumption of alcoholic beverages or traditional foods such as smoked fish, or of varieties of fish known to be toxic). Note that in all those instances a different ALOP is justified by the divergence in the risk perception (eg resulting from traditional consumption patterns).

There are also pragmatic reasons that support the liberal interpretation of the second condition of the test provided by Article 5.5. Note that a broad reading of the first condition, combined with an undemanding threshold for making a prima facie case gives a considerable authority to a panel and leads to a de facto shift in the burden of proof. A defendant is expected to provide explanations for identified differences which, in turn, are assessed and judged by a panel as to the underlying rationality. This expansive reading of Article 5.5 may be contrasted with the more restricted approach that would require a close similarity between situations that are compared. Such narrow conception of the first condition would increase the national autonomy by limiting the scope of the consistency requirement. Although the current interpretation definitely has a sufficient textual basis (after all Article 5.5 speaks about different and not similar situations), the panel should use its authority very carefully. Deciding on such delicate issues as rationality of domestic risk responses requires granting considerable discretion to WTO Members who appear to be best equipped in making judgments over the need to make regulatory distinctions on the basis of differences in risk perception. At the same time, recognizing that such differences may serve as a possible

[72] See eg, David Vogel, 'Ships Passing in the Night: The Changing Politics of Risk Regulation in Europe and the United States', EUI RSC Working Paper No 2001/16.

[73] Cf Chapter 1, section 1.2.

justification does not mean that WTO dispute settlement bodies are required to unconditionally accept such assertions in every case. This would probably lead to the inoperability of disciplines provided by Article 5.5 as it is relatively easy to manipulate fears of citizens (eg by directing the attention of media to specific risk issues). A middle ground between those two extremes where the panel considers the source and reasons standing behind the popular perception of risk probably provides the best solution.

The distinction between ALOPs in different but comparable situations needs to result in discrimination or a disguised restriction on international trade. As elucidated by the Appellate Body, the third condition refers 'to the measure embodying or implementing a particular level of protection... resulting, in its application, in discrimination or a disguised restriction on international trade'.[74] On the textual level, the third condition appears to provide two distinctive requirements, each having a specific legal meaning. On the one hand, the examination of a disguised restriction arguably involves the question of intent of the party imposing the measure. As noted by Hurst, 'while one can objectively assess whether a measure creates a "restriction on international trade," whether or not such a restriction was "disguised" is a matter of intent'.[75] If the real purpose of a measure is to protect domestic producers at the expense of foreign suppliers rather than to achieve specific health policy goals, this would amount to a disguised restriction on international trade. On the other hand, the concept of discrimination does not have such a subjective aspect and is more concerned with objective consequences resulting from the application of a measure.[76]

The approach of the SPS case law is somewhat ambiguous. It did not settle whether Article 5.5 indeed requires identification of both objective and subjective factors. It also failed to establish a clear distinction between two elements of the alternative (discrimination or a disguised restriction). For example, the Appellate Body in *EC – Hormones* considered both elements in one analysis without distinguishing between the objective dimension of discrimination and subjective aspect which is implied in the concept of disguised restriction. Nevertheless, both elements were present. In particular, the question of intent appeared in the analysis of the rationales that stood behind the EC measure. The Appellate Body admitted in this context that the measure was at least partially motivated by a need to protect the health and safety of the European population and to harmonize rules of the EC internal market.[77] On the other hand, the WTO dispute settlement bodies in *Australia – Salmon* concentrated only on the notion of 'disguised restriction' and the problem of discrimination was not addressed at all. Neither the panel nor the Appellate Body specifically discussed the issue of the intent that underlied

[74] Appellate Body Report, *EC – Hormones*, para 214; remember that the Appellate Body characterized that element as the most important under Article 5.5 (ibid, para 240).

[75] Hurst ('Hormones') 26.

[76] Cf Panel Report, *EC – Asbestos*, para 8.226.

[77] Appellate Body Report, *EC – Hormones*, para 245.

the Australian sanitary regime for salmon. This confusion is well reflected in the submissions of the parties in *EC – Biotech*. Note that all complainants argued the existence of both elements (discrimination and a disguised restriction on international trade), without making any distinction between them.[78] The panel did not ultimately examine the compatibility of the EC measures with Article 5.5.

One would intuitively look to GATT 1994 rules for additional guidelines. As Article 5.5 and Article XX of GATT 1994 use very similar language, it seems natural to incorporate into the SPS Agreement some of the interpretations that were developed in the context of Article XX. Surprisingly, the Appellate Body took a different stance and after emphasising the structural difference between those two provisions, explicitly warned against the casual importation of GATT 1994 case law into Article 5.5.[79] The reasons which stand behind the finding of the Appellate Body are not entirely clear. Of course, both provisions differ as Article XX constitutes an exception from general obligations such as national treatment, most-favoured nation principles, or prohibition of quantitative restriction. Consequently, the examination of a measure under Article XX presupposes the violation of other provisions of GATT 1994.[80] This is not the case for Article 5.5 which is itself a positive obligation and not the exception. Moreover, Article XX uses the concept of discrimination with respect to the application of measures while Article 5.5 is concerned with the distinctions in ALOPs. It is not clear to what extent those differences may be relevant when incorporating GATT 1994 jurisprudence into the SPS Agreement. At the same time, the Appellate Body's enigmatic statement does not really clarify the nature of the structural difference between those two provisions.[81] Irrespective of the rationale which stood behind the approach of the Appellate Body, it seems that its main concern was a verbatim incorporation of the Article XX case law into the SPS Agreement, rather than the rejection of possible interpretative dialogue between those two provisions.[82]

Bearing this reservation in mind, let us look into GATT 1994 case law. First of all, it recognizes that discrimination and disguised restriction are distinct, although related, concepts.[83] While discrimination has received a rather

[78] See eg, First Submission of the US in *EC – Biotech Products*, WT/DS293, 21 April 2004, paras 114–27; First Submission of Canada in *EC – Biotech Products*, 21 April 2004, WT/DS292, paras 215–22.

[79] Appellate Body Report, *EC – Hormones*, para 239.

[80] This fact led the Appellate Body to observe that 'the provisions of the chapeau [of Article XX] cannot logically refer to the same standard(s) by which a violation of the substantive rule has been determined to have occurred' (Appellate Body Report, *US – Gasoline*, 25).

[81] Interestingly, a closer examination of the SPS case law also reveals a number of similarities in the approach of case law (eg overlapping character of the relationship between discrimination and disguised restriction).

[82] Note that the Appellate Body used the expression 'casually imported', which indicates that importation is in principle possible and only calls for careful examination.

[83] Appellate Body Report, *US – Gasoline*, 25 (finding two elements as interrelated obligations and defining disguised restriction as very broad concept); Appellate Body Report, *US – Shrimp*, para 184.

consistent interpretation (less favourable treatment between foreign products as compared to like domestic products or between like products from different countries),[84] there is no uniform reading of disguised restriction. In some cases, the analysis under Article XX was based on objective factors,[85] in other instances the intent of the party was also a subject of examination.[86] In particular, the panel in *EC – Asbestos*, referring to Article 31 of the VCLT, found that the verb 'disguise' requires the inquiry into the intent of the party. Consequently, restriction is disguised if a measure is adopted with the intent to conceal the pursuit of trade-restrictive objectives.[87] According to the same panel, the protective application of a measure can be discerned from its design, architecture, and revealing structure.[88] The more recent case law tends to conceptualize a disguised restriction in terms of both the purpose of a measure and its effects. Determination of the purpose requires examination of the intent in order to answer the question whether a measure is only a disguise to conceal the pursuit of trade-restrictive objectives.[89] At the same time, the objective dimension (whether the restriction results in protection of domestic industry from the foreign competitors) seems to dominate the entire analysis. Consequently, the existence of a disguised restriction may be found solely on the basis of the effects produced by a measure.

Although the general case law is not without ambiguity, it suggests two things for the SPS Agreement. First, discrimination and disguised restriction are distinct concepts having their own legal meaning (despite the existence of some overlap). This requires, for the purpose of clarity, conducting a separate legal analysis with respect to each part of the alternative. Secondly, the more recent case law (ie *Brazil – Retreaded Tyres*) indicates that although the intent of the WTO Member is a relevant factor, its establishment is not essential for finding disguised restriction. When the intent is actually sought, it may be determined from the design, architecture, and revealing structure of a measure as well as other sources such as statements of the officials.

The SPS case law has developed in the context of Article 5.5 a rather casuistic test which relies on the concept of warning signals and additional factors. As noted by the Appellate Body, the warning signals operate as indirect proof that 'a Member may actually be applying an SPS measure in a manner that discriminates between Members or constitutes a disguised restriction on international trade'.[90]

[84] Under the SPS Agreement this would be translated into less favourable treatment between the Members where identical or similar SPS conditions prevail (cf Article 2.3 of the SPS Agreement).

[85] Appellate Body Report, *US – Gasoline*, 29 (finding the existence of a disguised restriction without any analysis of the intent of the Member).

[86] Panel Report, *US – Shrimp (Article 21.5 – Malaysia)*, para 5.143.

[87] Panel Report, *EC – Asbestos*, para 8.236.

[88] Ibid; cf also Appellate Body Report, *Japan – Alcoholic Beverages*, 31.

[89] Panel Report, *Brazil – Retreaded Tyres*, para 7.330. In another passage of the report the panel also noted that 'we agree that certain statements, in particular those made in an official capacity, may provide useful indications on the intent of the legislator or regulator in adopting a specific measure' (ibid, para 7.332).

[90] Appellate Body Report, *EC – Hormones*, para 240.

The additional factors perform the same function but, unlike the warning signals, they are very case-specific and cannot be easily generalized. The case law also clarified that none of the warning signals (or additional factors) has a dominant role and all of them should be considered cumulatively.[91] In contrast to the test provided by Article 5.5, it is not necessary to identify in a specific case the existence of all warning signals. The catalogue of warning signals is definitively not a closed list. Neither the panels nor the Appellate Body ever used a language which would exclude the identification of new signals in the future. That is actually reflected in the practice of the dispute settlement bodies, since the initial list that was introduced in *EC – Hormones*, was subsequently supplemented in *Australia – Salmon* with the new items.[92] Nonetheless, complainants in more recent disputes tended to present new circumstances as additional factors, rather than warning signals.[93]

The first warning signal identified by the case law is 'the arbitrary or unjustifiable character of the differences in levels of protection considered by a Member as appropriate in differing situations'.[94] Interestingly, this signal also constitutes the first element of the test provided by Article 5.5. The Appellate Body, while accepting this consideration as relevant, rightly refused to follow the approach of the panel which called for the deduction of discrimination/disguised restriction from the difference in levels of protection (the first element of the test under Article 5.5) and the arbitrariness (the second element).[95] According to the Appellate Body, such an inference was not justified by the language of the provision and the third element merited a separate analysis. The second warning signal is the existence of a rather substantial difference between ALOPs in situations that were identified for the purpose of the first signal.[96] This obviously excludes *de minimis* divergences. Note also that the qualification ('substantial') is additionally strengthened by the inclusion of the word 'rather' which arguably provides for an even higher threshold. Absence of scientific justification as required by Articles 2.2 and 5.1 constitutes the third warning signal.[97] The rationale behind this is simple. A lack of risk assessment (no risk assessment, insufficient assessment, or no rational relationship between such assessment and an SPS measure) or scientific evidence which would support a measure, may indicate that it is not really concerned with the protection of the life and health of human, animals, and plants but 'is instead a trade-restrictive measure taken in the guise of an SPS measure'.[98] The third warning signal is

[91] Ibid, para 240; Appellate Body Report, *Australia – Salmon*, para 177.

[92] Appellate Body Report, *Australia – Salmon*, para 166.

[93] See eg, First Submission of Canada in *EC – Biotech Products*, para 219.

[94] Appellate Body Report, *EC – Hormones*, para 215.

[95] Ibid, paras 238–9.

[96] Panel Report, *Australia – Salmon*, para 8.150. See also, Appellate Body Report, *Australia – Salmon*, para 164.

[97] Panel Report, *Australia – Salmon*, para 8.151.

[98] Appellate Body Report, *Australia – Salmon*, para 166.

also the only one which has a clear subjective dimension, as it aims to reveal the intent of the party.

The catalogue of warning signals is supplemented by the list of additional factors. These are very specific circumstances of particular dispute which are helpful in establishing discrimination and/or disguised restriction. Despite their highly context-dependent character, some generalization is possible. Thus, a substantial but unexplained change in the conclusions of risk assessments in the absence of new scientific evidence may be relevant.[99] Similarly, a difference between the measures applied internally (eg absence of internal control) as opposed to measures applied to the imported products (eg prohibition)[100] is also one of the additional factors. Apart from that it was argued in one of the disputes that a disproportion between the effects of an SPS measure on the producers located on the territory of the complainant, as compared to the producers of the defendant, could be regarded as an additional factor.[101] The panel did not address this argument.

(e) Guidelines to further the practical implementation of Article 5.5

The SPS Committee, on the basis of the second sentence of Article 5.5, adopted Guidelines that explain some practical issues relating to the implementation of the consistency requirement. These Guidelines are formulated in very general language and are built upon the existing SPS case law. As with any other decision of the SPS Committee, they also contain the explicit reservation as to their legal effect.[102] In line with this pronouncement, the Guidelines use rather soft language ('should' and 'may' instead of 'shall'), emphasising their non-binding character.[103]

The Guidelines address two issues: application of the concept of ALOP and its practical implementation. The Guidelines particularly prescribe that levels of protection should be stated in a clear manner, either in quantitative or qualitative terms. However, due to pragmatic reasons (easiness of comparison), the Guidelines express a preference for the quantitative approach and for the use of common terms or units across different SPS measures.[104] Countries, when setting ALOP (either in a specific situation or as an overall policy objective), should compare it with the existing levels as reflected in other SPS measures. If necessary, the new level should be adjusted or, alternatively, already existing levels of protection

[99] Panel Report, *Australia – Salmon*, para 8.154 (however, only if the subsequent conclusions are more stringent than those included in the initial risk assessment).

[100] Appellate Body Report, *Australia – Salmon*, para 176 (noting that '[w]e consider that [this factor does]... not carry much weight, but we agree with the Panel that [it] can nevertheless be taken into consideration').

[101] First Submission of Canada in *EC – Biotech Products*, para 219.

[102] The Guidelines specifically provide that they 'do not add to nor detract from the existing rights and obligations of Members under the SPS Agreement nor any other WTO Agreement'.

[103] Cf Chapter 2, section 2.2.5.

[104] SPS Committee, Guidelines (5.5), para A.4.

should be re-evaluated and modified.[105] In line with the three-prong test developed in the case law, countries are expected to consider differences between levels of protection in comparable situations and guarantee that these differences are not arbitrary or unjustifiable.[106] WTO Members should also identify those situations when they accept a lower level of protection, particularly with respect to those risks to which people expose themselves voluntarily.[107]

The Guidelines also prescribe a number of technical issues relating to the process of determination of an ALOP as well as selection and implementation of corresponding SPS measures. This is particularly ensured through establishment of proper ('clear and effective') communication channels between the competent authorities responsible for determining and implementing ALOPs.[108] The Guidelines advocate 'common approaches' and 'consistent procedures' for assessing risks and evaluating SPS measures. A common approach should be adopted with respect to particular categories of risk such as risks affecting human life or health, risks to animal life or health, and risks to plant life or health.[109] Finally, the Guidelines recommend considering, in determination of an ALOP or implementation of an SPS measure, the relevant international standards, guidelines or recommendations together with measures that are applied by other WTO Members facing similar risks.[110]

6.4.2 Options evaluation

The SPS Agreement contains very limited obligations with respect to the third phase of risk management. This is not surprising since the Agreement is more concerned with the final regulatory outcome rather than with the process itself.

In principle, the SPS Agreement does not impose one general and direct obligation which would require evaluation of different risk management options. There is an indirect requirement that applies to quarantine risks. Remember that one of the components of quarantine risk assessment is the evaluation of the likelihood of entry, establishment, or spread of diseases or pests according to the SPS measures which might be applied (Annex A(4) first sentence). This language was interpreted in the case law as requiring not only a risk evaluation for a measure actually applied, but also for other measures that constitute potential alternatives.[111] Consequently, risk managers need to be supplied with the information on different available quarantine measures. Another reference to option evaluation may be found in Article 5.3 which mandates, when assessing risk to animal and plant life and health, to take into account the relative

[105] Ibid. [106] Ibid, para A.2. [107] Ibid, para A.8.
[108] Ibid, para B.1. [109] Ibid, para B.2.
[110] Ibid, paras B.5 and B.6.
[111] Panel Report, *Japan – Apples*, para 8.283; that finding was subsequently upheld by the Appellate Body.

cost-effectiveness of alternative approaches to limiting risk.[112] Some other provisions of the SPS Agreement may also be indirectly relevant in this context. This particularly includes necessity (Article 2.2) and the least-trade-restrictive alternative requirements (Article 5.6) which although being concerned with final measure also affects the selection process. For instance, Article 5.6 implies that WTO Members assess available alternatives, taking into account their technical and economical feasibility. At the same time, the violation of Article 5.6 is of substantive and not procedural character. It is not important whether a WTO Member actually identified and considered other alternatives as long as its measure constitutes the least-trade-restrictive option.

The SPS Agreement also requires supplying other Members as well as the WTO Secretariat with information on proposed SPS regulation, including the product coverage, its objective, and rationale.[113] Other Members may present their comments to the implementing Member and discuss the content of a contemplated measure. The results of the discussion together with the comments need to be taken into the account when deciding on the final measure.[114] Although these obligations fall into the risk communication domain, they also have an impact on the options evaluation process. In particular, objections raised by trading partners may result in the additional assessment of available options and be followed by adjustment of the initial proposal or selection of a different measure. Although the SPS Agreement uses rather soft language ('take into account') which indicates that a final SPS measure does not have to reflect the position of other WTO Members, in practice their concerns frequently find a way into the national regulatory process. Consequently, in many instances the final measure is a form of compromise between the initial proposal and suggestions made by other WTO Members.[115]

6.4.3 Implementation of SPS decisions

Articles 2.2 and 5.1 are core science-based provisions of the SPS Agreement. Nevertheless, they also have a risk management dimension which is reflected in the necessity requirement, rational relationship test, and determination of legitimate subjects of SPS regulation (ie situations may be regulated).[116]

112 On the relationship between the definition of Annex A(4) and Article 5.3, see Chapter 4, section 4.4.1(i).

113 Annex B(5)(b) of the SPS Agreement.

114 Ibid, Annex B5(d). Note that in case of emergency all these obligations can be waived by WTO Members.

115 For example, the EC has changed its initial proposal for setting the maximum levels of aflatoxin in food products, following the objections raised by several WTO Members; cf Scott (*Commentary*) 57.

116 As will be discussed later, Article 2.2 also plays an important role in the last phase of risk management (monitoring and review) by prohibiting the maintenance of SPS measures which lack sufficient scientific backup.

These obligations are supplemented by Article 2.3, which establishes a general requirement of non-discrimination and prohibits the application of SPS measures in a manner which would constitute a disguised restriction on international trade, as well as Article 5.6, which introduces the requirement of the least-trade-restrictive alternative. Elaborating on the last obligation, Article 6 requires risk managers to adapt measures to the specific SPS characteristics of the area of the product's origin and its destination, while Article 4 points to the concept of equivalence as a specific example of least-trade-restrictive alternative. All these standards set up the limits for discretion of risk managers when selecting SPS measures.

(a) Necessity (Article 2.2)

Article 2.2 requires that WTO Members 'ensure that any SPS measure is *applied only to the extent that is necessary* to protect human, animal or plant life or health' (emphasis added). This language, and in particular the expression 'applied to', suggests that Article 2.2 is more concerned with the application of a measure than the measure itself. Consequently, even if the adoption of an SPS measure is necessary in general, its application in a specific case may go beyond what is needed. If one subscribes to such an interpretation, an analysis of the panel should concentrate not so much on the content of a measure but rather on the manner in which that measure is applied. Interestingly, this would distinguish the scope of Article 2.2 from that of Article 5.6, as the latter refers to a measure and not its application.[117] Consequently, two provisions may be seen as complementary with Article 2.2 covering the application of a measure and Article 5.6 applying to a measure itself. This conclusion also appears to correspond closely with the distinction made by the Appellate Body, in the context of the chapeau of Article XX of GATT 1994, between a measure and its application.[118]

The SPS case law went in a different direction. The *Agricultural* panel identified Article 5.6 as a specific formulation of a general rule provided by Article 2.2 and not as two complementary provisions.[119] The subsequent case law has followed this interpretation.[120] Despite the above objections, the approach of the case law seems to find some support in the structure of the SPS Agreement and SPS jurisprudence. Article 2.2 is clearly included in the provision that establishes basic rights and obligations under the SPS Agreement. Moreover, different subparagraphs of Article 5 were described by the case law as specific obligations of WTO Members that correspond with general rules of Article 2. It is natural to follow the same approach in the case of Article 5.6.

[117] For detailed discussion on disciplines of Article 5.6, see section 6.4.3(d) below.

[118] Appellate Body Report, *US – Gasoline*, para 22; see also, Appellate Body Report, *US – Shrimp*, para 115.

[119] Panel Report, *Japan – Agriculture Products*, para 8.71 (observing that the more specific language of Article 5.6 should be read in light of the more general language in Article 2.2).

[120] Panel Report, *EC – Biotech Products*, para 7.1433.

There is no case law which would address the risk management dimension of Article 2.2. Again, one may look for some guidance to the jurisprudence related to Article XX(b) of GATT 1994 which uses the very similar concept of necessity.[121] The relevant test is conventionally conceptualized as a process of weighing and balancing a series of factors,[122] including (i) the relative importance of the values furthered by the challenged measure; (ii) the contribution of this measure to the realization of the pursued ends; and (iii) the restrictive impact of the measure on international trade.[123] Thus, the more important the values at stake, a higher contribution of the measure to attain the objective, and a lesser impact on international commerce, the easier it will be to find the necessity. The test also asks for comparison of an examined measure with possible alternatives. The aim is to determine whether 'a WTO-consistent alternative measure, or a less WTO-inconsistent measure which the Member concerned could reasonably be expected to employ, is available'.[124] The three factors identified above play a role in this determination. At the same time, an alternative that is merely theoretical or imposes an undue burden on a Member, such as prohibitive costs or substantial technical difficulties, is not considered to be reasonably available.[125]

Does the necessity under Article 2.2 require weighing and balancing different factors? The answer is presumably no. Note that, as far as the first element is concerned, the protection of human life or health is recognized in the WTO case law as vital and important in the highest degree.[126] Similar high value is attached to animal or plant life and health.[127] Thus, SPS measures may be presumed to serve important values and there is no need for an additional assessment of their importance under Article 2.2. The contribution analysis (to what extent a measure contributes to pursued ends) also appears to be unnecessary as it is already done under the second part of Article 2.2 of the SPS Agreement (sufficiency of scientific evidence). It is also interesting to note that pursuant to the GATT 1994 case law such contribution is expected to be 'significantly closer to the pole of "indispensable" than to the opposite pole of simply "making a contribution to" the end'.[128] This seems to closely correspond with the rational relationship that is required between scientific evidence and an SPS measure. The third element of the necessity test, at least in theory, also appears to be irrelevant as the Agreement 'assumes that there is no compromising of authentic health values'.[129] Even if

121 Cf Scott (*Commentary*) at 160.

122 Appellate Body Report, *Korea – Various Measures on Beef*, para 164.

123 Panel Report, *Brazil – Retreaded Tyres*, para 7.104; Appellate Body Report, *US – Gambling*, para 306.

124 Panel Report, *Brazil – Retreaded Tyres*, para 7.104.

125 Appellate Body Report, *EC – Asbestos*, para 308.

126 Ibid, para 172.

127 Panel Report, *Brazil – Retreaded Tyres*, para 7.112.

128 Appellate Body Report, *US – Gambling*, para 310.

129 Gabrielle Marceau and Joel P Trachtman, 'The Technical Barriers to Trade Agreement, the Sanitary and Phytosanitary Measures Agreement, and the General Agreement on Tariff and Trade' (2002) 36 Journal of World Trade 811, 833.

there is some balancing under the SPS Agreement between restrictiveness and the extent of risk, the relevant analysis is performed under the third sentence of Article 2.2.[130] Consequently, one may think that the necessity test in the SPS Agreement is only limited to the examination of available alternatives in order to assess whether any WTO-compatible or less restrictive option exists. Taking into account the *quasi* absolute right of WTO Members to establish their own level of protection, the alternative has to achieve the same ALOP as the contested measure.[131]

(b) Scientific disciplines and risk management (Articles 2.2 and 5.1)

As extensively discussed in Chapter 4, an SPS measure needs to be based on risk assessment (Article 5.1) and cannot be maintained without sufficient scientific evidence (Article 2.2). The SPS Agreement also requires the existence of a rational relationship between risk assessment/scientific evidence and an SPS measure. Moreover, risks that may be regulated have to be ascertainable. This excludes merely theoretical risks as a legitimate subject of regulation. Remember also that risk managers can base their decisions on both majority scientific opinions as well as those opinions that are produced by minority circles. All those standards set certain limits for the discretion of risk managers as they remove some options from the national SPS inventory. A measure that is finally selected as an appropriate risk management strategy will have to comply with all the above requirements.

(c) Non-discrimination (Article 2.3)

Article 2.3 contains two separate obligations. The first sentence deals with a qualified form of discrimination and provides that WTO Members shall ensure that their SPS measures do not arbitrarily or unjustifiably discriminate between Members where identical or similar conditions prevail (including between their own territories and that of other Members). The second part of Article 2.3 requires SPS measures to be applied in a manner which would not constitute a disguised restriction on international trade. The rationale behind the provision is two-fold. First, it helps to eliminate those measures which serve a protectionist purpose rather than genuine health concerns (in theory if there are similar or identical conditions in two countries there are no reasons to establish a different SPS status). Secondly, it introduces an element of technocratic rationality in the national risk regulation, as WTO Members are expected to respond to the same (or similar) risk in a consistent fashion.

Article 2.3 may be seen as a variation of the basic GATT 1994 principles (the national treatment combined with the most-favoured-nation principle), supplemented by some elements of the chapeau of Article XX. Of course, due to the specific character of SPS measures, the conceptualization of these rules under the

[130] See Chapter 4, section 4.2.

[131] Cf section 6.4.3(d) below.

SPS Agreement is significantly different. The central element under Article 2.3 is not a product but the SPS situation in the exporting Member. This makes the like products analysis, which is a basic element of the GATT rules, completely irrelevant.[132] Consequently, discrimination under Article 2.3 may also include discrimination between different products.[133] What is prohibited under Article 2.3 is discrimination between WTO Members where identical or similar conditions prevail (external dimension) and between a Member's own territory and that of other WTO Members (internal dimension). In this regard, the scope of Article 2.3 is wider than that of Article I:1 and III:4 of GATT 1994, as it aims to account for the fact that different products can pose the same or similar health risks.[134] On the other hand, unlike the above GATT 1994 provisions, Article 2.3 addresses only a qualified form of discrimination that needs to be arbitrary or unjustifiable. This narrower formulation may again be explained by the nature of SPS measures which sometimes precisely discriminate between countries on the basis of their pest or disease status. Furthermore, under GATT 1994, discrimination may also be justified within Article XX. The SPS Agreement does not contain an equivalent provision, and therefore it was necessary to qualify its non-discrimination obligation.

The case law on Article 2.3 is very limited. It only identified, in very general terms, three elements that are required to find a violation of the first sentence of Article 2.3. This includes the following conditions: (i) the SPS measure needs to discriminate either between the WTO Members or alternatively between an importing Member's own territory and that of other WTO Members; (ii) the discrimination has to be arbitrary or unjustifiable; and (iii) identical or similar conditions prevail in the territory of the Members compared.[135] Due to similarities between the disciplines of Articles 2.2 and 5.5, case law decided under the latter may be relevant here. This would mean that the arbitrary and unjustifiable character of discrimination will be considered in one analysis on a case-by-case basis. The possible justifications are most probably also the same.[136]

The notion of identical or similar conditions under Article 2.3 refers to the SPS status of relevant countries. This includes both bioecological factors such as the existence of the same or similar hazard (ie pest or disease) as well as the quality of SPS enforcement (eg observance of good veterinary practice). Although Article

[132] For details concerning like product analysis under GATT 1994 see, Donald H Regan, 'Regulatory Purpose and "Like Products" in Article III:4 of the GATT (With Additional Remarks on Article III:2)' (2002) 36(3) Journal of World Trade 443; Robert Hudec, '"Like Product": The Differences in Meaning in GATT Articles I and III' in Thomas Cottier and Petros Mavroidis (eds), *Regulatory Barriers and the Principle of Non-discrimination in World Trade Law* (2000).

[133] Cf Panel Report, *Australia – Salmon (Article 21.5 – Canada)*, para 7.112.

[134] Denise Prévost and Peter Van den Bossche, 'The Agreement on the Application of Sanitary and Phytosanitary Measures' in Patrick FJ Macrory, Arthur E Appleton, and Michael G Plummer (eds), *The World Trade Organization: Legal, Economic and Political Analysis* (2005) 265 (giving as an example different fruits which transmit the same disease).

[135] Panel Report, *Australia – Salmon (Article 21.5 – Canada)*, para 7.111.

[136] Cf section 6.4.1(d) below.

2.3 refers to the territory of a country as a whole, its interpretation also needs to reflect the regionalization principle as provided by Article 6. Accordingly, a territory of an exporting country cannot be treated as a monolith and determination of similarities should be based on the SPS status of a particular region within a country. At the same time, the case law has not defined what is required to find that conditions in two countries are similar and this has to be ascertained on a case-by-case basis.[137]

The second sentence of Article 2.3 prohibits the application of SPS measures in a manner which constitutes a disguised restriction on international trade. This part of the obligation had not yet been addressed in the SPS jurisprudence. The Appellate Body only observed that the second sentence of Article 2.3 is closely related to the third element of the Article 5.5 test.[138] However, the standards that were developed under the latter provision are not easily applicable in the context of Article 2.3. Note that under Article 5.5 the examination of a disguised restriction is combined with the assessment of discrimination into one analysis. Consequently, it is difficult to identify those findings of the case law that relate to disguised restrictions, the only element that is referred to in the second sentence of Article 2.3. The relevance of the intent of the party introducing a measure also remains an open question here.[139] What was at least clarified by the case law is the existence of a negative presumption with respect to Article 2.3 for those measures that are incompatible with Article 5.5.[140] The nature of the presumption, as in the case of many other instances under the SPS Agreement, is far from clear. Although the language used by the Appellate Body indicates that the presumption is rebuttable,[141] in practice it may operate as an irrefutable one. Finding that a measure does not arbitrarily or unjustifiably discriminate against the countries with similar SPS conditions, even if it fails to meet the requirements of Article 5.5, seems to be very difficult. Given the more general character of Article 2.3, the positive presumption is, of course, not available, as Article 5.5 does not exhaust the whole meaning of Article 2.3. Consequently, an additional examination of the measure under Article 2.3 is not excluded, even if such a measure is found to be consistent with Article 5.5.

(d) Least-trade-restrictive alternative (Article 5.6)

Article 5.6 establishes the least-trade-restrictive alternative requirement[142] by compelling WTO Members to ensure that SPS measures are not more

[137] Appellate Body Report, *EC – Hormones*, para 240. [138] Ibid, para 238.

[139] Cf section 6.4.1(d) below.

[140] Panel Report, *Australia – Salmon*, para 8.109; Appellate Body Report, *Australia- Salmon*, para 240.

[141] Note that the Appellate Body simply identified the existence of the presumption without introducing any qualification, which could indicate its irrefutable character.

[142] Scott (*Commentary*) 157 (describing that provision as a 'weak proportionality requirement' since it is only concerned with means and does not require balancing the value of the objective being pursued against the trade-restrictive effects of a measure).

trade-restrictive than required in order to achieve their ALOP. In addition, when assessing the availability of alternatives, Members are expected to take into account their technical and economic feasibility. The text of Article 5.6 is supplemented by a footnote which clarifies the meaning of the phrase 'not more trade-restrictive than required'. The footnote to Article 5.6 reads as follows:

> [f]or purposes of paragraph 6 of Article 5, a measure is not more trade-restrictive than required unless there is another measure, reasonably available taking into account technical and economic feasibility, that achieves the appropriate level of sanitary or phytosanitary protection and is significantly less restrictive to trade.

Regrettably, the case law relating to Article 5.6 is again very limited. It only succeeded in identifying a three-prong test that needs to be passed in order to establish a violation of Article 5.6. The test calls for (i) identification of alternative measures that are reasonably available, taking into account technical and economic feasibility; and which (ii) achieve the Member's ALOP; and (iii) are significantly less restrictive to trade than the contested measure.[143] The Appellate Body stressed that, as in the case of Article 5.5, the test is of cumulative character and all its elements need to be demonstrated before finding the violation.[144]

The SPS case law does not explain what is to be understood by reasonable availability or technical and economic feasibility. It is probably safe to conclude that alternatives that are merely theoretical or impose an undue burden on a Member, for example in the form of high economic costs or substantial technical difficulties, are not considered to be reasonably available.[145] However, it seems impossible to establish any general standard that would identify when an alternative becomes reasonably unavailable and each situation will have to be assessed on a case-by-case basis.

The second element of the test requires the determination of whether alternatives identified by a complainant may reach the defendant's ALOP. Note that the question here is not whether an alternative meets the level of protection currently achieved by the measure under examination, but the ALOP as determined by a particular WTO Member. As already mentioned, those two levels are not necessarily the same and ALOP may be higher or lower as compared to the level of protection reflected in an SPS measure. As far as ALOP is concerned, a panel is expected to rely on the determination of a defendant. It may ascertain such a level on its own only if ALOP was not determined at all or a defendant did it without sufficient precision. As a subsequent part of the analysis, the panel should assess identified alternatives in terms of their effectiveness (ie what level of protection is actually achieved by these measures).

The conceptualization of ALOP as a category which is external from a measure and which can be identified by a WTO Member in a panel proceeding gives,

[143] Panel Report, *Australia – Salmon*, para 8.167. See also Appellate Body Report, *Australia – Salmon*, para 194.

[144] Appellate Body Report, *Australia – Salmon*, para 194.

[145] Appellate Body Report, *EC – Asbestos*, para 308.

however, unsatisfactory results. In particular, such an interpretation encourages a defendant to claim that its ALOP is higher than the level of protection that is reflected in a measure. Remember that Article 5.6 is concerned with the ALOP and not with the level of protection that is manifested in a measure. Consequently, a complainant may identify an alternative which is reasonably available and significantly less restrictive to trade and which also reflects the same level of protection as a contested measure, and still fail the applicable test because such an alternative does not achieve a Member's ALOP. A better solution would be to accept that a level of protection reflected in a measure is equal to or at least as high as ALOP. Alternatively, the analysis under Article 5.6 should be limited to a level of protection reflected in a contested measure. Note that such an interpretation would also be in line with the function of Article 5.6, which is to remove those measures that have less trade-restrictive alternatives.

Under the third condition of the test, a panel must examine the extent to which an alternative measure(s) is less restrictive than a measure under the examination. Article 5.6 introduces a *de minims* threshold as only significant difference is relevant. This is a sensible approach as it leaves a margin of additional discretion to domestic governments with respect to the selection of appropriate SPS measures. The provision gives a power to a panel to remove an ostensibly restrictive measure without interfering excessively with the right of a WTO Member to decide on its risk policy. Again, taking into account the variety of different SPS measures, it seems impossible to establish here any abstract threshold and each case will require an individual assessment.

The case law identifies Article 5.6 as closely related to Article 2.2, constituting a specific application of the necessity requirement.[146] In the words of one of the panel 'Article 5.6 must be read in context... [a]n important part of the context of Article 5 is Article 2. We consider that Article 5.6 should, in particular, be read in light of Article 2.2'.[147] This may suggest that a breach of Article 5.6 activates a presumption of inconsistency of a measure with the first sentence of Article 2.2 (similarly to the case of Articles 2.3 and 5.5). On the other hand, both provisions may also be seen as having a complementary character and applicable to different situations. If one accepts the complementary character between Article 5.6 and the first sentence of Article 2.2, the necessity analysis under Article 2.2 should also incorporate a *de minimis* threshold. If these two provisions are complementary, there is no reason to establish a different threshold for the measure itself and the measure as it is applied. This would mean that the violation of Article 2.2 is only possible if the difference between the original measure and its alternatives, in terms of restrictiveness, is significant. This claim is valid, even if we reject the complementary character of both provisions. If any difference in restrictiveness between a measure and its alternatives constitutes a violation of

[146] Panel Report, *EC – Biotech Products*, para 7.1433.
[147] Panel Report, *Australia – Salmon*, para 8.165.

Article 2.2, while under Article 5.6 only a significant difference is relevant, the latter would be deprived of any legal meaning. A measure that is found compatible with Article 5.6 could violate Article 2.2 anyway even if the difference in restrictiveness between a measure and its alternative is minimal.

(e) Regionalization (Article 6)

State borders frequently do not correspond with SPS risks. The same disease or pest may be present in different countries. At the same time, some parts of a single country can be free from certain diseases or pests, while products from other areas of the same country may pose significant SPS risks. There are diverse bioecological or regulatory factors which explain these differences. Variations in climatic, environmental, or geographic conditions within a country may affect the existence and occurrence of pests or diseases (eg prevalence in a particular area of specific species that may serve as a host for a disease). State regulatory activities are also relevant in this context. Programmes that aim at the eradication and control of pests and diseases may radically affect the SPS status of regions. Consequently, two geographically identical areas may have entirely different SPS status.

The SPS Agreement recognizes this heterogeneity and requires WTO Members to adapt their SPS measures to regional conditions (Article 6). The provision, which is referred to in the SPS practice as a regionalization requirement, may be seen as a specific reflection of the necessity test provided by Article 2.2 and the least-trade-restrictive obligation of Article 5.6. Since WTO jurisprudence has not yet addressed disciplines of Article 6, the subsequent analysis is predominantly based on the textual analysis of the provision. Where appropriate, the discussion also refers to relevant international standards and SPS practice.

Article 6.1 introduces a general substantive obligation to adapt national SPS measures to SPS characteristics of the area from which a particular product is originated and destined. That area may include a country as a whole, a part of a country, or all or parts of several countries. The characteristics of the destination place are also relevant. That is understandable as the climatic or geographical conditions of the territory of the importing WTO Member (or some parts of it) may be such as to physically prevent a particular risk from materialization. If, for example, there is no suitable host for a disease in an importing country there will simply be no SPS risk. Having said this, one also needs to recognize that this part of the regionalization requirement may have only limited importance. While the need to take the whole territory of the importing Member is indisputable, the relevance of the SPS status of a particular region within such a country implies the existence of a control system for the internal movements of goods. In practice, internal SPS restrictions, although not completely absent, are not very common. The SPS Agreement as such does not impose any obligations that would require implementing such a control system and the issue is left entirely to WTO Members. Thus, it seems that the obligation to adapt measures to SPS conditions of destination place is relevant only in those rare cases where an importing country is able to control the internal movement of goods.[148]

The second sentence of Article 6.1 supplements the general substantive rule with a procedural requirement. It specifically provides that when assessing the SPS characteristics of a region (both in the exporting and importing country), a WTO Member needs to take into account a number of factors such as the level of prevalence of specific diseases or pests, the existence of eradication or control programmes, and appropriate criteria or guidelines which may be developed by the relevant international organizations. This list is not exhaustive and other elements may also be considered. It is also worth remembering that WTO Members are also obliged to take into account in their risk assessment the existence of pest- or disease-free areas as well as the prevalence of specific diseases or pests. Both obligations use the rather soft language of 'taking into account'. Pursuant to the *Apples* panel, 'taking into account' indicates a softer relationship than would be provided by the expressions 'in conformity with' or 'based on'. Paraphrasing the panel, the 'failure to respect each and every aspect of them (elements that are enumerated by Article 6.1) would not necessarily, *per se*, signal'[149] the violation of this provision; however it may provide a useful guidance as to whether conduct of a WTO Member is indeed in conformity with Article 6.1.

There seems to be some overlap between the disciplines of Article 6.1 and the obligation of Article 3.1. The first one requires criteria or guidelines developed by the relevant international organization to be taken into account when assessing an SPS characteristic of a region. Simultaneously, Article 3.1 obliges WTO Members to base their SPS measures on international standards, guidelines, and recommendation, a category which also includes documents relating to regionalization. One reading of these two provisions would be to see them as setting parallel obligations which are equally applicable to SPS measures. This would mean that when assessing the SPS characteristics of a region WTO Members have to take into account criteria and guidelines developed by relevant international bodies, while a final measure needs to be based on international standards, guidelines, and recommendations (including those which are relevant for regionalization) unless there is scientific justification for deviation. This approach, although textually correct, seems to be unsatisfactory. The notion of standards, guidelines, and recommendations is sufficiently broad to cover the expression 'criteria or guidelines which may be developed by the relevant international organizations'.[150] Consequently, applying disciplines of Article 3.1, as far as regionalization is concerned, would deprive a part of Article 6.1. If a measure is based on

148 For example, some internal SPS control exists between different states of the US.

149 Panel Report, *Japan – Apples*, para 8.241. See also, Panel Report, *US – Continued Suspension*, para 6.82.

150 The expression 'criteria' is nowhere defined in the SPS Agreement; a contextual reading would suggest that it refers to the relevant requirements set by international standards, guidelines, and recommendations.

international standards, guidelines, or recommendations, a WTO Member must have taken into account relevant criteria or guidelines. A more appropriate reading would be to conceptualize the second sentence of Article 6.1 as *lex specialis*[151] to a general obligation of Article 3.1. In practical terms, this would mean that WTO Members are only obliged, when adapting their measures to regional conditions, to take into account relevant criteria or guidelines developed by international organizations, without a need to base measures on those documents.

Although the SPS Agreement does not explain what is to be considered as a relevant international organization in the context of Article 6.1, this clearly covers the sister organizations. The pertinent IPPC documents include four standards on requirements for the establishment of pest-free areas, pest-free places of production and production sites, areas of low pest prevalence, and recognition of pest-free areas and areas of low pest prevalence.[152] The IPPC has also adopted complementary standards relating to eradication of pests from particular areas and subsequent monitoring and surveillance.[153] As far as animal diseases are concerned, the OIE has developed extensive general guidelines in its Terrestrial Animal Health Code and Aquatic Animal Health Code. Both Codes also provide detailed recommendations with respect to particular diseases.[154] These documents are discussed below in more detail.

The substantive obligation of Article 6.1 finds its elaboration in Article 6.2, which specifically requires WTO Members to recognize the concepts of pest- or disease-free areas and areas of low pest or disease prevalence. According to the SPS Agreement, determination of such areas has to be based on several factors, including geography, ecosystems, epidemiological surveillance, and the effectiveness of sanitary or phytosanitary controls. The pest- or disease-free area is defined as an area which could be composed of several countries, only one country, or some part of the country or countries as identified by the competent authorities, in which a specific pest or disease does not occur at all.[155] A pest- or disease-free area may surround, be surrounded by, or be adjacent to, an area in which a specific pest or disease is known to occur but is subject to regional control measures, including protection, surveillance, and buffer zones. An area of low pest or disease prevalence is defined as an area (again all of a country, part of a country, or all or parts of several countries), as identified by the competent authorities,

151 *Lex specialis* in its strong form, meaning that Article 6.1 excludes the applicability of Article 3.1 as far as regionalization is concerned.

152 ISPM No 4, Requirements for the Establishment of Pest Free Areas (1995), ISPM No 10, Requirements for the Establishment of Pest Free Places of Production and Pest Free Production Sites (1999), ISPM No 22, Requirements for the Establishment of Areas of Low Pest Prevalence (2005), ISPM No 29, Recognition of Pest Free Areas and Areas of Low Pest Prevalence (2007), ISPM No 6, Guidelines for Surveillance (1997).

153 ISPM No 9, Guidelines for Pest Eradication Programmes (1998).

154 OIE, *Terrestrial Animal Health Code*, Article 10.4.2 (avian influenza) and Article 15.3.2 (classic swine fever).

155 Annex A(6) of the SPS Agreement.

in which a specific pest or disease occurs at low levels and which is subject to effective surveillance, control, or eradication measures.[156] Thus, the main difference between a pest- or disease-free area and an area of low prevalence is that the first class excludes a presence of pest or disease in a particular area, while in the second case such presence is possible albeit it needs to be below a specific level. In practice, the concept of a low pest prevalence area is used when the ALOP of the exporting country is not so high as to require the products to originate from disease- or pest-free areas.

Does Article 6.2 exhaust the whole meaning of Article 6.1? Its language indicates that this is not necessarily the case. Note that Article 6.2 uses the expression 'in particular' which suggests that recognition of pest- and disease-free areas (and areas of low pest or disease prevalence) constitutes only one possible way of applying the general requirement to adapt SPS measures to regional conditions. This conclusion is, however, not supported by SPS practice where regionalization is equated with the concepts of pest- and disease-free areas as well as areas of low pest or disease prevalence. Consequently, one may assume that the scope of Articles 6.1 and 6.2 is the same.

The obligation to adapt measures to regional conditions is not translated under the SPS Agreement into automatic recognition of specific SPS status of a part of the territory of the exporting Member (pest- or disease-free area or area of low pest or disease prevalence). This can be done only after conducting a special investigation. Article 6.3, elaborating on this issue, details the obligations which rest on the exporting Member in the recognition process. An exporting country bears a burden of proof as to the existence of a relevant area, which includes the obligation to provide the importing Member with all the necessary evidence objectively demonstrating that a particular area is indeed pest- or disease-free (or of low pest or disease prevalence) and that it is likely to retain such status in the future. Article 6.3 also specifies that for the purpose of demonstration, reasonable access needs to be given to the importing Member for inspection, testing, and other relevant procedures. This should allow an importing Member to verify the information provided by the exporting Member and assess the actual operation of its domestic SPS system. Apart from that, Article 6.3 remains silent on many important issues. As noted by Scott, it does not specify the time horizon which should be relevant in the evaluation of a pest- or disease-free area or area of low pest or disease prevalence status, and simply refers to likelihood ('likely to *remain*').[157] The documents of the relevant international standard-setting bodies are also silent on that issue. Logic suggests that the time horizon cannot be too short as this may impair the right of an importing Member to establish its ALOP. A pest- or disease-free status is not granted for a particular shipment of goods but has quasi-permanent character. Consequently, the SPS safety of an importing Member requires assurance that the exporting country will maintain

[156] Annex A(7) of the SPS Agreement.

[157] Scott (*Commentary*) 185.

such status in the future. On the other hand, if a temporal horizon is too long, Article 6.2 may lose some of its significance as it is normally very difficult (if not impossible) to make an accurate SPS prediction on the distant future. Bearing in mind these concerns, it is clear that the SPS Agreement gives considerable discretion to WTO Members in this respect. Probably the sensible way of dealing with this imprecision is to apply a standard of reasonableness where a panel assesses all relevant circumstances such as historical records of maintenance of disease- or pest-free areas on the territory of an exporting Member, the seriousness of a particular disease, the time horizon applicable to other Members, and biological factors such as the incubation of the disease.[158]

The second problem, which arises in the context of Article 6.3, is the required probability threshold. The language of the provision (*likely* to remain) resembles the definition of quarantine risk assessment which requires the evaluation of the likelihood of entry, establishment, or spread of a pest or disease within the territory of an importing Member. As noted by Scott, this may suggest that 'theoretical risks that the area in question will not remain pest- or disease-free will not suffice to reject a regionalization request'.[159] However, apart from this observation, it is difficult to say anything more general on applicable threshold.

The concept of pest- and disease-free areas, as well as areas of low pest or disease prevalence, although exhausting the principle of regionalization, does not dilute the idea of adaptation of measures to specific SPS conditions in an exporting country. SPS practice also recognizes the concept of compartmentalization (or pest-free places of production and pest-free production sites) which defines a population as pest- or disease-free through reference to management practices. In opposition to regionalization, which relies on geographical factors (areas that can be separated from each other using natural, artificial, or legal boundaries), compartmentalization concentrates on management and husbandry practices as a mean of ensuring biosecurity.

The central element of compartmentalization is the concept of the compartment. The OIE defines this as a single or group of establishments (any premises where animals are kept) under a common biosecurity management system. The compartment contains an 'animal subpopulation with a distinct health status with respect to a specific disease for which required surveillance, control and biosecurity measures have been applied for the purpose of international trade'.[160] The relevant biosecurity measures consist, for example, of limiting non-essential traffic on the premises (including access of non-essential people), keeping wildlife and other domestic animals confined to the premises, and providing disinfectant and appropriate disposable footwear for visitors to the premises. The aim of this

[158] In practice, WTO Members often grant time limited pest- or disease-free (low pest or disease prevalence) status, subject to the review.

[159] Scott (*Commentary*) 185.

[160] OIE, *Terrestrial Animal Health Code*, Glossary.

is to limit or eliminate epidemiological links with external populations. The same is true for IPPC pest-free places of production or pest-free production sites.[161] As in the case of compartmentalization, a place and production site is established on the basis of management practices and not geographical characteristics. The IPPC sets out criteria that can be used in evaluating the status of a particular establishment (eg characteristic of the pest, characteristic of production site or place) and suggests certain management practices (ie preventive, exclusion, and pest control measures).[162]

To what extent are these concepts relevant for the SPS Agreement? Note that Article 6 only requires adapting SPS measures to the characteristic of the area. This language clearly removes the idea of compartmentalization and place/production sites from the scope of Article 6 as they are based on management and not geographical criteria. However, it would go too far to conclude that these concepts are completely irrelevant under the SPS Agreement. One option is to apply them through Article 3.1, which requires WTO Members to base their SPS measures on international standards. Since those rules are part of international standards, one may argue that deviation from them is possible if supported by a scientific justification. Again, this interpretation, although textually correct, cannot be accepted. Remember that under the SPS Agreement standards (or guidelines and recommendations) pertaining to regionalization are only to be taken into account (Article 6.1). Applying more demanding requirements with respect to compartmentalization and pest-free place/production sites, concepts that are even not explicitly referred to by the SPS Agreement, appears to be counter-intuitive. A WTO Member would only be obliged to take into account criteria or guidelines relevant for regionalization, but it will be required to base its SPS measures on the concept of compartmentalization and pest-free place/production sites. This in turn suggests that the disciplines of Article 3.1 do not apply here. On the other hand, it is uncontroversial that both concepts may play a role in the assessment of necessity of a measure (Article 2.2) and be relevant under the test of the least-trade-restrictive alternative (Article 5.6). Consequently, they will act not as standards (guidelines or recommendations) but rather as factual indicators which show available alternatives. A measure that achieves the required ALOP and implements the concept of compartmentalization and place/production sites may be a less restrictive alternative than a measure which restricts trade for all establishments in the particular area. This interpretation also seems to be supported by the official position of the OIE. In one of the documents submitted

[161] A pest-free place of production is defined as a 'place of production in which a specific pest does not occur as demonstrated by scientific evidence and in which, where appropriate, this condition is being officially maintained for a defined period', while a pest-free production site is defined as a portion of a place of production that can be managed as a separate unit within a place of production, and which satisfies the above mentioned conditions (ISPM No 10, Requirements for the Establishment of Pest Free Places of Production and Pest Free Production Sites (1999) 108).

[162] Ibid, 110.

to the SPS Committee, the OIE clarified that the application of both concepts (compartmentalization and zoning) is not mandatory and should only be used when appropriate, taking into account the epidemiology of the disease and the particular factors applying in each situation.[163] This language closely corresponds to the interpretation proposed above.

As already mentioned, both the IPPC and the OIE adopted a number of standards and guidelines which are relevant in the context of Article 6. The IPPC provides general guidelines on the establishment and maintenance of pest-free areas. Such areas may be based either on artificial or natural boundaries such as administrative (eg country, province, or commune borders), physical (eg rivers, seas, mountain ranges, roads) or property borders.[164] The IPPC standards are formulated in very general language. This should be not surprising since the IPPC itself admits that the nature of the process varies according to the biology of the pest, types and characteristics of pest-free areas, and the level of required phytosanitary security.[165] Nevertheless, the IPPC identifies three general components in establishing and maintaining pest-free areas. This particularly includes systems to establish freedom (ie collection of data to identify free areas), phytosanitary measures to maintain freedom (ie regulatory actions such as restriction of the movement of certain products between areas of a country), and verification activities (ie monitoring and surveillance of the area). A separate set of substantive requirements is provided with respect to areas of low pest prevalence.[166] As in the case of pest-free areas, the areas of low pest prevalence may occur naturally or be established through the development and application of phytosanitary measures aimed at controlling a pest(s).[167] A separate IPPC standard sets out the procedural requirements with respect to recognition by the importing country of pest-free areas and areas of low pest prevalence.[168]

The OIE standards, both of substantive and procedural character, establish similar set of guidelines for animal diseases. The substantive provisions can be divided into two categories. The first group relates to general issues such as principles for defining the zone (the SPS Agreement instead of 'a zone' uses the notion of 'an area'), the basic obligations of the veterinary service in the exporting country, and the surveillance and traceability of live animals. The second group consists of specific requirements with respect to particular diseases (eg trichinellosis,

163 SPS Committee, OIE Activities on Zoning and Compartmentalization, Communication from the World Organization for Animal Health (OIE), 23 January 2006, G/SPS/GEN/625.

164 ISPM No 4, Requirements for the Establishment of Pest Free Areas (1995) 38.

165 Ibid, 39.

166 This includes elements such as determination of specific pest level, surveillance activities, or reducing the risk of entry of a particular pest, etc.

167 ISPM No 22, Requirements for the Establishment of Areas of Low Pest Prevalence (2005) 266.

168 ISPM No 29, Recognition of Pest Free Areas and Areas of Low Pest Prevalence (2007). The standard details the basic principles of the process such as non-discrimination, avoidance of undue delay, transparency, allocation of responsibilities between the parties and prescribes particular steps of the recognition process.

foot and mouth disease). As far as procedural provisions are concerned, the Terrestrial Animal Health Code specifies the recommended steps for the recognition of disease-free areas.[169] The OIE has not adopted any special standards or guidelines concerning the establishment and maintenance of areas of low disease prevalence.

Apart from development of standards, the OIE also has a right to recognize officially a country or a part of it as being free from certain specific diseases.[170] The process of official recognition is initiated by an interested OIE member, whose application is subsequently examined by the Scientific Commission advised by ad hoc groups of experts. On that basis, the Commission may issue the recommendation to grant a disease-free status (with or without vaccination). Such recommendation is subsequently distributed to all OIE members which may submit their objections; however, these may be based only on scientific or technical grounds. If no objection is raised or the objection is unfounded, the International Committee includes a particular country or a list of countries (or zones) that are free from particular diseases.[171] The maintenance of a disease-free status by the OIE member depends on the observance of the OIE rules and in case of their violation may be revoked. The IPPC does not have comparable provisions with respect to pest-free areas.

What are the consequences under the SPS Agreement of official OIE recognition? From a legal viewpoint, there are at least two possible approaches. First, official recognition may be considered as a relevant standard, guideline, or recommendation. If this is the case, WTO Members are obliged to base their SPS measures on such official recognition, unless they are able to present scientific evidence which justifies the deviation. Secondly, the official recognition may be considered as a distinctive act which does not qualify as a standard. Its relevance would again be limited to the evaluation of necessity (Article 2.2) as well as the least-trade-restrictive alternative requirement (Article 5.6). Since the SPS Agreement does not define standards, guidelines, and recommendations, their meaning needs to be established within the OIE. The list of countries which are officially recognized as free from certain diseases is adopted in the form of a resolution of the International Committee. The resolution does not mention (explicitly or implicitly) that it should be considered as an international standard, guideline, or recommendation. Nevertheless, due to differences in the nomenclature used in the OIE and the SPS Agreement, this lack of reference cannot be conclusive.[172] There are, however, other arguments which suggest that

[169] OIE, *Terrestrial Animal Health Code*, Article 4.3.3.

[170] The OIE may recognize a disease status for the following diseases: foot and mouth disease (FMD), rinderpest, contagious bovine pleuropneumonia (CBPP) and bovine spongiform encephalopathy (BSE).

[171] See eg, OIE, Resolution No XII of the OIE International Committee, 'Procedure for the Recognition of the Foot and Mouth Disease Status of Member Countries', 15–19 May 1995.

[172] Cf Chapter 3, section 3.4.

official recognition cannot be regarded as an international standard, guideline, or recommendation. First, note that despite some similarities between the procedure for standard-setting and official recognition, the voting requirements are entirely different. Standards are adopted by consensus (if consensus is not achieved, then only if no more than 10 countries do not oppose the proposal). Official recognition is subject to a less stringent threshold as a simple majority is sufficient for the adoption of a resolution.[173] Moreover, the compilation of the countries that are included on the OIE list is predominantly based on the questionnaires submitted by OIE Members. Although the Commission (or ad hoc group) may request permission for a technical visit to be made to the particular country in order to verify submitted data, this is an option rather than a general practice. Recognizing limited mandate, the OIE's resolution always includes a disclaimer to indemnify the OIE for any inaccuracies in the publication of country disease status if they result from imprecise or incomplete information, or changes in epidemiological conditions.[174] This also corresponds with the practice of WTO Members. The official recognition is never automatically translated into bilateral recognition[175] and is only considered as one of the relevant factors. There is also a pragmatic reason that supports such an approach. The OIE members, unwilling to lose discretion in the process of recognition, would probably block the adoption of any official list of disease-free countries. This in turn may have additional trade distorting effects since many countries require such official recognition as a prerequisite for bilateral talks.

In addition to official recognition, the Terrestrial Animal Health Code and Aquatic Animal Health Code also provide for self-declaration of disease-free status. That alternative is available to all OIE members with respect to any disease listed in one of the Codes and for which there is no official OIE procedure of recognition. Declaration may relate either to the whole or part of the territory of the OIE member. A member wishing to take advantage of such an option needs to observe the requirements that are imposed by the Code with respect to particular diseases. The procedure for self-declaration of disease-free status is simple: an OIE member submits an appropriate declaration to the OIE which is subsequently published by the OIE without any prior verification. As noted by the OIE 'the self-declaration for freedom from an OIE listed disease is on the responsibility of the Member Country concerned and does not have official OIE recognition status as for diseases'.[176] The value of self-declaration for the purpose of the SPS Agreement is rather limited. Obviously, this document cannot be classified

[173] OIE, General Rules of the Office International des Epizooties adopted by the International Committee on 24 May 1973, Article 6.3.

[174] See eg, OIE, Resolution XX of the OIE International Committee, List of Foot and Mouth Disease Free Countries, 24 May 2005, para 10.

[175] See eg, SPS Committee, Foot and Mouth Disease Free Zone—Brazilian Experience on Regionalization, Statement by Brazil at the meeting of 29–30 June 2005, G/SPS/GEN/584.

[176] Information on the webpage of the OIE (<http://www.oie.int/eng/info/en_procedures.htm?e1d6>, last visited 20 July 2009).

as a standard (guidelines or recommendation) promulgated by an international standard-setting body. At best it may serve as one of the relevant factors in the process of national recognition.

In SPS practice, the application of the concept of regionalization is far more important than it may seem when analysing only the provisions of the SPS Agreement. Regionalization has significant potential to reduce the transaction costs for the exporting country through minimizing administrative burdens connected with export. Sometimes, regionalization is even a *sine qua non* condition for international trade in agricultural products. The eradication of pest or disease from the whole territory of the country may be impossible or prohibitively costly. This is particularly true for large developing countries such as Brazil or China. On the other hand, regionalization strengthens the quality of the veterinary or phytosanitary infrastructure of the exporting country in general by requiring establishment of operable systems that would guarantee the maintenance of disease- and pest-free status.

Despite some success, the overall experience with the implementation of the regionalization principle is rather unsatisfactory. This is reflected in the activities of the SPS Committee where regionalization has been a standing item on the agenda since 2003.[177] Although the SPS Committee was able to identify the most problematic issues, this knowledge was not translated into substantial improvements. The main problem is the length and costs of the bilateral recognition process. The reasons for this vary, ranging from inefficient and complex administrative procedures applied in the importing countries to a lack of trust in the capacities of the exporting country to guarantee the achievement of domestic ALOP. Another concern is excessive data requirements which sometimes do not find scientific support. Administrative procedures required by importing Members are often not transparent, and in some cases are not even clearly defined.[178] An additional problem is created by the lack of consistency in procedures required by different WTO Members. In consequence, an exporting Member needs to comply with different requirements depending on the importing Member. Finally, there is an almost complete lack of predictability in the bilateral recognition process. In some cases, the process takes years to be completed, while in others it results in the unexpected rejection of recognition. This additionally discourages exporting WTO Members from committing to investment in the development and maintenance of pest- or disease-free areas.

[177] See eg, SPS Committee, Summary of the Meeting Held on 19–20 March 1997, Note by the Secretariat, G/SPS/R/7, paras 18–19 (Chilean restrictions on imports of wheat and fruit—issue raised by the US); SPS Committee, Summary of the Meeting Held on 11–12 October 2006, Note by the Secretariat, G/SPS/R/43, paras 17–21 (Indonesian legislation on importation of live animals and meat products—issue raised by Brazil); SPS Committee, Summary of the Meeting Held on 27–28 June 2006, Note by the Secretariat, G/SPS/R/42, paras 25–26 (Japanese import suspension of heat-processed straw and forage for feed—issue raised by China).

[178] SPS Committee, Submission of New Zealand, Monitoring of International Harmonization: Regionalization, 29 September 2004, G/SPS/W/151, para 5.

The efforts undertaken by the SPS Committee have only partially resolved the above problems. During the consultations it appeared that there was strong opposition among some WTO Members (mainly developed countries) to adopt any specific procedure, including time frames, for the recognition process.[179] Those countries particularly argued that it was availability and quality of scientific information which dictated the length of process and not so much the complexity of administrative proceedings. Some of the importing countries also pointed out that in many cases the process was affected by the unwillingness of the exporting country to collect and share information. Moreover, since WTO Members differ in their capacities and resources which can be dedicated to the recognition process, it could be difficult for Members to implement a uniform timescale.[180] These practical concerns are also supported by more systemic problems. Some Members argue that adoption of any general time frames may not be possible due to the diverse epidemiological and ecological characteristics of diseases and pests. In consequence, a case-by-case approach provides a better solution.

Without going into details, since this problem deserves a separate and comprehensive study, it is worthwhile noting that although some of these concerns are valid, probably the main reason behind the opposition by some importing countries to introduce time frames or common procedures is not so much technical, but rather political. Without unequivocal requirements in the SPS Agreement (or through the decision of the SPS Committee), countries are not really eager to lose their unrestricted discretion. Note that this discretion relates not only to the final decision on the recognition of the SPS status of a country (or part of it) but also to mediating issues, such as the distribution of financial and technical resources between different governmental tasks in the SPS area. Since foreign producers have limited opportunities to influence the allocation of such resources, this particular task is frequently underinvested.

Despite this disagreement, WTO Members have recently reached a general consensus on the procedure applicable to regionalization. The compromised text, which was developed in the process of informal consultation with the key WTO Members, entered into force on 19 May 2008.[181] The guidelines, formulated in very general language, set out the basic elements of the recognition process. In particular, they ask WTO Members to publish the basis for recognition of pest- or disease-free areas together with its description (including, for

[179] See eg, SPS Committee, Summary of the Meeting Held on 11–12 October 2006, Note by the Secretariat, 3 January 2007, G/SPS/R/43, 15.

[180] SPS Committee, Communication from the Republic of Korea, The Implementation of Article 6 of the Agreement on the Application of Sanitary and Phytosanitary Measures (Regionalization), 31 May 2006, G/SPS/W/195, para 12.

[181] Information on the webpage of the WTO (<http://www.wto.org/english/news_e/news08_e/sps_may08_e.htm>, last visited 20 July 2009).

example, information required for evaluation of the request).[182] The guidelines do not set any specific time frames and only generally stipulate that WTO Members should proceed with a recognition process without undue delay. This stipulation is supplemented with the instructive term of 90 days (from the date of request) for entering into the initial discussion on recognition.[183] The guidelines also provide a description of typical administrative steps in the recognition process.[184] This part of the guidelines has more of a narrative than normative character as WTO Members retain a sovereign right to determine their own recognition process with regard to pest- or disease-free areas (and areas of low pest or disease prevalence).[185] Although the guidelines fall short of providing any specific time frames or detailed procedure, they should be regarded as a positive development. The guidelines themselves recognize that they only constitute a first step and will be reviewed and developed as the new consensus emerges.

(f) Equivalence (Article 4)

The concept of equivalence is rooted in the idea that ALOP required by an importing WTO Member may be achieved in different ways. An example of SPS measures applied against the risks of fruit fly species may serve as a good illustration. Measures such as heat treatments (eg high temperature forced air, vapour heat treatment, hot water dip), cold treatments (eg cold storage—both in-transit and in country of origin), chemical treatments (eg fumigants, dips, dustings, flood spraying) or irradiation, although different, may provide the equivalent levels of protection. Thus, at least in theory, it is possible for an importing country to rely on the SPS measures, control, and inspection mechanisms which are in place in an exporting country, even if such measures (mechanisms) are different from its own. This is also the essential point of equivalence; it does not require the sameness of the measures but rather the same level of protection that may be achieved in different ways.

As mentioned in Chapter 3, section 3.2, mutual recognition (equivalence) is one way of addressing regulatory heterogeneity between different jurisdictions. Consequently, the concept of equivalence may be seen as an alternative approach to harmonization. In some instances, international harmonization

[182] SPS Committee, Guidelines to Further the Practical Implementation of Article 6 of the Agreement on the Application of Sanitary and Phytosanitary Measures, 25 February 2008, G/SPS/W/218, para 4.

[183] Ibid, para 15.

[184] The guidelines recognizes the following steps: (i) information request of an exporting Member with respect to recognition process; (ii) explanation of the requirements; (iii) provision of the documentation by an exporting Member; (iv) evaluation of the documentation and, if necessary, requesting additional data; (v) provision by an exporting Member requiring clarifications and additions; (vi) evaluation of additional information; (vii) on-site verification by an importing Member; (viii) response of an exporting Member to inspection report; and (ix) final determination.

[185] Ibid, para 19.

could be simply impossible due to political, ecological, epidemiological, or other reasons. The concept of equivalence may help in such cases to maintain the benefits of harmonization (ie easier market access), without a need to agree on the same standard. As noted by one scholar, 'equivalence is a key to permitting the maintenance of regulatory diversity, while at the same time promoting market integration'.[186] It preserves a process of mutual learning between different regulatory regimes and supports policy innovations. Contrary to the traditional non-discrimination disciplines, mutual recognition helps to eliminate unnecessary duplication of enforcement activities and encourage the use of collective resources in a more efficient and effective way.[187] There are also obvious economic benefits for private actors (ie exporters), which need to comply with only one set of rules instead of meeting the requirements of two regulatory systems—its own domestic system and that of the importing WTO Member. The equivalence may also contribute to the overall improvement of SPS protection as it 'decrease[s] reliance on routine end-product testing (eg at port of entry), which may be of limited effectiveness'.[188] At the same time, it should be noted that in the SPS practice the equivalence is not as widely spread as one would expect. It requires the existence of mutual trust between Members and may involve high initial costs.

In technical terms, the equivalence may relate to (i) inspection and control systems (in the language of the SPS Agreement, conformity assessment systems); (ii) processing techniques such as HACCP;[189] and (iii) product standards.[190] In practice, the most common is the determination of equivalence with respect to process standards. As far as the form is concerned, equivalence may be used on an ad hoc basis, or through the conclusion of more or less formal arrangements or agreements between interested states. Ad hoc applications usually relate to a specific SPS measure aimed at the protection against particular risk(s) in a specific product. Informal arrangements have a wider scope and are applied for longer periods of time. Formal agreements, which can be either unilateral or reciprocal,[191] have the broadest character. They may be concluded as agreements

[186] Scott (*Commentary*) 165.

[187] Codex Alimentarius Commission, Guidelines for the Development of Equivalence Agreements Regarding Food Imports and Export Inspection and Certification Systems (CAC/GL34) ('Guidelines on Equivalence Agreements'), para 2.

[188] Digby Gascoine, 'Harmonization, Mutual Recognition and Equivalence—How and What Is Attainable?', Paper presented at the Conference on International Food Trade Beyond 2000: Science-Based Decisions, Harmonization, Equivalence and Mutual Recognition, Melbourne Australia, 11–15 October 1999.

[189] HACCPs (hazard analysis and critical control points) can be defined as a systematic preventative approach to food safety which concentrates on critical points in the production and distribution process, rather than on the final product.

[190] SPS Committee, Equivalence: Consideration of Article 4 of the SPS Agreement, 1.

[191] In the first case, only one country recognizes a measure(s) of the other country as equivalent, while in the second case both countries reciprocally bind themselves to allow each other's products to enter their respective markets.

recognizing the equivalence of the whole SPS system or limited to specific group of products.[192]

The SPS Agreement establishes only very general obligations with respect to equivalence. Article 4, which is a central provision in this context, is supplemented with the guidelines of the SPS Committee that explain specific implementation details.[193] The disciplines of the SPS Agreement are also accompanied by a set of international standards. In particular, the Codex adopted two standards describing the process of developing bilateral and multilateral equivalence agreements in the area of food, including equivalence of food inspection and certification systems.[194] The OIE dedicates to equivalence a separate part of its *Terrestrial Animal Health Code* which establishes guidelines on the process of reaching the judgment of equivalence of sanitary measures.[195] The IPPC standards have similar provisions with respect to phytosanitary measures.[196] The subsequent section will present a more elaborate discussion of these rules.

The SPS Agreement imposes on WTO Members a substantive requirement to accept as equivalent measures of other Members, even if such measures differ from their own or from those used by other Members trading in the same product. This acceptance is conditional, as the exporting Member needs to demonstrate that its measure achieves relevant ALOP. Consequently, the burden of showing the equivalence lies with the exporting Member. On the other hand, the importing Member is responsible for determining that the conditions for recognizing a measure as equivalent are met. Seen from a different angle, equivalence may be conceptualized as an exception from the principle of most-favoured nation as it allows making special privileges to some WTO Members without the need to extend them to others.

The Equivalence Decision, elaborating upon the very enigmatic language of the SPS Agreement, explains that the demonstration of equivalence has a two-dimensional character. First, it consists of the evaluation of scientific and technical information underlying a measure of an exporting Member.[197] The aim of such an evaluation is to ensure that a measure is scientifically sound and able to achieve the ALOP that was identified by the importing Member (scientific

[192] SPS Committee, Equivalence: Consideration of Article 4 of the SPS Agreement, Summary of Informal Discussions on Equivalence, Second Report by the Chairman, 21 March 2001, G/L/445, 1.

[193] SPS Committee, Decision on Implementation of Article 4 of the Agreement in the Application of Sanitary and Phytosanitary Measures, 24 October 2001, G/SPS/19, last time revised by the decision of 23 July 2004, G/SPS/19/Rev ('Equivalence Decision').

[194] Codex Alimentarius Commission, Guidelines on Equivalence Agreements; Codex Alimentarius Commission, Guidelines on the Judgment of Equivalence of Sanitary Measures associated with Food Inspection and Certification Systems (CAC/GL/53) ('Guidelines on the Judgment of Equivalence').

[195] OIE, *Terrestrial Animal Health Code*, Chapter 5, section 5.3.

[196] ISPM No 24, Guidelines for the Determination and Recognition of Equivalence of Phytosanitary Measures (2005).

[197] SPS Committee, Equivalence Decision, para 4.

level). This information may include, amongst other things, reference to relevant international standards, or to relevant risk assessments undertaken by the importing Member or by another WTO Member.[198] A separate part of determination relates to the practical application of a measure in an exporting country within the framework of its overall SPS system (factual level). Although a measure as such may constitute an effective means to reduce particular risk, the quality of its enforcement can actually make it inoperable. The SPS Agreement recognizes that fact and grants to an importing Member a right to request reasonable access to an exporting Member for inspection, testing, and other relevant procedures. The Codex identifies some elements which are relevant in this context, such as a broadly understood infrastructure, including the legislative basis (eg food and enforcement law), design of programme, implementation and monitoring, documentation of systems, performance, decision criteria and action, laboratory capabilities, transportation infrastructure and provisions for certification and audit, and requirements applicable to individual facilities, equipment, processes, procedures, tests, and methods of sampling and inspection.[199] The assessment of these factual issues obviously gives a wide margin of discretion to the authorities of the importing country.

Demonstration of equivalency evidently requires cooperation of an importing Member. That is implied in the provision of Article 4 and explicitly recognized by the Equivalence Decision, which requires importing Members to supply the exporting Member with relevant information, including an explanation of the objective and rationale of its SPS measure, and identification of relevant risks and applicable ALOP. The Equivalence Decision goes even further and instructs an importing Member to provide a copy of its risk assessment on which an SPS measure is based and/or a technical justification based on a relevant international standard, guideline, or recommendation.[200] The above list is only indicative and the exporting Member may also request other information that can help it to objectively demonstrate the equivalence of its own measure.

The SPS Agreement does not set any time frames for the recognition process. Particularly, such a deadline is not provided by Annex C which requires initiating and completing some SPS procedures without undue delay as the recognition process cannot be qualified as a control, inspection, and approval procedure.[201] The Equivalence Decision is more specific on this issue and introduces an instructive term of six months during which the importing Member is expected to provide a response to the request for the initiation of the recognition process.[202] The Equivalence Decision also recommends an accelerated procedure for

[198] Ibid. [199] Codex, Guidelines on Equivalence Agreements, para 13.

[200] SPS Committee, Equivalence Decision, para 2.

[201] The footnote to Annex C provides that 'control, inspection and approval procedures include, inter alia, procedures for sampling, testing and certification'.

[202] SPS Committee, Equivalence Decision, para 3.

those products that were historically traded between Members.[203] A long trade relationship between Members with respect to a particular product or group of products provides useful data on the actual operation of the SPS system in an exporting country (including its credibility), and may shorten the recognition process. In practice, the length of the procedure is considered to be one of the main obstacles to the broader reliance on the concept of equivalence.

The SPS Agreement also promotes bilateral and multilateral recognition agreements by urging WTO Members to enter into a consultation process leading to the conclusion of such agreements. Although a WTO Member may not refuse to start negotiations if so requested, the obligation remains weak as the SPS Agreement does not provide for any specific procedure or deadline. In practice, most of the equivalence agreements are concluded on a bilateral basis. The SPS Agreement does not contain any direct obligation which would require WTO Members to notify the conclusion of equivalence agreements or inform on ad hoc recognition. Nevertheless, the Equivalence Decision, through reference to transparency provisions of the Agreement, insists on notification of equivalence agreements.[204] A special procedure, which was subsequently developed by the SPS Committee, asks for the notification of any equivalence agreements or arrangements (note that it is not clear whether this obligation also covers ad hoc recognition) as well as any significant variations to existing equivalence arrangements, including their suspension or rescission.[205] Leaving aside the question of whether the SPS Committee acted within its mandate, it is worth noting the relatively low record of notifications. As observed by Scott, this reluctance may result from the lack of any interest in making notifications. An importing Member, which usually bears the high financial costs of the recognition process, is not really interested in encouraging other Members to initiate the process. The exporting country, on the other hand, is not interested in publicizing the successful outcome of negotiations, since that may push other countries to seek the same recognition and jeopardize its competitive advantage.[206]

The formulation of Article 4.1 raises problems similar to those which were identified under Article 5.6. Note that Article 4 requires an objective demonstration that a measure of the exporting country achieves the ALOP identified by the importing Member. Remember also that the SPS case law distinguishes ALOP from the level of protection that is reflected in a specific measure and accepts that these two may differ (eg ALOP may be higher than the level of protection reflected in a measure).[207] In the context of Article 4.1, this would mean that a measure which is claimed to be equivalent, needs to achieve ALOP, irrespective of whether a measure in an importing country does so. As in the case of Article

[203] Ibid, para 5.

[204] Ibid, para 11.

[205] SPS Committee, Decision on Notification of Determination of the Recognition of Equivalence of Sanitary or Phytosanitary Measures, Addendum, 25 July 2002, G/SPS/7/Rev.2/Add.1.

[206] Scott (*Commentary*) 176.

[207] See section 6.4.1(a) above.

5.6 this leads to unacceptable results. A measure of an exporting Member may guarantee the same level of protection as its counterpart in an importing country and still fail the test of Article 4. Interestingly, the Equivalence Decision takes a different approach and speaks about the level of protection reflected in a measure. The Decision specifies that an importing Member should process data with the view of 'determining whether ... measures [of the exporting country] achieve the level of protection provided by its own relevant sanitary or phytosanitary measures'.[208] The Codex Guidelines on the Judgment of Equivalence take a similar stance. Although the Guidelines provide that a sanitary measure applied in an importing country should in practice achieve the ALOP of this country[209] they also stipulate that an exporting country may prove equivalency 'by demonstrating that the measures it proposes ... have the same effect, relative to the achievement of the importing country's ALOP, as the corresponding sanitary measures applied by the importing country'.[210] This approach is definitely preferable as it allows the avoidance of questionable interpretative results. Nonetheless, it is not clear how future case law will address this issue.

The Equivalence Decision also emphasizes the special obligations borne by WTO Members towards developing countries.[211] In particular, Members shall give full consideration to requests by another Member for appropriate technical assistance to facilitate the implementation of Article 4. The specific examples include assistance in identification and implementation of measures which can be recognized as equivalent, and provision of scientific and technical information that helps to objectively demonstrate that a measure achieves ALOP. In line with the requirement of Article 9 of the SPS Agreement, the Equivalence Decision uses soft language and requires WTO Members to consider granting assistance rather than actually doing so.

In practice, the operation of equivalence obligations appears to be disappointing. The reluctance of importing Members to rely more extensively on this concept can probably be explained by the distrust that remains with respect to the operability of other SPS systems and high costs related to the recognition process. There is also an additional reason which explains why some countries are so reluctant to initiate and conclude equivalence agreements. As noted in the literature, 'recognizing the equivalence of an alternative regulatory regime may require national regulators to offer the same alternative to domestic producers, requiring new domestic regulations'.[212] In consequence, importing WTO Members tend to require from exporting countries duplication of their own measures.[213] Interestingly, in some cases exporting countries also prefer to adopt the same measure rather than rely on the concept of equivalence. This should not come as

[208] SPS Committee, Equivalence Decision, para 7.
[209] Codex Alimentarius Commission, Guidelines on the Judgment of Equivalence, para 7.
[210] Ibid, para 15. [211] Ibid, para 8.
[212] Prévost and Van den Bossche ('Agreement') 332.
[213] SPS Committee, Equivalence, Note by the Secretariat, 4 July 2001, G/SPS/W/111, 1.

a surprise as the recognition process is resource intensive and time consuming,[214] while its results are highly uncertain. Sometimes the implementation of exactly the same measure may be regarded as a more feasible option. The existing initiatives both on the level of the SPS Committee and within international standard-setting bodies are probably insufficient. Despite the fact that equivalence is one of the issues which has been continually discussed in the SPS Committee, this increased attention has not been translated into any meaningful improvements. The process of capacity building in developing WTO Members (both with respect to infrastructure and development of human resources) is similarly unsatisfactory.

6.4.4 Monitoring and review

As mentioned in section 6.4.3(b), the requirement to base a measure on risk assessment and not maintain it without sufficient scientific evidence also has a risk management dimension that is relevant for the monitoring and review phase. The word 'maintain' clearly indicates that the obligation of Article 2.2 has a dynamic and continuous character. Such a formulation has potentially important implications as it requires WTO Members to track the latest scientific developments related to a particular risk. This also forces risk managers to review a measure in order to ensure that it is actually based on the latest science. As a consequence of such a review, risk managers may be obliged to withdraw a measure (if the new scientific evidence invalidates the scientific basis for the measure), to modify it (softening or tightening its requirements), to introduce a new measure(s), or to take no action if a state of scientific knowledge remains the same or developments do not materially change the previous determinations.

Article 5.1 does not contain comparable language as it only speaks about basing a measure on risk assessment without any reference to its maintenance. This may suggest that Article 5.1 is concerned not so much with the maintenance of a measure but rather with its adoption. If one accepts such a textual approach, a panel examination under Article 5.1 would be limited to risk assessment as it existed at the time when a measure was adopted. The case law chose, however, a different direction, with the *Biotech* panel holding that under Article 5.1 WTO Members are actually required to have updated risk assessment that reflects the latest scientific developments.[215] Although such an interpretation may be questioned from a textual point of view, it seems to be correct.[216] Article 5.1 is a 'specific application of the basic obligations contained in Article 2.2',[217] which 'may be seen to be marking out and elaborating a particular route leading to the same

[214] Ibid, 2.
[215] Panel Report, *EC – Biotech Products*, para 7.3034.; cf also Chapter 4, section 4.4.1(d).
[216] Cf Scott (*Commentary*) 125.
[217] Appellate Body Report, *Japan – Agriculture Products*, para 82.

destination set out in Article 2.2'.[218] A reading that limits the obligation of Article 5.1 only to the moment of adoption of a measure will hardly fit into that description. Moreover, such an approach is in line with the practice of international standard-setting bodies. For example, the OIE explicitly provides that the '*risk assessment* should be amenable to updating when additional information becomes available'.[219]

6.5 Conclusions

This chapter argues that it is legitimate to speak about the risk management dimension of the SPS Agreement. The relevant passages from both *Hormones* reports indicate that what was actually rejected by the Appellate Body was a narrow, and in fact incorrect, conceptualization of risk assessment rather than the distinction itself. Simultaneously, the analysis of the relevant provisions clearly shows that risk management disciplines are easily identifiable in the SPS Agreement. Recognition of risk management also appears to be justified as it helps to reconcile the text of the Agreement with international and domestic practice in the field of human health and environmental protection. It may also facilitate the application of the Agreement by clearly identifying those SPS provisions which are relevant for individual steps in the domestic regulatory process.

As far as concrete risk management obligations are concerned, the chapter notes a moderate success of the case law in elucidating the normative content of different provisions. Some rules were addressed only on a very general level, leaving WTO Members without clear guidelines as to their legal meaning. This is particularly true for Articles 2.3 and 5.6. There are also those provisions, such as Articles 4 and 6, which have not yet come to the dispute settlement process. This probably results from the fact that both of them give a considerable level of discretion to importing Members, which makes establishment of a prima facie case of violation particularly difficult. At the same time, this chapter observes that in practice disciplines on equivalence as well as regionalization generate a number of concerns among WTO Members, in particular developing countries, with respect to their operability. The initiatives undertaken on the level of the SPS Committee, although partially successful, have not yet produced a satisfactory compromise.

At the same time, some developments of the case law appear to be disappointing. This observation applies to the strict proportionality test of Article 2.2 that prohibits WTO Members from addressing the so-called negligible risk with highly restrictive measures. Fortunately this particular drawback seemed to be properly addressed by the Appellate Body in one of its recent decisions. Equally unsatisfactory is the reading of Article 5.4 which reduces the provision to the

218 Panel Report, *Australia – Salmon*, para 8.52.

219 DIE, Aquatic Animal Health Code, Article 2.2.3.

expression of the objective without any real legal meaning. As an alternative, this chapter proposes to construe Article 5.4 as providing a procedural rule that would require consideration of the impact of ALOP on international trade. The chapter also notes the unprincipled approach of the Appellate Body to the evaluation of arbitrariness and a lack of justification under Article 5.5 which makes the regime of the SPS Agreement less predictable. Last but not least, conceptualizing ALOP as an external benchmark together with an undemanding approach to its determination has serious implications for the operation of Articles 4 and 5.6. In particular, it invites a defendant to manipulate its level of protection, as the comparison under the SPS Agreement is made between ALOP and the equivalent measure/least-trade-restrictive alternative. It seems that a better solution would be to accept that a level of protection reflected in a measure is equal to or at least as high as ALOP.

One of the particularly interesting issues in the context of risk management disciplines is the effect of Article 5.5 on trade liberalization. The case law shows that the successful challenge of an SPS measure under Article 5.5 may actually result in an additional restriction on trade rather than a freer flow of goods. Although this may appear to be at odds with the idea of free trade, it is probably an unavoidable consequence of the *quasi* absolute character of the right of WTO Members to establish ALOP. Consequently, Article 5.5 allows for the existence of false negatives rather than false positives. Another interesting issue is the relevance of risk perception as a possible justification for divergences in domestic risk regulatory responses. This chapter submits that since perceived risk needs to be defensible in technical terms (otherwise a measure will fail the scientific tests provided by Articles 2.2 and 5.1), there are no compelling reasons which would exclude the acceptance of such a justification. This seems to correspond with the function of Article 5.5 which intends to detect discrimination through identification of inconsistencies in risk regulation rather than to impose technocratic rationality under which legitimate responses to risk are solely determined on the basis of probability and adverse effects.

7

Concluding Remarks

The SPS Agreement constitutes an important development in the field of international economic law. It introduces a number of new mechanisms, such as requirements of scientific justification or consistency, which appear to go beyond the traditional trade disciplines. Simultaneously, it touches on the area that has been traditionally considered as a core of national sovereignty. This novelty of the Agreement combined with the delicacy of its subject matter has not surprisingly generated various concerns as to the appropriateness of its disciplines.

The SPS Agreement was drafted in very general and enigmatic language which leaves many questions open to competing interpretations. As observed by Scott, 'it is a framework agreement, laying down standards which are so loose [as] to be essentially unworkable in their own terms'.[1] The analysis of the SPS case law shows that the WTO dispute settlement bodies were rather successful in translating these general requirements into concrete legal obligations. The science-based provisions, different risk management obligations, and to a lesser extent harmonization disciplines, have been given a real normative content. The general trend of the SPS case law seems to be positive. At the same time, one also has to admit that some of the developments are questionable or incoherent.

Following the objectives that were set out in the Introduction, this concluding chapter summarizes the main findings of the book, arranging them into three problematic categories: (i) appropriateness of disciplines established by the Agreement; (ii) consistency of the SPS case law; and (iii) normative content of those substantive provisions that have not yet been addressed in the SPS jurisprudence.

7.1 SPS Agreement—An appropriate tool for management of SPS disputes?

Does the SPS Agreement strike an optimal balance between the competing objectives of international free trade and protection against environmental and human health risks? I believe that this is actually the case. As discussed in the

[1] Joanne Scott, *The WTO Agreement on Sanitary and Phytosanitary Measures: A Commentary* (2007) 44.

preceding chapters, a part of the earlier jurisprudence subscribed to a simplified and incorrect conception of science and an overly technical view of risk assessment. Many concepts, such as insufficiency of scientific evidence or risk assessment, were defined in purely scientific terms without proper regard for their socio-cultural dimension. Simultaneously, the panels experimented with different dubious standards such as the strict proportionality test, the concept of negligible risk, or overly high requirements with respect to specificity of the assessment. This was combined with a rather intrusive examination of science underlying domestic SPS measures. The panels, particularly in the *EC – Biotech Product* and *US/Canada – Continued Suspension* disputes, engaged in the inquiry into the quality of scientific evidence, its persuasive force, and the correctness of conclusions made by domestic regulators on the basis of such data. Although the case law has never gone so far as to openly admit its preference for mainstream science and has formally always accepted divergent scientific opinions, in practice their relevance was minimal. This approach could indeed be questioned on different levels. The WTO dispute settlement bodies seem to be poorly equipped to make complex scientific determinations as they simply lack relevant expertise. Such an approach also fails to recognize that assessment of risk is not a purely scientific task and may depend on the socio-cultural conditions specific to a particular country. Risk assessment inevitably involves subjective judgments of the assessors that reflect their attitudes toward risks, values of the relevant community in which experts are acting, and other normative elements, such as the required level of protection or the treatment of uncertainties. Imposing a monolithic and imperialist vision of science on all WTO Members obviously cannot correctly account for all those elements and will generate questions as to the appropriateness of such a supervision system.

The majority of the above concerns were, however, properly addressed by the Appellate Body in its recent *US/Canada – Continued Suspension* report. The Appellate Body was clear that a panel is not authorized to determine whether the risk is appreciable. The specificity requirement was mitigated by the recognition of methodological difficulties that may be posed by the nature and characteristics of the particular risk. In the same vein, complex and interrelated risks do not need to be separated for the purpose of risk assessment. The applicable standard of review requires panels to perform a rather limited task. A panel is not permitted to weigh different scientific evidence in order to establish which view is better and, once it determines that a particular claim is at least scientifically defensible, even if in the opinion of the panel other claims are better, the examination ends. The approach taken by the Appellate Body under Article 5.7 is also more nuanced than the previous case law and corresponds more closely with the reality of risk assessment. For instance, insufficiency of scientific evidence was recognized as a relational category which may depend on normative values, such as the level of protection or the attitude of risk assessors. This recognition permitted the Appellate Body to acknowledge that insufficiency is a multidimensional concept that is determined not only by scientific developments but may also result from differences in the appreciation of available scientific information. One of the consequences of such an approach was the denial of the critical mass test that was introduced by the *Suspension* panel. Therefore, other risk assessments, including those underlying international standards, may only serve as a source of scientific data and cannot be dispositive for the determination of insufficiency.

All these developments allow me to conclude that the SPS Agreement is actually able to provide a workable mechanism that seriously takes into account the complex nature of science and scientific risk assessment and does not compromise the legitimate regulatory choices of WTO Members. This also means that the deficiencies and failures of the case law that are identified below should not be seen as a criticism of the whole system but rather as an appeal for specific adjustments.

I also believe that even without deciding whether a proper role for international trade rules should be limited to the eradication of protectionism, or be extended to non-discriminatory measures which lack technical rationality, the scientific benchmark may also be appreciated as an imperfect but very useful proxy for the detection of protectionism. As noted in Chapter 4, the problem of over-inclusiveness of science (condemning measures that lack scientific rationality but are not pursued for protectionist reasons) may be mitigated by the proper application of specific scientific standards. Although some residual over-inclusiveness will always remain (this is also the case under any other criterion) in practice this may prove to be less problematic than one might imagine. Under the *Suspension* standards, the new EC measure concerning hormone-treated beef, the emblematic example of over-intrusiveness of the SPS Agreement, most probably will be found to be compatible with the SPS requirements. On the other hand, the *Biotech* report seems to be a relic of the earlier SPS case law which would have been decided in an entirely different way under the new approach. Moreover, one may also think of the TBT Agreement as a possible alternative for WTO Members that can be applied to all those cases where the regulatory measure has been adopted as a response to citizens' demands which, however, lack any scientific justification. Remember that the TBT Agreement contains an open-ended enumeration of legitimate objectives. Protection of consumers' expectations as to the quality of goods may definitively constitute one of them. A relevant measure will be examined as to its necessity (external legitimacy concerning availability of less-trade-restrictive alternatives) rather than through reference to its internal legitimacy (scientific justification).

In this context, it also worth recalling the observation of Hudec which, although made in the context of the GATT rules, also holds true for the SPS Agreement. According to Hudec:

> dispute settlement is only the tip of the GATT legal system. GATT law applies to a very wide range of day-to-day government conduct, very little of which ever comes before a

GATT legal proceeding. An analysis of the relative success of formal legal proceedings is manifestly not a description of this larger whole.[2]

While the impact of the SPS Agreement on the practice of WTO Members definitely merits a separate and detailed study, some general remarks seem to be justified. As noted in the literature, the adoption of the Agreement, particularly in the early years of its operation, provoked a broad review of national SPS measures that resulted in a number of adjustments of domestic risk regulations.[3] This arguably led to a greater overall convergence between different SPS regimes and reduced some of the tensions in the SPS area. An introduction of the regionalization principle by the United States is a good example of a significant departure from its long-standing practice of only recognizing entire countries as free of a particular pest or disease. Note also that the SPS Agreement operates as a dispute prevention device through its notification system, which requires reporting of the drafts of new SPS measures and engaging in a discussion with interested trading partners. In many cases this allows reaching a compromise that takes into account the concerns of other WTO Members and reduces the potential for future trade disputes. Yet another possibility of resolving controversies between WTO Members before they turn into disputes is provided by the SPS Committee. One of the functions of the Committee is to establish a regular forum for consultations where WTO Members may discuss their trade concerns. During the 14 years of the operation of the SPS Agreement (1995–2008), more than 277 such specific concerns were raised by the WTO Members.[4] Out of these, 76 have been reported as fully resolved, with 19 partially resolved.[5] Although no solution has been reported for the remaining 182 trade concerns, as observed by the Secretariat, some of them may have been resolved without the Committee having been informed.[6] These are all important elements which also highlight the value of the SPS Agreement.

7.2 Recommendations

As far as possible improvements of the SPS system are concerned, this book proposes two types of recommendations. The first intends enhancing the clarity of the SPS Agreement by introducing changes in the conceptualization of some of

[2] Robert E Hudec, Daniel LM Kennedy, and Mark Sgarbossa, 'A Statistical Overview of GATT Dispute Settlement Cases: 1948 –1989' (1993) 2 Minnesota Journal of Global Trade 1, 3.

[3] Cf Donna Roberts, 'Preliminary Assessment of the Effects of the WTO Agreement on Sanitary and Phytosanitary Trade Regulations' (1998) 1(3) Journal of International Economic Law 377, 396–401.

[4] SPS Committee, 'Specific Trade Concerns, Note by the Secretariat', Revision, 5 February 2009, G/SPS/GEN/204/Rev 9, para 8.

[5] Partial resolution means that trade was allowed for selected products or by some of the importing Members maintaining the measure in question.

[6] SPS Committee ('Specific Trade Concerns') para 10.

its disciplines. The second relates to the specific obligations or standards of the Agreement.

In particular, this book argues that the SPS Agreement should be interpreted as providing three distinctive modes of compliance. First, a WTO Member may decide to conform to the requirements of international standards. As a consequence, a measure will be refutably presumed to be compatible with all obligations of the SPS Agreement. If a WTO Member wishes to reach a different level of SPS protection than would be achieved by a measure that follows the relevant international standard (or if there is no standard at all), the second or third mode of compliance becomes available. The applicability of a particular mode is determined by the existence of the sufficiency/insufficiency of scientific evidence. Under the second mode, a WTO Member is required to base its SPS measure on scientific evidence organized in the form of risk assessment (Articles 2.2/5.1 to 5.3) and observe all other provisions of the SPS Agreement (apart from Article 5.7). The third mode could be activated in those cases where scientific evidence is insufficient to perform an adequate risk assessment. A WTO Member that nevertheless decides to act will need to conform to the specific requirements of Article 5.7 as well as those obligations of the Agreement that are applicable to provisional measures. The relationship between the second and third mode of compliance is one of exclusion, in the sense that inconsistency of an SPS measure with the provisions that are relevant for a particular mode does not result in their inapplicability. Therefore, a measure that fails to meet some of the requirements of Article 5.7 (apart from insufficiency) is not sent back to Article 2.2.

Another general recommendation relates to the nature of Article 3.3. The book specifically submits that the SPS Agreement does not establish any general requirement which would force WTO Members to maintain certain minimum SPS standards. Although one may find some indications (eg the text of Article 3.3 that explicitly refers only to a higher level of protection) that could support the opposite interpretation, there are a number of important reasons against such an approach. In particular, it would go against the spirit of the Agreement, which is the promotion of free trade rather than the protection of human, animal, and plant life and health, as well as the existing case law, which clearly refuses to recognize international standards as legally binding norms. Moreover, this would also produce a disequilibrium between the rights and obligations of WTO Members by establishing a double benchmark, where Members may deviate upwards from international standards but cannot depart downwards. The practice shows that this will primarily affect Members from developing countries. In this context, Chapter 3 also notes that a tendency of the SPS case law to describe Article 3 as providing a possibility to adopt measures with a higher, rather than a different, level of protection may be explained by the fact that all disputes decided until now were concerned with upward deviation. Consequently, there was no need for the WTO dispute settlement bodies to articulate any rules that would be applicable to measures deviating from international standards in a downward direction. The textual argument equally does not appear to be persuasive as the second sentence of Article 3.3 actually refers to measures which result in a *different* level of protection.

This book also suggests reintroducing the concept of risk management to the SPS Agreement. In this context, the book submits that the Appellate Body in its reports was more concerned with the specific consequences resulting from the restrictive approach of the panel (which anyway were erroneous) rather than the distinction between risk assessment and management. Although the book admits that the strict separation of these two phases of the risk analysis model is not possible, it also notes the value in maintaining the distinction that consists in a continued attempt to control political aspects involved in the risk assessment to ensure, to the greatest possible extent, the integrity and quality of scientific advice. Moreover, the analysis of the SPS Agreement shows that the risk management dimension can be rather easily identified in the Agreement. These provisions establish a number of different requirements on risk and options evaluation, implementation, monitoring, and review—activities that are conventionally described as elements of the risk management. At the same time, Chapter 6 also identifies a number of additional advantages that seem to justify the recognition of the risk management dimension of the SPS Agreement. First of all, such an approach allows the text of the Agreement to be reconciled with the practice of international and domestic risk regulators. This may also facilitate compliance with the Agreement by clearly identifying SPS provisions that are relevant for individual stages in the domestic regulatory process. A Member that relies on a risk analysis model will be able to determine more easily which obligations are applicable to specific steps that are normally taken within its risk management process.

Another issue of a general nature relates to the over-textual approach of the WTO panels to the interpretation of the SPS provisions. The employment of strict textual methods may be seen (wrongly as I believe) as a way of ensuring the legitimacy of decisions adopted by WTO dispute settlement bodies. Deciding disputes on a technical level with text playing the central (or even exclusive) role in the interpretative process allows, at least on its face, the controversy to be depoliticized. Such an approach denies relevance to any policy considerations in the dispute settlement process since it assumes that a particular issue is decided on the basis of a neutral text and does not require the dispute settlement bodies to take difficult normative decisions. The logical consequence of strict textualism is the denial of any policy-making role for dispute settlement bodies.[7] This in turn allows them to defend against charges of having exceeded their acting powers (particularly if a dispute is highly politicized, as was the case in *EC – Biotech Products*). The sad part of the story is that the

[7] Claus-Dieter Ehlermann, 'Six Years on the Bench of the "World Trade Court". Some Personal Experience as Member of the Appellate Body of the World Trade Organization' (2002) 36(4) Journal of World Trade 605, 617.

interpretation is never value free, while over-reliance on the text may actually lead to wrong results. This problem is well exemplified in the discussion of the *Biotech* panel on the applicability criteria of the SPS Agreement. As noted in Chapter 2, a number of different technical terms such as 'food' or 'additive' were interpreted almost solely on the basis of dictionaries. As a consequence, the meanings that were produced as an outcome of such analysis seem to be rather artificial and somehow disconnected with the rest of the SPS area. This should not come as a surprise as interpreting terms that have a specific technical sense only on the basis of dictionaries can hardly produce any satisfactory results. A far more preferable approach would be to look for guidance in the relevant technical or scientific discipline where a term is conventionally used. One of the possible sources which the panel may consult are documents produced by relevant international bodies that specialize in the SPS field. Such documents can be regarded as a kind of glossary that reveals the common meaning ascribed to specific technical terms by the international community.

This book also identifies number of specific problems relating to the interpretation of particular provisions of the SPS Agreement. For instance, I criticize the approach of the case law which conceptualizes ALOP as an external benchmark and accepts a statement of a defendant on its ALOP as sufficient for its determination. This method seems to be particularly problematic for the operation of Articles 4 and 5.6. A complainant may identify an alternative that is reasonably available, significantly less restrictive to trade, and reflects the same level of protection as a contested measure, but it will lose if such an alternative does not reach an ALOP (even if this is not the case for the contested measure). The same problem arises under Article 4 since the equivalent measure needs to achieve ALOP irrespective of whether a domestic measure does so. Another example of erroneous approach is the confusion between the issue of applicability and consistency under Article 5.7. According to the case law, Article 5.7 is applicable only if all its requirements are met. In other words, violation of any requirement of Article 5.7 removes a measure from its scope and results in applicability of the regime provided by Articles 2.2/5.1. As a result, a WTO Member violates the requirement not to maintain an SPS measure without sufficient scientific evidence (Article 2.2) in a situation where there may be no such evidence. This also creates confusion as to what should be considered as a right or exception under the SPS Agreement. In the current practice the provision that is correctly described as a right actually operates as a form of justification.

7.3 Consistency of the SPS jurisprudence

The SPS case law appears to lack overall consistency. This can probably be explained by the fact that the Agreement is relatively new and its disciplines are still evolving. On the other hand, the recent decision of the Appellate Body in

US/Canada – Continued Suspension seems to be an important contribution to the greater uniformity of SPS standards.

As noted above, the applicable standard of review proved to pose particular difficulties for the WTO dispute settlement bodies. Consequently, the history of SPS jurisprudence may be seen as a constant endeavour to identify a proper standard of review that could be applied for examination of complex technical determinations. The Appellate Body report in *EC – Hormones* established a rather differential approach that left WTO Members with a considerable degree of discretion. The case law that followed went, however, in the direction of a *de novo* review, reaching its culminating point with panels' decisions in *EC – Biotech Products* and *US/Canada – Continued Suspension*. The Appellate Body report in *US/Canada – Continued Suspension* is a new turning point, with an approach which is much closer to deferential extreme. Under this new standard, a panel is only obliged to verify whether the scientific basis comes from respected and qualified sources and assess whether the reasoning articulated on the basis of the scientific evidence is objective and coherent.

The treatment of insufficiency of scientific evidence under Article 5.7 was equally problematic. The panels tended to interpret it in absolute and purely scientific terms. This approach was translated into the excessive importance that was attached to risk assessment existing on a supranational and international level. Similarly, panels recognized only a temporal and not spatial dimension of insufficiency. The absolute character of insufficiency also resulted in the marginalization of minority scientific opinions. They were frequently underestimated or even disregarded in clear violation of standards developed by the Appellate Body under Articles 2.2/5.1. The Appellate Body in *US/Canada – Continued Suspension* reintroduced a normative dimension in the assessment of insufficiency by recognizing that ALOP and values of specific community may have an impact on the final determination.

The approach of the dispute settlement bodies to the allocation of burden of proof under Article 5.7 may serve as an example of another inconsistency. The early panels rightly recognized that provision as a positive obligation, with the burden of making a prima facie case of violation on a complainant. This position was changed by the *Apples* case, where the panel held that it was for the party invoking a particular provision to bear the burden of making a prima facie case. In other words, the panel saw Article 5.7 as a form of justification for an otherwise WTO-inconsistent measure, rather than a provision which establishes a positive obligation. The *Biotech* panel again recognized Article 5.7 as a positive obligation (in the panel's nomenclature as 'qualified right') and not the exception to other obligations of the SPS Agreement. Consequently, it was the complainant who bore the burden of showing a prima facie case of inconsistency of provisional measures. This decision also seems to finally settle the problem of allocating burden of proof under Article 5.7.

The incontinency that still persists relates to the identification of a relevant time for assessment of insufficiency. The *Apples* panel, when assessing sufficiency/

insufficiency of scientific evidence, correctly refused to limit the scope of its examination to insufficiencies existing at the time the Japanese measure was adopted. Consequently, such an assessment required the consideration of insufficiencies as of the time of the panel proceeding. On the other hand, the *Biotech* panel conceptualized the first sentence of Article 5.7 as referring to the time when an SPS measure was taken. Regrettably, the same approach was followed by the *Suspension* panel. However, as explained in Chapter 5, such an approach, although defendable on the textual level, does not establish a workable standard. Such an interpretation also seems to create systemic problems for the relationship between Articles 5.7 and 2.2. As the relevant moment for the assessment of sufficiency under Article 2.2 is the time when the panel's terms of reference are fixed, a single measure may be simultaneously compatible with the requirement of sufficiency under Article 2.2 and insufficiency under Article 5.7. If one conceptualizes both provisions as mutually exclusive it is not clear which of them should apply.

Another inconsistency which remains unsettled concerns not so much the SPS Agreement but the more general problems relating to the relevance of international treaties in the context of WTO law. Remember that on the one hand, the *Biotech* panel held that provisions of other international treaties can be relevant for the purpose of interpretation only if all WTO Members are also parties to such a treaty. The panel came to this conclusion based on a textual argument and the need to ensure consistent application of WTO disciplines. On the other hand, the panel did not exclude the possibility of considering rules of other treaties which are applicable only to the parties of a particular dispute, if all of them argued that a particular WTO agreement should be interpreted in the light of such rules. These contradicting conclusions do not establish any clear standard.

7.4 Interpreting the uninterpreted

This book also identifies a number of SPS provisions that have not yet been addressed by the case law and attempts to clarify their normative content. It is worth recalling here some of the most important findings. As far as the harmonization disciplines are concerned, this includes the presumption of Article 3.2 for those SPS measures which conform to international standards. Chapter 3 particularly submits that there are good reasons for such reading of the presumption of Article 3.2, which makes a measure immune under all provisions of the SPS Agreement. This seems to provide a clear and unambiguous standard that is easy to apply and does not prevent proof to the contrary (as the presumption is conceptualized as rebuttable). Moreover, such an approach also helps to maintain the consistency of the results under both the SPS Agreement and GATT 1994. A measure that is presumed to be compatible with the disciplines of GATT 1994 will enjoy the same status under the SPS Agreement.

In the context of Article 3.1, I argue that in order to be considered as based on international standards, a measure needs to reflect the same level of protection as provided by the relevant standards. This conclusion seems to be justified by the text of Article 3 which unequivocally links the obligation to base a measure on international standards with the level of protection (by establishing a dichotomy between measures that are based on international standards and those that aim for a different level of protection). Chapter 3 also recognizes that in legal terms potential gains from basing a measure on international standards are rather limited and are exhausted by Article 5.8 and Annex B(5). At the same time, the practical advantages appear to be substantial. Inapplicability of both provisions increase the difficulties that are faced by a complainant when making a prima facie case of inconsistency, while a defendant may rely on evidence and scientific data that were used as a basis for a standard. This should make the task of showing that its measure is supported by sufficient scientific evidence much easier.

The science-based provisions were extensively addressed by the SPS case law. Nevertheless some specific aspects still remain unclear. In particular, the relationship between the first sentence of Annex A(4) and Article 5.3 appears to be problematic. The textual interpretation indicates that both provisions establish two separate sets of rules that apply to different situations. On the other hand, a number of other considerations seem to suggest that they should be applied cumulatively, with Article 5.3 only specifying which economic factors must be taken into account in the quarantine risks related to animal or plant life or health. This book subscribes to the second position and submits that such an approach allows conceptualizing both obligations in a more harmonious way and guarantees the consistent treatment of the same categories of risk. Moreover, as noted above, such a reading of the two provisions also preserves the right of WTO Members to establish their ALOP.

The disciplines of Article 5.7 are definitively less developed by the case law. First, the scope of the exclusion provided by Article 2.2 remains unclear. Chapter 5 argues for a narrow interpretation under which the exclusion covers only the last obligation of Article 2.2 (not to maintain an SPS measure without sufficient scientific evidence). This seems to be justified by the character of Article 5.7 as it does not address the question of necessity of a particular SPS measure. Moreover, since pertinent information is a category of scientific character, there are no reasons to reject the applicability of the second sentence of Article 2.2 to measures that fall into the ambit of Article 5.7. As a consequence, both the selection and the evaluation of pertinent information should be compatible with scientifically sound methodological standards. Secondly, the relationship between Article 5.7 and other substantive provisions of the SPS Agreement is also obscure. On the basis of textual and pragmatic arguments I submit that disciplines other than the second sentence of Article 2.2 and Article 5.1 to 5.3 remain applicable to provisional measures. At the same time, I agree that their application may require a degree of flexibility compared to measures that are based on risk assessment and scientific evidence. Thirdly, it seems that the category of pertinent information (albeit not the quality) is principally of the same nature as scientific evidence. Having said this, one also needs to recognize that pertinent information may also include general studies on a particular risk or mere suggestive and indirect evidence but not necessary data relating to public values or consumer preferences. Fourthly, the case law is silent on the standard applicable to the fourth condition of Article 5.7 (obligation to review a measure in a reasonable period of time). In particular, it is not clear whether a specific situation of a WTO Member such as its development status, available financial resources, complexity of a domestic regulatory process, or national priorities in addressing SPS risks, may have any relevance. In this context, the book proposes applying a flexible standard of reasonableness which takes into account both the situation of a Member as well as actual availability of information. Moreover, the assessment of what constitutes a 'reasonable period of time' seems to ask not so much for specific time limits but is rather concerned with the scientific developments. Consequently, slow progress of scientific research in a particular field that results from objective difficulties may justify a longer period of time.

With regard to risk management obligations, Chapter 6 particularly notes that the SPS Agreement is silent on the significance of the public perception of risk as a potential justifying factor for differences in ALOP. I conclude that even without deciding whether perception of risk by laypersons is a psychological deficiency or rather a rival rationality to scientific discourse, it is advisable to recognize it as a possible justification. This observation seems to be supported by pragmatic reasons as well as the function of Article 5.5. In this context, Chapter 6 particularly submits that Article 5.5 intends to detect discrimination and protectionism through identification of inconsistencies in national risk regulation rather than to impose a technocratic rationality under which legitimate responses to risk are determined on the basis of probability and adverse effect. Secondly, Chapter 6 notes that neither the disciplines of Article 4 nor those of Article 6 have been addressed by the SPS case law. In this context, I examine the relationship between the requirements of Articles 6.1 and 3.1 and conclude that Article 6.1 should be seen as *lex specialis* to the general obligation of Article 3.1. As far as the specific obligations of Article 6 are concerned, two particularly problematic issues include the time horizon that is relevant for acquiring such status and the required threshold of probability. The book proposes a case-by-case approach where a panel assesses all relevant circumstances such as historical records of maintenance of disease- or pest-free areas on the territory of an exporting Member, the seriousness of a particular disease, and biological factors. Chapter 6 also recognizes that the concepts of compartmentalization and pest-free place/production sites as developed by the OIE and IPPC have limited utility under the SPS Agreement and may only be helpful in the evaluation of necessity of a measure (Article 2.2) or corresponding obligation of the least-trade-restrictive alternative (Article 5.6). In the same vein, Chapter 6 submits that the official OIE recognition of disease-free status cannot be considered as a type of international standard and may be relevant only as one of consideration in a domestic determination process.

7.5 A final word

The history of the SPS Agreement has definitively not come to its end. Its disciplines are still evolving, with the WTO dispute settlement bodies constantly trying to find an appropriate balance between the need to guarantee unconstrained trade in agriculture products and regulatory freedom of WTO Members in the domain of environmental and human health risk.

The most recent decision of the Appellate Body in *US/Canada – Continued Suspension* tends to favour the second element, presumably recognizing the limits of international supervision in this area. Nonetheless, it still remains unclear how those new standards will be applied by the prospective panels. The new *quasi* procedural standard of review which defers to scientific determination made on the domestic level, may quite easily turn into an intrusive examination of underlying methodology. The evalution of insufficiency of scientific evidence, despite accepting the existence of normative elements in such an assessment, may in practice attach a higher probative value to 'objective' scientific data. It is possible to continue this list without end. The first test for the new standards will come with the panel report in *Australia – Measures Affecting the Importation of Apples from New Zealand*.

What seems to be certain is that in the future one may expect an increasing number of SPS disputes. Regardless of the outcome of the Doha Round, further liberalization of international trade in agricultural sectors appears to be inevitable. This will probably provoke domestic governments to make a more frequent recourse to regulatory measures which, although taken in the guise of human health and environmental protection, are primarily concerned with the defence of interests of domestic producers. This will be accompanied, at least in the developed countries, by the growing demand from citizens for effective and highly protective measures that will guard against smaller and smaller risks. How will the international trading system react to those developments? Will it be able to maintain the appropriate balance? Past experience shows that one may be optimistic. The Appellate Body learns quickly and seems to respond to the challenges of the new globalized world in a reasonable manner.

Bibliography

Alemanno, Alberto, *Trade in Foods. Regulatory and Judicial Approaches in the EC and the WTO* (Cameron May, London: 2007)

Applegate, John S, 'The Taming of the Precautionary Principle' (2002–2003) 27(13) William and Mary Environmental Law & Policy Review 12

Arcuri, Alessandra and Dari-Mattiacci, Giuseppe, 'Centralization versus Decentralization as a Risk–Return Trade-off' Amsterdam Center for Law & Economics Working Paper No 2007–06

Arnoldi, Jakob, *Risk: An Introduction* (Polity Press, Cambridge: 2009)

Atik, Jeffery, 'The Weakest Link: Demonstrating the Inconsistency of "Appropriate Levels of Protection" in Australia – Salmon' (2004) 24(2) Risk Analysis 483

——'Science and International Regulatory Convergence' (1996–97) 17 Northwestern Journal of International Law & Business 736

Baldwin, Robert, Hutter, Bridget, and Rothstein, Henry, 'Risk Regulation, Management and Compliance, A Report to the BRI Inquiry', available at <http://www.bristol-inquiry.org.uk>

Barceló III, John J, 'Burden of Proof, Prima Facie Case and Presumption in WTO Dispute Settlement' (2009) 42(1) Cornell International Law Journal 23

——'Product Standards to Protect the Local Environment—the GATT and the Uruguay Round Sanitary and Phytosanitary Agreement' (1994) 27 Cornell International Law Journal 755

Bechtel, William, 'Scientific Evidence: Creating and Evaluating Experimental Instruments and Research Techniques', PSA: Proceedings of the Biennial Meeting of the Philosophy of Science Association, Vol 1 (1990)

Beck, Ulrich, *Risk Society: Towards a New Modernity* (Sage, Newbury Park, CA: 1992)

Bergkamp, Lucas, *European Community Law for the New Economy* (Intersentia, Antwerp, New York: 2003)

Bhuiyan, Sharif, *National Law in WTO Law. Effectiveness and Good Governance in the World Trading System* (Cambridge University Press, Cambridge: 2007)

Birnie, Patricia W and Boyle, Alan E, *International Law and the Environment* (Oxford University Press, London: 2002)

Bloche, M Gregg, 'WTO Deference to National Health Policy: Toward an Interpretative Principle' (2002) 5(4) Journal of International Economic Law 825

Blomkvist, Anna-Christina, 'Psychological Aspects of Values and Risks' in Lennart Sjöberg (ed), *Risk and Society* (Allen & Unwin, London, Boston: 1987)

Boehmer-Christiansen, Sonja, 'The Precautionary Principle in Germany—Enabling Government' in Timothy O'Riordan and James Cameron (eds), *Interpreting the Precautionary Principle* (Cameron May, London: 1994)

Bohanes, Jan, 'Risk Regulation in WTO Law: A Procedure-Based Approach to the Precautionary Principle' (2002) 40 Columbia Journal of Transnational Law 323

Bossche, Peter van den, *The Law and Policy of the World Trade Organization: Text, Cases, and Materials* (Cambridge University Press, Cambridge: 2005)

Boudia, Sonya and Jas, Nathalie, 'Introduction: Risk and Risk Society in Historical Perspective' (2007) 23(4) History and Technology 317

Bradbury, Judith A, 'The Policy Implications of Differing Concepts of Risk' (1989) 14(4) Science, Technology & Human Values 380

Broude, Tomer, 'Genetically Modified Rules: The Awkward Rule-Exception-Right Distinction in EC-Biotech' (2007) 6(2) World Trade Review 215

Button, Catherine, *The Power to Protect. Trade, Health and Uncertainty in the WTO* (Hart Publishing, Oxford, Portland: 2004)

Byrd, Daniel M III and Cothern, C Richard, *Introduction to Risk Analysis: A Systematic Approach to Science-Based Decision Making* (Government Institutes, Rockville, MD: 2000)

Cameron, James and Abouchar, Juli, 'The Status of the Precautionary Principle in International Law' in David Freestone and Ellen Hey (eds), *The Precautionary Principle in International Law* (Kluwer Law International, The Hague: 1996)

Carmody, Chios, 'WTO Obligations as Collective' (2006) 17(2) European Journal of International Law 419

Chang, Howard F, 'Risk Regulation, Endogenous Public Concerns, and the Hormones Dispute: Nothing to Fear but Fear Itself' (2004) 77 Southern California Law Review 743

Charnovitz, Steve, 'The Supervision of Health and Biosafety Regulation by World Trade Rules' (2000) 13 Tulane Environmental Law Journal 271

Cole, Stephen, *Making Science: Between Nature and Society* (Harvard University Press, Cambridge, Mass: 1992)

Conrad, Christiane R, 'The EC-Biotech Dispute and Applicability of the SPS Agreement: Are the Panel's Findings Built on Shaky Ground?' (2007) 6 World Trade Review 23

Covelli, Nick and Hohots, Viktor, 'The Health Regulation of Biotech Foods under the WTO Agreement' (2003) 6(4) Journal of International Economic Law 773

Covello, Vincent T and Merkhofer, Miley W, *Risk Assessment Methods: Approaches for Assessing Health and Environmental Risk* (Plenum Press, New York: 1993)

—— and Mumpower, Jeryl, 'Risk Analysis and Risk Management: An Historical Perspective' (1985) 5(2) Risk Analysis 103

—— and Detlof von Winterfeldt, Vincent T, and Slovic, Paul, 'Communicating Scientific Information about Health and Environmental Risks: Problems and Opportunities from a Social and Behavioral Perspective' in Vincent T Covello, Lester B Lave, Alan Moghissi, and VR Uppuluri (eds), *Uncertainty in Risk Assessment, Risk Management and Decision Making* (Plenum Press, New York: 1984)

Cranor, Carl F, 'Science Courts, Evidentiary Procedures, and Mixed Science Policy Decisions' (1993) 4(2) Risk: Issue in Health and Safety 113

Crawford-Brown, Douglas, Pauwelyn, Joost, and Smith, Kelly, 'Environmental Risk, Precaution and Scientific Rationality in the Context of WTO/NAFTA Trade Rules' (2004) 24(2) Risk Analysis 461

Douglas, Mary, 'Risk as a Forensic Resource' (1990) 119(4) Daedalus 1

—— and Aaron Wildavsky, *Risk and Culture: An Essay on the Selection of Technical and Environmental Dangers* (University of California Press, Berkeley: 1982)

Echols, Marsha A, *Food Safety and the WTO. The Interplay of Culture, Science and Technology* (Kluwer Law International; The Hague: 2001)

Editors of The American Heritage Dictionaries (ed), *The American Heritage Dictionary of the English Language*, 4th edn (Houghton Mifflin, Boston: 2000)

Ehlermann, Claus-Dieter and Lockhart, Nicolas, 'Six Years on the Bench of the "World Trade Court". Some Personal Experience as Member of the Appellate Body of the World Trade Organization' (2002) 36(4) Journal of World Trade 605

Epps, Tracey, *International Trade and Health Protection. A Critical Assessment of the WTO's SPS Agreement* (Edward Elgar Publishing, Cheltenham: 2008)

European Commission, Health & Consumers Protection Directorate-General, *First Report on the Harmonisation of Risk Assessment Procedures, Part 1: The Report of the Scientific Steering Committee's Working Group on Harmonisation of Risk Assessment Procedures in the Scientific Committees Advising the European Commission in the Area of Human and Environmental Health* (2000)

Food and Agriculture Organization and World Health Organization, *Report of the Joint FAO/WHO Expert Consultation, Application of Risk Analysis to Food Standard Issues*, WHO/FNU/FOS/95.3 (WHO, Geneva: 1995)

—— *Understanding the Codex Alimentarius*, 3rd edn (Rome: 2005)

—— *Expert Consultation on Food Safety, Science and Ethics* (Food and Agriculture Organization of the United Nations, Rome: 2003)

—— *Report of the Evaluation of the Codex Alimentarius and Other FAO and WHO Food Standards Work* (Rome: 2002).

—— *Report of a Joint FAO/WHO Consultation on Risk Management and Food Safety* (Rome: 1997)

Fisher, Elizabeth, Jones, Judith, and von Schomber, Rene (eds), *Implementing the Precautionary Principle: Perspectives and Prospects* (Edward Elgar Publishing, Cheltenham: 2006)

—— 'Beyond the Science/Democracy Dichotomy: The World Trade Organisation Sanitary and Phytosanitary Agreement and Administrative Constitutionalism' in Christian Joerges and Ersnt-Ulrich Petersmann (eds), *Constitutionalism, Multi-level Trade Governance and Social Regulation* (Hart Publishing, Oxford, Portland: 2006)

Fraiberg, Jeremy and Trebilcock, Michael J, 'Risk Regulation: Technocratic and Democratic Tools for Regulatory Reform' (1998) 43 McGill Law Journal 835

Freestone David and Hey, Ellen (eds), *The Precautionary Principle in International Law* (Kluwer, Hague: 1996)

—— 'The Road to Rio: International Environmental Law After the Earth Summit' (1994) 6 Journal of Environmental Law 193

Gardiner, Richard K, *Treaty Interpretation* (Oxford University Press, Oxford: 2008)

Gascoine, Digby, 'Harmonization, Mutual Recognition and Equivalence—How and What Is Attainable?', Paper presented at the Conference on International Food Trade Beyond 2000: Science-Based Decisions, Harmonization, Equivalence and Mutual Recognition, Melbourne Australia, 11–15 October 1999

Giddens, Anthony, *Modernity and Self Identity: Self and the Society in the Late Modern Age* (Politi Press, Cambridge: 1991)

Godard, Olivier, 'The Precautionary Principle, the Environment and International Trade: Sovereignty and Collective Preferences in Question', Ecole Polytechnique, Centre National de la Recherche Scientifique, Cahier No 2006–05

Graham, John D, Chang, Bei-Hung, and Rhomberg, Lorenz, 'Challenges in Risk Assessment and Risk Management' (1996) 545 Annals of the American Academy of Political and Social Science 15

Graubard, Stephen R, 'Preface to the Issue "Risk"' (1990) 119(4) Daedalus i

Guruswamy, Lakshman D, 'Sustainable Agriculture: Do GMOs Imperil Biosafety' (2002–03) 9 Indiana Journal of Global Legal Studies 461

Gutteling, Jan and Kuttschreuter, Margot, 'The Role of Expertise in Risk Communication: Laypeople's and Expert's Perception of the Millennium Bug Risk in the Netherlands' (2002) 5(1) Journal of Risk Research 35

Guzman, Andrew T, 'Dispute Resolution in SPS Cases' in Dan Horovitz, Daniel Moulis and Debra Steger (eds), *Ten Years of WTO Dispute Settlement* (International Bar Association, London: 2007)

Hathcock, John N, 'The Precautionary Principle—An Impossible Burden of Proof for New Products' (2000) 3(4) AgBioForum 255

Hohmann, Harald, *Precautionary Legal Duties and Principles of Modern International Environmental Law* (Graham & Trotman, London: 1994)

Hood, Christopher, Rothstein, Henry, and Baldwin, Robert, *The Government of Risk: Understanding Risk Regulation Regimes* (Oxford University Press, Oxford: 2004)

Horn, Henrik and Weiler, Joseph HH, 'European Communities—Trade Description of Sardines: Textualism and its Discontent' in Henrik Horn and Petros C Mavroidis (eds), *The WTO Case Law of 2002* (New York: Cambridge University Press, 2005)

Howse, Robert, 'The Use and Abuse of Other "Relevant Rules of International Law" in Treaty Interpretation: Insight from WTO Trade/Environment Litigations', IILJ Working Paper 2007/1

—— 'Democracy, Science and Free Trade—Risk Regulation on Trial at the WTO' (1999–2000) 98 Michigan Law Review 2343

Hudec, Robert E, 'Science and Post-discriminatory WTO Law' (2003) 26 Boston College International and Comparative Law Review 185

—— '"Like Product": The Differences in Meaning in GATT Articles I and III' in Thomas Cottier and Petros Mavroidis (eds), *Regulatory Barriers and the Principle of Non-Discrimination in World Trade Law* (University of Michigan Press, Ann Arbor: 2000)

—— Kennedy, Daniel LM, and Sgarbossa, Mark, 'A Statistical Overview of GATT Dispute Settlement Cases: 1948–1989' (1993) 2 Minnesota Journal of Global Trade 1

Hunter, Paul R and Fewtrell, Lorna, 'Acceptable Risk' in Lorna Fewtrell and Jamie Bartram (eds), *Water Quality: Guidelines, Standards and Health* (IWA Publishing, London: 2001)

Hurst, David R, 'Hormones: European Communities—Measures Affecting Meat and Meat Products' (1998) 1(9) European Journal of International Law

International Association for Impact Assessment & Institute for Environmental Assessment, *Principles of Environmental Impact Assessment Best Practice* (Fargo, Lincoln: 1990)

International Law Commission, *Fragmentation of International Law: Difficulties Arising from the Diversification and Expansion of International Law, Report of the Study Group of the International Law Commission*, A/CN.4/L.682 (2006)

Jaeger, Carlo, Webber, Thomas, Rose, Eugene A, and Renn, Ortwin, *Risk, Uncertainty, and Rational Action* (Earthscan Publications, Sterling: 2001)

Jasanoff, Sheila, 'The Songlines of Risk' 8(2) Environmental Values, Special Issue: Risk 135

Johnson, Renée, *Potential Farm Sector Effects of 2009 H1N1 'Swine Flu': Questions and Answers*, CRS Report for Congress (Congressional Research Service, Washington, DC: 2009),

Kastner, Justin and Powell, Douglas, 'The SPS Agreement: Addressing Historical Factors in Trade Dispute Resolution' (2002) 19 Agriculture and Human Values 283

Klinke, Andreas and Renn, Ortwin, A New Approach to Risk Evaluation and Management, Risk-Based, Precaution-Based, and Discourse-Based Strategies (2002) 22(6) Risk Analysis 1071

Kremlis, Georges, Balias, Yiorgos, and Sifakis, Antonis (eds), *The Precautionary Principle* (Sakkoulas, Athens: 2004)

Kuhn, Thomas S, *The Structure of Scientific Revolutions* (University of Chicago Press, Chicago, London: 1996)

Lagadec, Patrick, *Major Technological Risk : An Assessment of Industrial Disasters* (Oxford [Oxfordshire], New York: Pergamon Press, 1982)

Lang, Andrew, 'Provisional Measures under Article 5.7 of the WTO's Agreement on Sanitary and Phytosanitary Measures: Some Criticism of the Jurisprudence So Far' (2008) 42(6) Journal of World Trade 1085

Lave, Lester B, *Quantitative Risk Assessment in Regulation* (Brookings Institution, Washington, DC: 1982)

Leebron, David W, 'Lying Down with Procrustes: An Analysis of Harmonization Claims' in Jagdish Bhagwati and Robert Hudec (eds), *Fair Trade and Harmonization: Prerequisites for Free Trade? Vol. 1: Economic Analysis* (MIT Press, Cambridge, Massachusetts: 1996)

Leiss, William, 'Three Phases in the Evolution of Risk Communication Practice' (1996) 545 Annals 85

Lennard, Michael, 'Navigating by the Stars: Interpreting the WTO Agreement' (2002) 5(1) Journal of International Economic Law 17

Lowrance, William W, *Of Acceptable Risk: Science and the Determination of Safety* (W Kaufmann, Los Altos, California: 1976)

Lupton, Deborah, *Risk* (Routledge, London, New York: 1999)

MacGarvin, Malcolm, 'Precaution, Science and the Sin of Hubris' in Timothy O'Riordan and James Cameron (eds), *Interpreting the Precautionary Principle* (Cameron May, London: 1994)

Majone, Giandomenico, 'What Price Safety? The Precautionary Principle and Its Political Implications' (2002) 40(1) Journal of Common Market Studies 89

Marceau, Gabrielle and Trachtman, Joel P, 'The Technical Barriers to Trade Agreement, the Sanitary and Phytosanitary Measures Agreement, and the General Agreement on Tariff and Trade' (2002) 36 Journal of World Trade 811

Margolis, Howard, *Dealing with Risk: Why the Public and the Experts Disagree on Environmental Issues* (University of Chicago Press, Chicago: 1996)

Maruyama, Warren H, 'A New Pillar of the WTO: Sound Science' (1998) 32 International Lawyer 651

Mayeda, Graham, 'Developing Disharmony? The SPS and TBT Agreements and the Impact of Harmonization on Developing Countries' (2004) 7(4) Journal of International Economic Law 737

McGinnis, John O, 'The Appropriate Hierarchy of Global Multilateralism and Customary International Law: The Example of the WTO' (2003–04) 44 Virginia Journal of International Law 229

McGrady, Benn, 'Fragmentation of International Law or Systemic Integration of Treaty Regimes: EC—Biotech Products and the Proper Interpretation of Article 31(3)(c) of the Vienna Convention on the Law of Treaties' (2008) 42(4) Journal of World Trade 589

McIntyre, Owen and Mosedale, Thomas, 'The Precautionary Principle As a Norm of Customary International Law' (1997) 9 Journal of Environmental Law 221

McLachlan, Campbell, 'The Principle of Systemic Integration in Treaty Interpretation and Article 31(3)(c) of the Vienna Convention' (2005) 54(2) International and Comparative Law Quarterly 279

McMahon, Joe and Young, Margaret A, 'The WTO's Use of Relevant Rules of International Law: An Analysis of the Biotech Case' (2007) 56(4) International and Comparative Law Quarterly 907

McNelis, Natalie, 'The Role of the Judge in the EU and WTO: Lessons from the BSE and Hormones Cases' (2001) 4(1) Journal of International Economic Law 189

McNiel, Dale E, 'The First Case Under the WTO's Sanitary and Phytosanitary Agreement: The European Union's Hormone Ban' (1998) 39 Virginia Journal of International Law 89

Merkhofer, Miley W, *Decision Science and Social Risk Management: A Comparative Evaluation of Cost-Benefit Analysis, Decision Analysis, and Other Formal Decision-Aiding Approaches* (Kluwer Academic Publisher, Dordrecht, Netherlands: 1987)

Molak, Vlasta, *Fundamentals of Risk Analysis and Risk Management* (Lewis Publishers, Boca Raton: 1997)

Motaal, Doaa A, 'The "Multilateral Scientific Consensus" and the World Trade Organization' (2004) 38(5) Journal of World Trade 855

National Research Council, Committee on the Institutional Means for Assessment of Risks to Public Health, *Risk Assessment in the Federal Government: Managing the Process* (National Academy Press, Washington, DC: 1983)

National Research Council, Committee on Risk Assessment of Hazardous Air Pollutants, Commission on Life Sciences, *Science and Judgment in Risk Assessment* (National Academy Press, Washington, DC: 1994)

Neugebauer, Regine, 'Fine-Tuning WTO Jurisprudence and the SPS Agreement: Lessons from the Beef Hormone Case' (2000) 31 Law and Policy in International Business 1255

OECD Joint Working Party on Trade and Environment, *Uncertainty and Precaution: Implications for Trade and Environment*, COM/ENV/TD(2000)114/FINAL

Oesch, Matthias, *Standards of Review in WTO Dispute Resolution* (Oxford University Press, Oxford, New York: 2003)

Ore, Oystein, 'Pascal and the Invention of Probability Theory' (1960) 67(5) The American Mathematical Monthly 409

O'Riordan, Timothy and Cameron, James (eds), *Interpreting the Precautionary Principle* (Cameron May, London: 1994)

Ortino, Federico, 'Treaty Interpretation and the WTO Appellate Body Report in US-Gambling: A Critique' (2006) 9(1) Journal of International Economic Law 117

Palmeter, David and Mavroidis, Petros C, *Dispute Settlement in the World Trade Organization: Practice and Procedure* (Cambridge University Press, New York: 2004)

Pauwelyn, Joost, 'Expert Advice in WTO Dispute Settlement' in George A Bermann and Petros C Mavroidis (eds), *Trade and Human Health and Safety* (Cambridge University Press, Cambridge, New York: 2006)

—— 'Non-Traditional Patterns of Global Regulation: Is the WTO "Missing the Boat"' in Christian Joerges and Ernst-Ulrich Petersmann (eds), *Constitutionalism, Multilevel Trade Governance and Social Regulation* (Hart Publishing, Oxford, Portland: 2006)

—— *Conflict of Norms in Public International Law: How WTO Law Relates to Other Rules of International Law* (Cambridge University Press, New York: 2003)

—— 'Does the WTO Stand for "Deference to" or "Interference with" National Health Authorities When Applying the Agreement on Sanitary and Phytosanitary Measures (SPS Agreement)' in Thomas Cottier and Petros C Mavroidis (eds), *The Role of the Judge in International Trade Regulation: Experience and Lessons for the WTO* 175 (University of Michigan Press, Ann Arbor: 2003)

—— 'A Typology of Multilateral Treaty Obligations: Are WTO Obligations Bilateral or Collective in Nature?' (2003) 14(5) European Journal of International Law 907

—— 'The Use of Experts in WTO Dispute Settlement,' (2002) 51(2) International and Comparative Law Quarterly 325

—— 'The Role of Public International Law in the WTO' (2001) 95 American Journal of International Law 538

—— 'The WTO Agreement on Sanitary and Phytosanitary (SPS) Measures as Applied in the First Three Disputes' (1999) 4 Journal of International Economic Law 641

Peel, Jacqueline, 'A GMO by any Other Name... Might be an SPS Risk!: Implications of Expanding the Scope of the WTO Sanitary and Phytosanitary Measures Agreement' (2007) 17(5) European Journal of International Law 1009

—— Nelson, Rebecca, and Godden, Lee, 'GMO Trade Wars: The Submissions in the EC–GMO Dispute in the WTO' (2005) 6 Melbourne Journal of International Law 141

—— 'Risk Regulation under the WTO SPS Agreement: Science as an International Normative Yardstick?', Jean Monnet Working Paper No 2002/04

Petersmann, Ernst-Ulrich, 'Trade Policy as a Constitutional Problem. On the "Domestic Policy Functions" of International Trade Rules' (1986) 41 *Aussenwirtschaft* 405

Pfitzer, James Headen and Sabune, Shiela, 'Burden of Proof in WTO Dispute Settlement: Contemplating Preponderance of the Evidence', ICTSD Dispute Settlement and Legal Aspects of International Trade, Issue Paper No 9 (2009)

Presidential/Congressional Commission on Risk Assessment and Risk Management, *Final Report, Volume 1: Framework for Environmental Health Risk Management* (National Academy Press, Washington, DC: 1997)

Presidential/Congressional Commission on Risk Assessment and Risk Management, *Final Report, Volume 2: Risk Assessment and Risk Management in Regulatory Decision-Making* (National Academy Press, Washington, DC: 1997)

Presley, Mari, 'Sovereignty and Delegation Issues Regarding U.S. Commitment to the World Trade Organization's Dispute Settlement System' (1998) 8 Journal of Transnational Law & Policy 173

Prévost, Denise and Van den Bossche, Peter, 'The Agreement on the Application of Sanitary and Phytosanitary Measures' in Patrick FJ Macrory, Arthur E Appleton, and Michael G Plummer (eds), *The World Trade Organization: Legal, Economic and Political Analysis* (Springer, New York: 2005)

Raffensperger, Carolyn and Tickner, Joel A, *Protecting Public Health and the Environment: Implementing the Precautionary Principle* (Island Press, Washington, DC: 1999)

Regan, Donald H, 'What are Trade Agreements for? Two Conflicting Stories Told by Economists, with a Lesson for Lawyers' (2006) 9(4) Journal of International Economic Law 951

—— 'Regulatory Purpose and "Like Products" in Article III:4 of the GATT (With Additional Remarks on Article III:2)' (2002) 36(3) Journal of World Trade 443

Renn, Ortwin, *Risk Governance. Coping with Uncertainty in a Complex World* (Earthscan Publications, Sterling: 2008)

—— 'Concepts of Risk: A Classification' in Sheldon Krimsky and Dominic Golding (eds), *Social Theories of Risk* (Praeger, Westport, Conn: 1992)

Ritzer, George (ed), *The Blackwell Companion to Major Contemporary Social Theorists* (Blackwell Publishing, Oxford: 2003)

Roberts, Donna, 'Preliminary Assessment of the Effects of the WTO Agreement on Sanitary and Phytosanitary Trade Regulations' (1998) 1(3) Journal of International Economic Law 377

Rosa, Eugene A, 'Metatheoretical Foundations for Post-Normal Risk' (1998) 1(1) Journal of Risk Research 15

Rowe, Gene and Wright, George, 'Differences in Expert and Lay Judgments of Risk: Myth or Reality' (2001) 21(2) Risk Analysis 341

Sadeleer, Niolas de, *Implementing the Precautionary Principle: Approaches from the Nordic Countries, EU and USA* (Earthscan Publications, Sterling: 2006)

Sands, Philippe, *Principles of International Environmental Law* (Cambridge University Press, Cambridge: 2003)

Schepel, Harm, *The Constitution of Private Governance. Product Standards in the Regulation of Integrating Markets* (Hart Publishing, Oxford, Portland: 2005)

Scott, Joanne, *The WTO Agreement on Sanitary and Phytosanitary Measures: A Commentary* (Oxford University Press, New York: 2007)

—— 'European Regulation of GMOs: Thinking about Judicial Review in the WTO' in Jane Holder, Colm O'Cinneide, and Michael Freeman (eds), *Current Legal Problems 2004: Vol 57* (Oxford University Press, Oxford: 2005)

—— 'On Kith and Kine (and Crustaceans): Trade and Environment in the EU and WTO' in Joseph HH Weiler (ed), *The EU, the WTO and the NAFTA—Towards a Common Law of International Trade* (Oxford University Press, New York: 2000)

Segger, Marie-Claire Cordonier and Gehring, Markus W, 'The WTO and Precaution: Sustainable Development Implications of the WTO Asbestos Dispute' (2003) 15(3) Journal of Environmental Law 289

Siegrist, Michael and Cvetovich, George, (2000) 'Perception of Hazards: the Role of Social Trust' (2000) 20(5) Risk Analysis 712

Silbergeld, Ellen K, 'Risk Assessment and Risk Management: An Uneasy Divorce' in Deborah G Mayo and Rachel Hollander (eds), *Acceptable Evidence: Science and Values in Risk Management* (Oxford University Press, New York: 1991)

Slovic, Paul, *The Perception of Risk* (Earthscan Publications, Sterling: 2000)

—— 'Trust, Emotion, Sex, Politics, and Science: Surveying the Risk-Assessment Battlefield' (1999) 19(4) Risk Analysis 689

—— Fischoff, Baruch, and Lichtenstein, Sarah, 'Facts versus Fears: Understanding Perceived Risk' in Daniel Kahneman, Paul Slovic, and Amos Tversky (eds), *Judgment under Uncertainty: Heuristics and Biases* (Cambridge University Press, Cambridge: 1982)

Smith, V Kerry, 'A Conceptual Overview of the Foundations of Benefit-Cost Analysis' in Judith D Bentkover, Vincent T Covello, and Jeryl Mumpower (eds), *Benefit Assessment: A State of the Art* (Reidel Publishing Co, Dordrecht: 1986)

Spangler, Miller B, 'Policy Issues Related to Worst Case Risk Analyzes and the Establishment of Acceptable Standards of De Minimis Risk' in Vincent T Covello, Lester B Lave, Alan Moghissi, and VR Uppuluri (eds), *Uncertainty in Risk Assessment, Risk Management and Decision Making* (Plenum Press, New York: 1984)

Stone, Christopher D, 'Is There a Precautionary Principle?' (2001) 31 Environmental Law Reporter 10790

Sunstein, Cass R, *Laws of Fear: Beyond the Precautionary Principle* (Cambridge University Press, New York: 2005)

Sykes, Alan O, 'Domestic Regulation, Sovereignty and Scientific Evidence Requirements: A Pessimistic View' (2002) 3 Chicago Journal of International Law 353

—— 'Regulatory Competition or Regulatory Harmonization? A Silly Question?' (2000) 3 Journal of International Economic Law 257

—— 'The (Limited) Role of Regulatory Harmonization in International Goods and Services Markets' (1999) 2 Journal of International Economic Law 49

Taniguchi, Yasuhei, 'Understanding the Concept of Prima Facie Proof in WTO Dispute Settlement in Merit E Janow, Victoria Donaldson, and Alan Yanovich (eds), *The WTO: Governance, Dispute Settlement, and Developing Countries* (Juris Publishing, Huntington: 2008)

Thirlway, Hugh, 'The Sources of International Law in International Law' in Malcolm D Evans (ed), *International Law* (Oxford University Press, Oxford: 2003)

Thomas, Ryan D, 'Where's the Beef? Mad Cows and the Blight of the SPS Agreement' (1999) 32 Vanderbilt Journal of Transnational Law 487

Trebilcock, Michael J and Howse, Robert, *The Regulation of International Trade* (Routledge, London, New York: 2005)

—— and Robert Howse, 'Trade Liberalization and Regulatory Diversity: Reconciling Competitive Markets with Competitive Politics' (1998) 6(1) European Journal of Law & Economics 5

Trettin, Lilliana and Musham, Catherine, 'Is Trust a Realistic Goal of Environmental Risk Communication' (2000) 32(2) Environment & Behaviour 410

Van Asselt, Marjolein BA and Rotmans, Jan, 'Uncertainty in Integrated Assessment Modelling' (2002) 54(1–2) Climatic Change 75

—— *Perspectives in Uncertainty and Risk. PRIMA Approach to Decision Support* (Kluwer Academic Publisher, Dordrecht, Netherlands: 2000)

Van Damme, Isabelle, *Treaty Interpretation by the WTO Appellate Body* (Oxford University Press, Oxford: 2009)

Veggeland, Frode and Borgen, Svein Ole, 'Negotiating International Food Standards: The World Trade Organization's Impact on the Codex Alimentarius Commission' (2005) 18(4) Governance: An International Journal of Policy, Administration, and Institutions 675

Victor, David G, 'The Sanitary and Phytosanitary Agreement of the World Trade Organization: An Assessment after Five Years' (1999–2000) 32 New York University Journal of International Law and Policy 890

Vogel, David, 'Ships Passing in the Night: The Changing Politics of Risk Regulation in Europe and the United States', EUI RSC Working Paper No 2001/16

Walker, Vern R, 'The Myth of Science as Neutral Arbiter for Triggering Precaution' (2003) 26 Boston College International and Comparative Law Review 197

—— 'Keeping the WTO from Becoming the "World Trans-Science Organization": Scientific Uncertainty, Science Policy, and Factfinding in the Growth Hormones Dispute' (1998) 31 Cornell International Law Journal 251

—— 'The Siren Songs of Science: Toward a Taxonomy of Scientific Uncertainty for Decisionmakers' (1990–91) 23 Connecticut Law Review 567

Wallace, Rebecca MM, *International Law* (Sweet & Maxwell, London: 2005)

Wiener, Jonathan B, 'Whose Precaution After All? A Comment on the Comparison and Evolution of Risk Regulatory Systems' (2003) 13 Duke Journal of Comparative & International Law 207

—— and Michael D Rogers, 'Comparing Precautions in the US and Europe' (2002) 5(4) Journal of Risk Research 317

Wilson, Norbert, 'Clarifying the Alphabet Soup of the TBT and the SPS in the WTO' (2003) 8 Drake Journal of Agriculture Law 703

Winickoff, David, Jasanoff, Sheila, Busch, Lawrence, Grove-White, Robin, and Wynne, Brian, 'Adjudicating the GM Food Wars: Science, Risk, and Democracy in World Trade Law' (2005) 30 Yale Journal of International Law 81

Wirth, David A, 'The Role of Science in the Uruguay Round and NAFTA Trade Disciplines' (1994) 27 Cornell International Law Journal 817

Wolfrum, Rüdiger, Stoll, Peter-Tobias, and Seibert-Fohr, Anja (eds), *WTO—Technical Barriers and SPS Measures, Max Planck Commentaries on World Trade Law, Vol 3* (Martinus Nijhoff Publishers, Leiden, Boston: 2007)

World Health Organizations, *Principles for the Assessment of Risks to Human Health from Exposure to Chemicals, IPCS Environmental Health Criteria* 210 (Geneva: 1999)

Zarilli, Simonetta, 'WTO Sanitary and Phytosanitary Agreement: Issues for Developing Countries', TRADE Working Papers No 3 (1999)

LEGAL MATERIALS

Agreement on the Application of Sanitary and Phytosanitary Measures, opened for signature 15 April 1994, 1867 UNTS 493 (entered into force 1 January 1995)

Agreement on the Implementation of Article VI of the General Agreement on Tariffs and Trade 1994, opened for signature 15 April 1994, 1868 UNTS 201 (entered into force 1 January 1995)

Agreement on Technical Barriers to Trade, opened for signature 15 April 1994, 1868 UNTS 120 (entered into force 1 January 1995)

Agreement on Technical Barriers to Trade, 12 April 1979, 1186 UNTS 276, GATT, BISD, 26th Supp 8 (1980)

Agreement on Trade-Related Aspects of Intellectual Property Rights, opened for signature 15 April 1994, 1869 UNTS 299 (entered into force 1 January 1995)

Codex Alimentarius Commission, Guidelines for the Development of Equivalence Agreements Regarding Food Imports and Export Inspection and Certification Systems, CAC/GL/34–1999

—— Criteria for the Consideration of the Other Factors Referred to in the Second Statement of Principles, Decision of the 24th Session of the Commission, 2001

—— Guidelines on the Judgment of Equivalence of Sanitary Measures associated with Food Inspection and Certification Systems, CAC/GL/53-2003
—— *Procedural Manual* (18th edn, 2008)
—— *Procedural Manual* (16th edn, 2006)
—— Report of the 21st Session, List of Standards and Related Texts Adopted by the 21st Session of the Codex Alimentarius Commission, ALINORM 95/37
Commission of the European Communities, Communication from the Commission on the Precautionary Principle, 2 February 2000, COM(2000)1 Final
Committee on Sanitary and Phytosanitary Measures, Working Procedures of the Committee, G/SPS/1 (1995)
—— Summary of the Meeting Held on 19–20 March 1997, Note by the Secretariat, G/SPS/R/7
—— Clarification of References to Codex Text, Draft Response to the Codex Alimentarius Commission, Note by the Chairman, 13 March 1998, G/SPS/W/86/Rev
—— Guidelines to Further the Practical Implementation of Article 5.5, 18 July 2000, G/SPS/15
—— Equivalence: Consideration of Article 4 of the SPS Agreement, 'Summary of Informal Discussions on Equivalence', Second Report by the Chairman, 21 March 2001, G/L/445
—— Equivalence, Note by the Secretariat, 4 July 2001, G/SPS/W/111
—— Decision on Implementation of Article 4 of the Agreement in the Application of Sanitary and Phytosanitary Measures, 24 October 2001, G/SPS/19, last time revised by the decision of 23 July 2004, G/SPS/19/Rev
—— Decision on Notification of Determination of the Recognition of Equivalence of Sanitary or Phytosanitary Measures, Addendum, 25 July 2002, G/SPS/7/Rev.2/Add.1
—— Submission of New Zealand, Monitoring of International Harmonization: Regionalization, 29 September 2004, G/SPS/W/151
—— Foot and Mouth Disease Free Zone—Brazilian Experience on Regionalization, Statement by Brazil at the meeting of 29–30 June 2005, G/SPS/GEN/584
—— OIE Activities on Zoning and Compartmentalization, Communication from the World Organization for Animal Health (OIE), 23 January 2006, G/SPS/GEN/625
—— Communication from the Republic of Korea, The Implementation of Article 6 of the Agreement on the Application of Sanitary and Phytosanitary Measures (Regionalization), 31 May 2006, G/SPS/W/195
—— Summary of the Meeting Held on 27–28 June 2006, Note by the Secretariat, G/SPS/R/42
—— Summary of the Meeting Held on 11–12 October 2006, Note by the Secretariat, G/SPS/R/43
—— OIE Dispute Mediation Process, Communication from the World Organization for Animal Health (OIE), 12 October 2006, G/SPS/GEN/731
—— Procedure to Monitor the Process of International Harmonization, Ninth Annual Report, 10 July 2007, G/SPS/45
—— IPPC Dispute Settlement Manual, Submission by the Secretariat of the International Plant Protection Convention, 22 June 2007, G/SPS/GEN/787
—— Guidelines to Further the Practical Implementation of Article 6 of the Agreement on the Application of Sanitary and Phytosanitary Measures, 25 February 2008, G/SPS/W/218

—— Specific Trade Concerns, Note by the Secretariat, Revision, 27 March 2008, G/SPS/GEN/204/Rev 8 Final Declaration of the First European 'Seas at Risk' Conference, Annex 1, Copenhagen 1994

—— Specific Trade Concerns, Note by the Secretariat, Revision, 5 February 2009, G/SPS/GEN/204/Rev 9

General Agreement on Tariffs and Trade, opened for signature 15 April 1994, 55 UNTS 194, 1867 UNTS 187 (entered into force 1 January 1995)

General Agreement on Trade in Services, opened for signature 15 April 1994, 1869 UNTS 183 (entered into force 1 January 1995)

International Plant Protection Convention, opened for signature 6 December 1951, 150 UNTS 67 (entered into force 3 April 1952), as subsequently amended

Marrakesh Agreement Establishing the World Trade Organization, opened for signature 15 April 1994, 1867 UNTS 154 (entered into force 1 January 1995)

Ministerial Declaration on the Uruguay Round: Declaration of 20 September 1986, GATT BISD 33S/19 (1987)

Montreal Protocol to the Vienna Convention on Substances that Deplete the Ozone Layer, Montreal, 16 September 1987, 30 ILM 541

Restatement (Third) of Foreign Relations Law of the United States (1986)

Rio Declaration on Environments and Development, 14 June 1992, UN A/Conf. 151/5/Rev 1 (1992), reprinted in (1992) 31 ILM 876

Secretariat of the International Plant Protection Convention, International Standards for Phytosanitary Measures (IPSM) No 2, Framework for Pest Risk Analysis (2007)

—— International Standards for Phytosanitary Measures (IPSM) No 4, Requirements for the Establishment of Pest Free Areas (1995)

—— International Standards for Phytosanitary Measures (IPSM) No 5, Glossary of Phytosanitary Terms (2007)

—— International Standards for Phytosanitary Measures (IPSM) No 6, Guidelines for Surveillance (1997)

—— International Standards for Phytosanitary Measures (IPSM) No 9, Guidelines for Pest Eradication Programmes (1998)

—— International Standards for Phytosanitary Measures (IPSM) No 10, Requirements for the Establishment of Pest Free Places of Production and Pest Free Production Sites (1999)

—— International Standards for Phytosanitary Measures (ISPM) No 22, Requirements for the Establishment of Areas of Low Pest Prevalence (2005)

—— International Standards for Phytosanitary Measures (ISPM) No 24, Guidelines for the Determination and Recognition of Equivalence of Phytosanitary Measures (2005)

—— International Standards for Phytosanitary Measures (IPSM) No 29, Recognition of Pest Free Areas and Areas of Low Pest Prevalence (2007)

Understanding on Rules and Procedures Governing the Settlement of Disputes, opened for signature 15 April 1994, 1869 UNTS 401 (entered into force 1 January 1995)

Vienna Convention on the Law of Treaties, opened for signature 23 May 1969, 1155 UNTS 331 (entered into force 27 January 1980)

World Organization for Animal Health, Resolution XII of the OIE International Committee, Procedure for the Recognition of the Foot and Mouth Disease Status of Member Countries, 15–19 May 1995

—— Resolution XX of the OIE International Committee, List of Foot and Mouth Disease Free Countries, 24 May 2005

—— *Terrestrial Animal Health Code* (18th edn, 2009)

—— *Aquatic Animal Health Code* (12th edn, 2009)

—— General Rules of the Office International des Epizooties adopted by the International Committee on 24 May 1973

World Trade Organization, Second Submission of the EC in *European Communities—Measures Affecting the Approval and Marketing of Biotech Products*, 19 July 2004, DS291, DS292, DS293

—— First Submission of the US in *European Communities—Measures Affecting the Approval and Marketing of Biotech Products*, 21 April 2004, WT/DS293

—— First Submission of Canada in *European Communities—Measures Affecting the Approval and Marketing of Biotech Products*, 21 April 2004, WT/DS292

—— Minutes of the DSU meeting of 10 December 2003, WT/DSB/M/160

—— First Submission of the US in *European Communities—Measures Concerning Meat and Meat Products*, 18 August 1997, WT/DS26/R/USA

Index

The manufacturer's authorised representative in the EU for product safety is Oxford University Press España S.A. of el Parque Empresarial San Fernando de Henares, Avenida de Castilla, 2 – 28830 Madrid (www.oup.es/en or product.safety@oup.com). OUP España S.A. also acts as importer into Spain of products made by the manufacturer.

www.ingramcontent.com/pod-product-compliance
Lightning Source LLC
LaVergne TN
LVHW011913200926
848484LV00002B/614

* 9 7 8 0 1 9 9 5 7 8 9 2 4 *